Oxford *Smart*

AQA GCSE BIOLOGY

TEACHER HANDBOOK

Katie Estruch

Editor: Lawrie Ryan

Curriculum Editor: Dr Andrew Chandler-Grevatt

Identity Editor: Jonathan Lansley-Gordon

OXFORD
UNIVERSITY PRESS

Contents

Introduction

About the series

This *Oxford Smart:* AQA GCSE Sciences (9-1) Biology Teacher handbook is part of a blended suite of resources from the UK's number 1 course for GCSE Science that has been brought up to date to meet the needs of today's science teachers and their students. The Student Book has been approved by AQA and this refreshed edition of the course has been developed in consultation with leading teachers and assessment and cognitive science experts to support all teachers and students to achieve more in their journey through GCSE Science. Student identity is a key focus – examples throughout the course use diverse scientists as well as relevant and contemporary ¬real-world contexts to help students see themselves reflected in the topics, promoting a sense of belonging and equity.

Oxford Smart: AQA GCSE Sciences (9-1) is designed to be flexible, enabling you to co-teach Foundation and Higher tiers, and Trilogy and Separate Sciences. Resources are clearly flagged throughout, helping you to identify the relevant content for your students. Assessments have been developed alongside the resources, ensuring coherence. It is an adaptive course, meaning that there are additional features and functionality on Kerboodle, including automated learning journeys for students and in-depth reporting, which offers data, analysis, and insight. In addition, there is plenty of support for teachers, including practical videos and a CPD and research area with links to current education research.

About your Teacher Handbook

This Teacher Handbook saves you time and effort by offering lesson plans and guidance on differentiation and assessment with a page-by-page match to the Student Book. Based on evidence-informed pedagogies, its clear guidance and explanations support both specialists and non-specialists; included are unit overviews, common misconceptions, tiered terminology, and ideas to help you to relate the content to relevant contexts and students' experiences. Curriculum narrative information is included, so that you can see the sequencing of topics, the preceding knowledge and what comes next.

About the *Oxford Smart* Curriculum for Science

The *Oxford Smart* Curriculum allows for a clear and coherent 11–16 curriculum journey and has been built to address what research shows to be the most pressing considerations for teachers and learners. It connects content, methods, assessment, insight, and evaluation to deliver a holistic learning experience. Drawing on data, insights, and research it equips teachers to deliver personalised learning for all students.

Designed with an explicit egalitarian purpose to enable teachers and students to experience the joy and wonder of learning whilst journeying towards exam readiness and success, the intent of the *Oxford Smart* Curriculum for Science is based on six founding pillars. More information on each pillar can be found in the CPD and Research Hub on Kerboodle.

Six Pillars

Provide clear and coherent pathways

Coherence

The well-researched progression and sequencing support transition from Key Stage 2, through Key Stage 3, and onto GCSE or equivalent qualifications. The pathways are described in the Curriculum narratives documents and embedded throughout Oxford Smart Activate. Unit and chapter openers clearly show the learning journey for students and highlight the common misconceptions identified by our research.

Hold high expectations, aspirations, and ambitions for all learners

High expectations

Oxford Smart Activate identifies strengths, opportunities for challenge, and addresses learning 'gaps'. Expectations are clearly defined using the Developing, Secure, and Extending assessment framework. Challenge and scaffolding are provided in the resources through opt-in, rather than task-based, differentiation.

Enable responsive teaching and learning that continually evolves and improves

Responsive teaching and learning

Regular opportunities for formative assessment are provided through in-book questions, Reactivate questions, mini-checkpoint tests, and checkpoint assessments, with embedded checkpoint follow-up lessons and recommended next steps allowing for easy intervention. Regular spaced practice is included to help students retain concepts.

Support engaged, self-regulated, and metacognitive learning

Metacognitive learning

The Oxford Smart Activate equips learners to succeed with the ability to understand and use learning strategies effectively, facilitating independence and resilience. Learners develop metacognitive skills with scaffolded resources and reflection plenaries for every lesson, while professional development and targeted resources support teachers in modelling metacognitive strategies and self-regulation in the classroom.

Promote development of learner identity and identification with science

Learner identity

Oxford Smart Activate reflects diversity and promotes a sense of belonging for students. The Trigger interest activities and unit openers demonstrate where science impacts our world, enabling learners to make connections with other subjects and their everyday lives.

Stimulate fascination, awe, and wonder in discovery of self and the world around us

Awe and wonder

Oxford Smart Activate promotes the awe and wonder of science through a wide range of hands-on practical activities and hooks for each topic. The unit openers and Trigger interest sections of lessons provide suggestions to spark curiosity and amazement.

Your Teacher Handbook

Unit opener

Introduction to unit
An overview of the unit.

Assessment
Suggested minimum guided learning hours and exam guidance.

Required practicals
This table indicates which required practicals are covered within this unit. It also gives a list of practical skills that could be assessed by that practical.

5 Ecology

Introduction to unit

This unit will look at how organisms are adapted to their environment and how they compete for mates and resources. It will explore how ecosystems are organised, how organisms play their role within the ecosystem, and how scientists monitor the population and spread of organisms. Students will also examine the impact of the human population on the planet and measures being taken to reduce the harm caused.

Assessment
Suggested minimum guided learning hours: 24 hours
Exam: Paper 2

Prerequisite knowledge from KS3 (chapter references from *Oxford Smart* Activate)

B2.3.3 Ecosystems
- A population is a group of a species in a habitat at a particular time.
- A community is composed of all the species in a habitat at one time.
- Changes to a population have an effect on other organisms in the community.
- Organisms are affected by factors that are living and non-living.

B2.3.4 Competition and adaptation
- Individuals will compete for resources.
- Adaptations help an organism survive in its habitat.

B3.4.4 Apply: Investigating competition
- The different types of average – mode, mean, median.
- How to add scales to axes on graphs and plot data using different formats.

B3.4.6 Reach: Maintaining biodiversity
- Biodiversity is the variety of all the different species of organisms.

B2.3.1 Food chains and webs
- Food chains show the flow of energy from one organism to another.
- Food webs are made of multiple food chains.

B2.3.2 Disruption to food chains and webs
- Humans are destroying habitats for land for building or agriculture.

AQA GCSE Biology specification links
4.7 Ecology
- 4.7.1 Adaptations, interdependence, and competition
- 4.7.2 Organisation of an ecosystem
- 4.7.3 Biodiversity and the effect of human interaction on ecosystems
- 4.7.4 Trophic levels in an ecosystem (biology only)

5 Ecology

Required practicals

AQA required practicals	Practical skills	Topic
Required practical 9: Field investigations Measure the population size of a common species in a habitat. Use sampling techniques to investigate the effect of a factor on the distribution of this species.	AT 1 – use appropriate apparatus to record length and area. AT 3 – use transect lines and quadrats to measure distribution of a species. AT 4 – safe and ethical use of organisms and response to a factor in the environment. AT 6 – application of appropriate sampling techniques to investigate the distribution and abundance of organisms in an ecosystem via direct use in the field. AT 8 – use of appropriate techniques in more complex contexts including continuous sampling in an investigation.	B16.3
Required practical 10: Decay Investigate the effect of temperature on the rate of decay of fresh milk by measuring pH change.	AT 1 – use appropriate apparatus to record temperature and pH. AT 3 – the use of appropriate apparatus to measure anaerobic decay. AT 4 – safe use of microorganisms. AT 5 – measurement of rate of decay by pH change.	B17.4

Working scientifically
There is high relevancy between the topics covered in this unit and a student's everyday life. This includes the impacts of human activity resulting in climate change and leading to extreme weather changes. This will be a good chance for students to practice their science communication skills and learn to present scientific data to a range of different audiences.

Students should also be able to monitor the distribution of plant species and organisms through a series of practical experiments by recording their observations and measurements using a range of apparatus and methods such as a quadrat or transect line.

Maths skills
In this unit, students will develop their numerical skills including being able to plot two variables on a graph from data they have collected and processed. Students should be able to choose a graph most suitable for the data (bar charts, scatter graphs, histograms) and handle data by calculating a mean, median, or mode.

Literacy links
There are a number of activities that focus on developing literacy skills in this unit. These include: B16.6 (WebQuest), B16.7, B18.6, and B18.7. Any of the practical investigations could be used to develop scientific and instructional writing skills such using a step-by-step method in B18.3.

Maths skills
This table provides an overview of the maths skills covered in the chapters of the unit.

Pre-requisite knowledge
This table outlines Key Stage 3 knowledge that is a pre-requisite for the unit along with the GCSE specification points that will be covered within the unit. There are also links which cover knowledge from *Oxford Smart* Activate.

Identity, awe and wonder
Ideas for how to provide context for the unit and make it relevant to students' lives.

Working scientifically skills
This table provides an overview of the Working scientifically skills covered in the chapters of the unit.

Literacy skills
Suggestions to where literacy skills could be covered within the unit and a list of topics that include literacy related activities.

Chapter closer

Student reflection
This section suggests how to help students to reflect on their learning over the chapter. Additional strategies to aid reflection can be found in the Metacognition chapter or in the Research hub on Kerboodle.

B2 Cell division

Introduction to checkpoint intervention

The checkpoint intervention provides suggestions for a lesson to follow up the *B2 Cell division* checkpoint assessment. Depending on the outcomes of the assessment, these suggestions can be used to move a class towards secure, to target specific learning outcomes, or to consolidate knowledge for students achieving secure outcomes. Use students' outcomes from the checkpoint assessment to address general misconceptions from the content covered. Be prepared to re-cover content as required; three key concepts from the chapter are covered in more detail below.

Checkpoint secure learning outcomes

- Describe the stages of the cell cycle, including the process of mitosis. Describe how this produces two identical cells and identify when cell division occurs in the life of an organism.
- Use the key words to describe the process of mitosis.
- Describe the production of a plant cell clone.
- Describe differences between embryonic and adult stem cells.
- Describe how stem cells can be used to treat medical conditions.
- Describe one ethical, one religious, and one medical issue with the use of stem cells in medicine.

Student reflection

Encourage students to reflect on whether there were any checkpoint questions they found difficult or straightforward as well as their preparation for the checkpoint assessment, including how they revised, what they revised, the time spent on revision, and what areas they could improve upon. For more strategies, see the Metacognition spread.

Intervention activity: Describe the stages of the cell cycle, including the process of mitosis. Describe how this produces two identical cells and identify when cell division occurs in the life of an organism.

Students perform a collective memory blurt. Provide students with a plain sheet of A4 paper and ask them to write their names on the sheet.

Give students two minutes to draw the first stage of the cell cycle, including key terms to describe what is happening.

When the time is up, students pass their A4 sheet on to the student next to them. Students then have another two minutes to add the second stage of the cell cycle to the new sheet, including key terms to describe what is happening. Repeat this so students can write the third stage of the cell cycle on a new sheet.

Return the A4 sheet to the original student and give them two minutes to add any missing details, or correct anything they think is incorrect.

Finally, project an image of the cell cycle and discuss the key terms they should include so that students can assess their collective diagram.

Answer:

Stage 1: Replication of DNA to form two copies of each chromosome and synthesis of new sub-cellular structures. This is the longest stage of the cell cycle.

Stage 2: Mitosis and nuclear division occurs.

Stage 3: Cell divides in two forming two identical daughter cells.

34

mitosis online. Students can complete the B2.1 Go Further: *Mitosis* activity sheet after the video has

Intervention activity: Describe one ethical, one religious, and one medical issue with the use of stem cells in medicine.

Present students with the six-mark question, 'Evaluate the use of stem cells from embryos instead of from adult bone marrow for treating human diseases'.

Students can split this six-mark question into a fish-bone scaffold as shown below. They should add two reasons why the use of embryonic stem cells would be an advantage over the use of adult bone marrow, and two disadvantages. The conclusion box is where they come to a judgment, but let students know their conclusions won't be shared with the class.

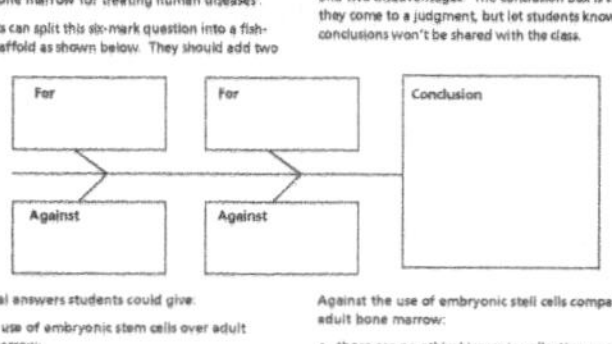

Potential answers students could give:

For the use of embryonic stem cells over adult bone marrow:
- can treat a wide variety / lots of diseases / problems
- many available / plentiful
- using them better than wasting them
- painless to the embryo
- collecting adult bone marrow requires an operation
- collecting adult bone marrow is painful so may deter donors.

Against the use of embryonic stell cells compared to adult bone marrow:
- there are no ethical issues in collection as adults can give permission
- there is a quick recovery after a bone marrow transplant
- collecting adult bone marrow is (relatively) safe
- using embryonic stem cells causes harm / death to embryo
- embryo can't be 'asked' / 'embryo rights' idea
- religious and ethical beliefs about right to life beginning at fertilisation.

35

Intervention activities
The intervention activities are suggested follow-up lessons after students have completed the relevant end of chapter Checkpoint Assessment on Kerboodle. The aim of the activities is to move all students from developing to secure on the key learning outcomes covered by the assessment.

Checkpoint secure learning outcomes
A list of all the secure learning outcomes covered in the checkpoint assessment.

Lesson

Suggested lesson plan
A suggested route through the lesson is provided; Trigger Interest, Exposition, Main activity, and Review and practice. The structure has been carefully designed using cognitive science research and is fully supported by resources on Kerboodle including lesson presentations. Suggested timings are given based on a 55-minute lesson.

Core concepts
This indicates the area of the AQA GCSE Biology (9-1) 2016 specification this lesson covers. The theme and sub-theme indicate the area of the *Oxford Smart* Curriculum for Science that this lesson focuses on. Students' understanding can be monitored with the curriculum reports. Relevant Working scientifically links are also provided, as well as the main activity for the lesson.

B1 Cell structure and transport

B1.5 Specialisation in plant cells

AQA GCSE Biology specification points
4.1.1.3 Cell specialisation

***Smart Curriculum* Narrative**
How organisms work

Theme/subtheme
Cells/Types of cell

Disciplinary knowledge
- WS 1.2, 3.1, 4.4

Practical
Observing root hair cells

Learning objectives	Learning outcomes: Developing Aiming for 3	Secure Aiming for 5	Extending Aiming for 8
Understand that many types of plant cells continue to differentiate throughout the life of the plant	State that cells differentiate and form specialised cells in plants as well as animals	**Understand that many types of plant cells continue to differentiate throughout the life of the plant**	Discuss the importance to a plant of cell differentiation continuing throughout life
Compare the structure of a specialised and a generalised plant cell	Identify specialised plant cells from diagrams	**Compare the structure of a specialised and a generalised plant cell**	Suggest the function of a specialised plant cell, familiar or unfamiliar, based on its structure
Describe the structural adaptations of three types of specialised plant cells	Name three types of specialised plant cells and describe their functions	**Describe the structural adaptations of three types of specialised plant cells**	Discuss how the structures seen in different examples of specialised plant cells are related to their functions within the organism
Observe, draw, and label a root hair cell using a light microscope, including a magnification scale	Observe and draw a root hair cell using a light microscope	**Observe, draw, and label a root hair cell using a light microscope, including a magnification scale**	Observe, draw, and label a root hair cell using a light microscope, including a magnification scale, and calculate the length of the root hair

Tier 2 vocabulary	Tier 3 vocabulary
cell, *organ*, organism, organ system, specialised, *tissue*	companion cell, phloem, phloem vessel, photosynthetic cell, root hair cell, sieve plate, xylem

Lesson plan

Reactivate knowledge

1 Name an adaptation of a leaf cell.
2 Name an adaptation of a root hair cell.
3 Which chemical reaction takes place in the mitochondria of cells?

Answers: 1 Lots of chloroplasts 2 Large surface area 3 Respiration

Common misconception
Students can think that *all* plant cells contain chloroplasts. Clarify that not all plant cells do. For example, root hair cells do not contain chloroplasts as they are underground, where there is no light to photosynthesise.

Trigger interest

Plant cell (5 mins)

Show (and describe if required) a diagram of a leaf cross-section and explain what it is showing. Ask students to identify a cell that is similar to the generalised plant cell previously studied (palisade cell). Discuss how the other cells have different structures and ask students to suggest why this might be. This is also an opportunity to show students a cross-section of a xerophyte as an example of how the structure can differ depending on a plant's needs.

Exposition

Plant cell differentiation (5 mins)

Show students an image of a spider plant where budding of 'mini' spider plants can be seen, or an actual example of such a plant, if available. Use this to begin a discussion of how plant cells can continue to become specialised throughout the life of the plant.

Specialised plant cells (20 mins)

Show (and describe if required) students images of a root hair cell, xylem, phloem, and palisade cell. Ask them to label any cell structures they recognise. (They should all be able to identify the cell membrane, cell wall, cytoplasm, nucleus, and vacuole. Some may also be able to identify the chloroplasts and mitochondria if they are visible in the images you provide.)

Inform the students of the function of each of these specialised cells. In pairs, they should discuss how the cell structures and cell shapes would help the cells perform their functions. If students need prompts, suggest they consider the quantity of mitochondria in the cell and the surface area of the cell.

Main activity

Observing root hair cells (20 mins)

Students study prepared slides of the root hair cells of mung beans using a light microscope. They should clearly be able to identify organelles but if students are struggling, they could be shown an example image with labels. They draw scientific drawings based on their observations and answer questions about the adaptations of these cells. If there is more time, students can make their own slides. They place the roots of sprouting mung bean seeds on slides and dye with iodine solution.

Review and practice

What am I? (5 mins)

Write the names of different specialised animal and plant cells on cards. Ask students to work in groups and give each student a card that they must hold in front of them without looking at it. Students then ask questions to determine which cell they are.

For example, students could ask:

- Do I have a cell wall?
- Do I have chloroplasts?
- Do I have a nucleus?
- Am I biconcave?

Learning outcomes
This table shows the lesson outcomes which are ramped and divided into three ability bands, Developing (Aiming for Grade 3), Secure (Aiming for Grade 5), and Extending (Aiming for Grade 8).

Vocabulary
Tier 2 vocabulary includes words that are more academic in nature and operate across different contexts. Tier 3 key words are subject-specific. Words in italics have multiple meanings and misconceptions may arise due to this.

Reactivate knowledge
The retrieval questions are included here, along with the answers and any suggestions for discussions relating to any common misconceptions.

Answers to the end-of-spread questions in the Student Book can be found from page 312.

Identity

What is diversity and inclusion?

Diversity, in its simplest form, is different-ness. In the context of curriculum, diversity calls for the presence of different-ness, or variety, in all its forms – a variety of content, people, stories, narratives, perspectives and voices. In seeking a diverse curriculum, the children and young people that we serve are likely to be and feel included – which in turn means they are likely to be engaged and inspired by what they learn. Inclusion, in this sense, is the ideal that all learners, irrespective of their individual identities or backgrounds, are active participants in the curriculum. An inclusive curriculum is integral to a quality education; one that ensures students are equipped to succeed now as well as in their future lives.

The importance of a science identity for GCSE students

A 'science identity' can be thought of as how you see yourself with respect to science. For instance, expressions like 'I could be a scientist', or 'I'm good at science', or even 'I think science is important', are a few of the ways that a strong science identity is articulated.

Students who have strong science identities are likely to:

- be interested in science
- feel competent at science
- be validated by their peers and teachers
- feel like they belong in science.

A strong science identity is a significant predictor of student outcomes as well as the likelihood of persistence with science beyond GCSE. In considering the intent and design of a science curriculum, therefore, it is crucial to consider the ways in which it establishes and reinforces the science identity of all students.

The Identity review process

The Identity review process centred the concept of a science identity; considering content and delivery across each of the sciences and ensuring that young people's diverse personal identities, skills, knowledge and experiences were valued, respected, visible, and represented. Recommendations from best practice in the sector, alongside evidence-based research on the views and opinions of young people doing GCSE science informed our guiding principles, which included:

- Harnessing and valuing individuals' cultural capital and identity, e.g., eliciting and valuing what the student already knows from prior knowledge, news, media, and social media
- Supporting learners to understand how content is relatable and meaningful to themselves at various levels of citizenship – as individuals, communities, and society
- Highlighting scientific elements in broader aspects of everyday life, that students can connect to now as well as in the future (i.e., hobbies, jobs, and careers)
- Considering areas of the curriculum in which diverse knowledge, stories, histories, and experiences are relevant and can be authentically explored
- Considering ways in which both science content as well as the nature of science can be understood by learners

- Supporting all learners to value and behave equitably and inclusively to each other, e.g., encouraging listening, respecting opinions, and ensuring active participation
- Challenging stereotypes about science and scientists by showcasing stories and achievements of people from historically underrepresented backgrounds.

Examples of building science identity in Biology

It was important at the start of the review process to acknowledge that all themes would not be relevant or appropriate to all spreads. This is something we encourage all teachers to consider, both in terms of if/how they choose to implement/tweak the suggestions enclosed, but also in terms of considering the nuances of their own settings and contexts – i.e., the needs of individual students. Alternative activities to approach identity have been provided in the Trigger Interest and Review and practice sections of the lesson so teachers can choose to cover these depending on the needs of the class.

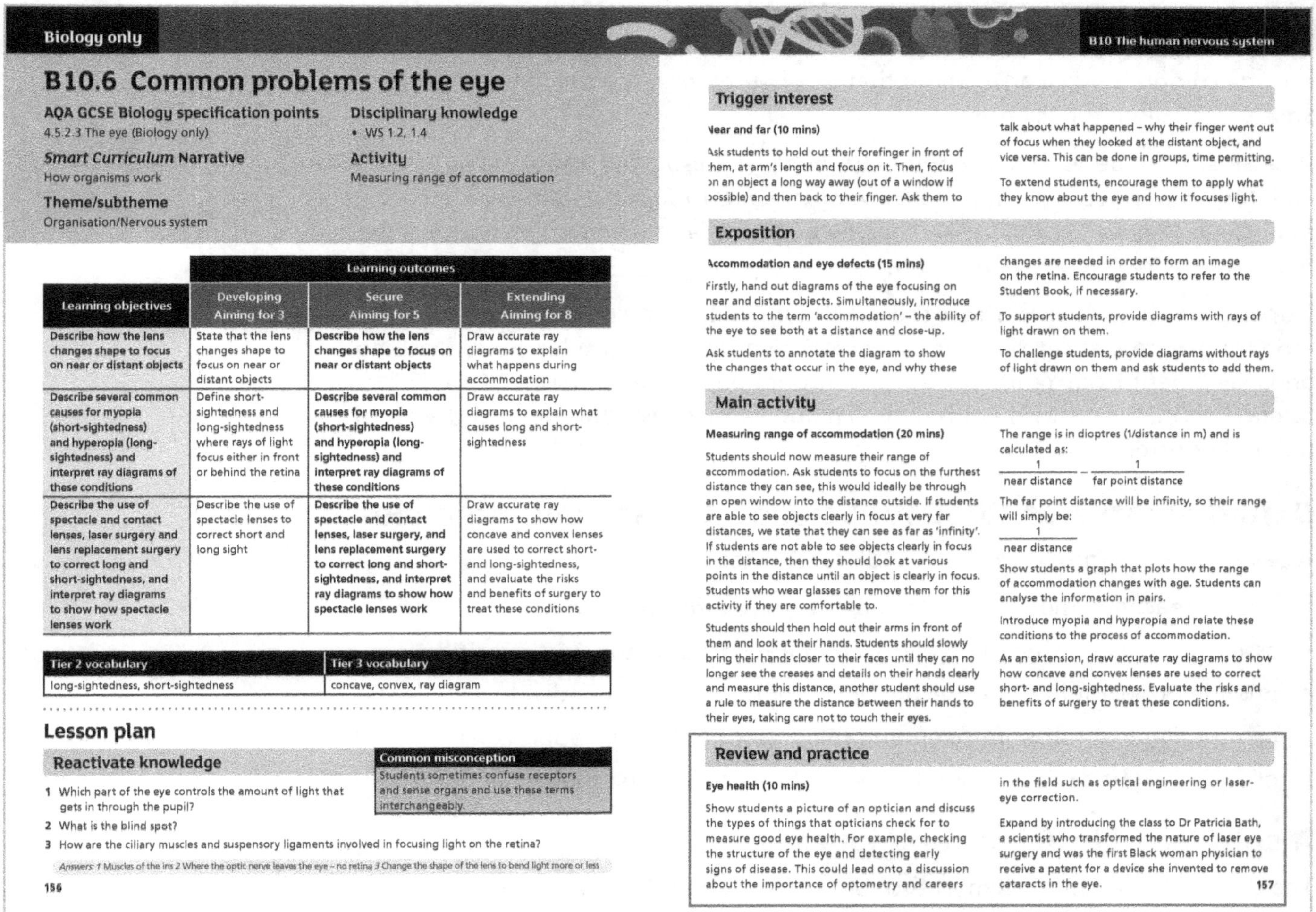

Biology only

B10 The human nervous system

B10.6 Common problems of the eye

AQA GCSE Biology specification points
4.5.2.3 The eye (Biology only)

***Smart Curriculum* Narrative**
How organisms work

Theme/subtheme
Organisation/Nervous system

Disciplinary knowledge
- WS 1.2, 1.4

Activity
Measuring range of accommodation

Learning objectives	Learning outcomes		
	Developing Aiming for 3	Secure Aiming for 5	Extending Aiming for 8
Describe how the lens changes shape to focus on near or distant objects	State that the lens changes shape to focus on near or distant objects	**Describe how the lens changes shape to focus on near or distant objects**	Draw accurate ray diagrams to explain what happens during accommodation
Describe several common causes for myopia (short-sightedness) and hyperopia (long-sightedness) and interpret ray diagrams of these conditions	Define short-sightedness and long-sightedness where rays of light focus either in front or behind the retina	**Describe several common causes for myopia (short-sightedness) and hyperopia (long-sightedness) and interpret ray diagrams of these conditions**	Draw accurate ray diagrams to explain what causes long and short-sightedness
Describe the use of spectacle and contact lenses, laser surgery and lens replacement surgery to correct long and short-sightedness, and interpret ray diagrams to show how spectacle lenses work	Describe the use of spectacle lenses to correct short and long sight	**Describe the use of spectacle and contact lenses, laser surgery, and lens replacement surgery to correct long and short-sightedness, and interpret ray diagrams to show how spectacle lenses work**	Draw accurate ray diagrams to show how concave and convex lenses are used to correct short- and long-sightedness, and evaluate the risks and benefits of surgery to treat these conditions

Tier 2 vocabulary	Tier 3 vocabulary
long-sightedness, short-sightedness	concave, convex, ray diagram

Lesson plan

Reactivate knowledge

1 Which part of the eye controls the amount of light that gets in through the pupil?
2 What is the blind spot?
3 How are the ciliary muscles and suspensory ligaments involved in focusing light on the retina?

Answers: 1 Muscles of the iris 2 Where the optic nerve leaves the eye – no retina 3 Change the shape of the lens to bend light more or less

Common misconception
Students sometimes confuse receptors and sense organs and use these terms interchangeably.

156

Trigger interest

Near and far (10 mins)

Ask students to hold out their forefinger in front of them, at arm's length and focus on it. Then, focus on an object a long way away (out of a window if possible) and then back to their finger. Ask them to talk about what happened – why their finger went out of focus when they looked at the distant object, and vice versa. This can be done in groups, time permitting.

To extend students, encourage them to apply what they know about the eye and how it focuses light.

Exposition

Accommodation and eye defects (15 mins)

Firstly, hand out diagrams of the eye focusing on near and distant objects. Simultaneously, introduce students to the term 'accommodation' – the ability of the eye to see both at a distance and close-up.

Ask students to annotate the diagram to show the changes that occur in the eye, and why these changes are needed in order to form an image on the retina. Encourage students to refer to the Student Book, if necessary.

To support students, provide diagrams with rays of light drawn on them.

To challenge students, provide diagrams without rays of light drawn on them and ask students to add them.

Main activity

Measuring range of accommodation (20 mins)

Students should now measure their range of accommodation. Ask students to focus on the furthest distance they can see, this would ideally be through an open window into the distance outside. If students are able to see objects clearly in focus at very far distances, we state that they can see as far as 'infinity'. If students are not able to see objects clearly in focus in the distance, then they should look at various points in the distance until an object is clearly in focus. Students who wear glasses can remove them for this activity if they are comfortable to.

Students should then hold out their arms in front of them and look at their hands. Students should slowly bring their hands closer to their faces until they can no longer see the creases and details on their hands clearly and measure this distance, another student should use a rule to measure the distance between their hands to their eyes, taking care not to touch their eyes.

The range is in dioptres (1/distance in m) and is calculated as:

$$\frac{1}{\text{near distance}} - \frac{1}{\text{far point distance}}$$

The far point distance will be infinity, so their range will simply be:

$$\frac{1}{\text{near distance}}$$

Show students a graph that plots how the range of accommodation changes with age. Students can analyse the information in pairs.

Introduce myopia and hyperopia and relate these conditions to the process of accommodation.

As an extension, draw accurate ray diagrams to show how concave and convex lenses are used to correct short- and long-sightedness. Evaluate the risks and benefits of surgery to treat these conditions.

Review and practice

Eye health (10 mins)

Show students a picture of an optician and discuss the types of things that opticians check for to measure good eye health. For example, checking the structure of the eye and detecting early signs of disease. This could lead onto a discussion about the importance of optometry and careers in the field such as optical engineering or laser-eye correction.

Expand by introducing the class to Dr Patricia Bath, a scientist who transformed the nature of laser eye surgery and was the first Black woman physician to receive a patent for a device she invented to remove cataracts in the eye.

157

Elevating the vital contributions of women in science, from historical figures such as Patricia Bath, and leading discussions to why their contributions were important.

Assessment and reporting

About the Curriculum Editor and Assessment Expert

Dr Andrew Chandler-Grevatt has a doctorate in school assessment and a passion for science teaching. Having worked as a teacher for 10 years, this stays at the front of his mind during all of his work in education. Alongside his work with schools and research in school assessment, Andy is a Senior Lecturer in Science education at the University of Brighton and a successful published assessment author.

Welcome from the Curriculum Editor

Welcome to your *Oxford Smart* AQA GCSE Biology Teacher Handbook. The Teacher Handbooks, together with Kerboodle, and the Student Books, provide comprehensive assessment that is coherent with the full 11–16 Oxford Smart Curriculum for Science.

Assessment can be used to set ambitious goals. [It] also plays a key role in enabling responsive teaching and learning needed to support as many learners as possible in reaching that goal. The power of awe and wonder: introducing the new Oxford Smart Curriculum, 2021.

Our assessment is built in across the curriculum and will help your school to monitor progress against a clear framework in order to inform learning, teaching, and assessment from Year 7 to Year 11. Throughout Oxford Smart AQA GCSE for Science, formative assessment has been made easy and we have followed a set of guiding principles.

Oxford Smart AQA GCSE for Science assessment principles

Assessment in *Oxford Smart* AQA GCSE for Science aims to:

- inform teaching and/or learning directly (have a formative function)
- assess agreed and shared objectives that are based on the AQA specifications
- provide opportunities for specific feedback to be given to individual students
- provide opportunities for peer- and self-assessment, when appropriate
- provide insightful data that informs teachers of class and individual progress

A coherent 11–16 learning journey

The *Oxford Smart* Curriculum assessment model is based on bands: the middle band indicates that students have a 'Secure' grasp of the knowledge or skills specified in the curriculum. The band below this is 'Developing' and the band above is 'Extending'. We use bands to help overcome the difficulty of predicting individual grades and should be used to decide on suitable feedback or interventions to help students reach their full potential.

This assessment model has informed the design of the learning outcomes thought the course. Learning outcomes are adapted, and there is a set of learning outcomes for every lesson for each band.

Band	Developing - Aiming for 3			Secure - Aiming for 5			Extending - Aiming for 8		
Grade	1	2	3	4	5	6	7	8	9
Demand	Low			Standard			High		

At GCSE, the framework is based on students' target grades in three bands. This has been informed by how GCSE questions are written at different demands: low, standard, and high demand.

- **Developing/Aiming for 3** targets the students who will have been Developing at KS3 and now have a target grade, aiming for Grade 3 at GCSE (this includes grades 1-3). In this band, students are developing core concepts, and should be able to demonstrate knowledge and understanding and carry out some application and analysis of familiar situations, with some scaffolding. These students will be able to answer most low demand questions and some standard demand questions.
- **Secure/Aiming for 5** is for students working at grades 4 to 6, who would have been secure at KS3. Their knowledge and understanding should be secure and they are able to apply the core concepts to familiar situations. This means that these students should be able to answer low and standard demand GCSE questions.
- **Extending/Aiming for 8** targets those students who have been secure or extending in GCSE and are predicted to get GCSE grade 7 or above. These students have secure knowledge and understanding or all the core concepts and can apply them to more complex or unfamiliar situations. These students will be able to answer GCSE high demand questions.

The checkpoint system

The formative assessment element comprises the checkpoints and mini-checkpoints on Kerboodle. Regular mini-checkpoint assessment tracks progress through the chapter and there is a checkpoint assessment following each chapter. These are all automarked and will help determine if students have a secure understanding of the chapter.

Following each checkpoint or mini-checkpoint, students are offered personalised next steps. They can consolidate their knowledge where they are at a developing level and challenge themselves where they have demonstrated secure knowledge.

Reporting and insights

The formative assessment data will feed online reporting on Kerboodle and live insights into students' strengths and weaknesses to identify misconceptions so you can adapt your teachings.

The *Oxford Smart* Curriculum breaks each science into the Curriculum narrative domains and then splits them into related themes that run through all key stages. The themes are divided into sub-themes to enable targeted interventions.

Kerboodle

Oxford Smart AQA GCSE Sciences (9-1) Kerboodle provides hundreds of engaging lesson resources, as well as a comprehensive assessment package. They are regularly updated, intuitive, and customisable. Resources include lessons and associated resources, automatically marked assessments, books, and the CPD and Research hub.

You can adapt many of the resources to suit your students' needs or your own scheme of work, with all non-interactive activities available as editable documents. You can also upload your existing resources so that everything can be accessed from one location.

Kerboodle is online, allowing you and your students to access the course anytime and anywhere. You can set homework and assessments through the Assessment system and track progress using the Markbook.

Kerboodle Book

The *Oxford Smart* AQA GCSE Sciences Kerboodle Book provides a digital copy of the Student Book for use at the front of the classroom. Teacher access to the Kerboodle Book is automatically available as part of the Lessons, Resources, and Assessment package. You can also purchase additional access for your students. A set of tools is available with the Kerboodle Book so you can personalise your book and make notes. Digital books are hot spotted to direct and relevant access to resources and quizzes on Kerboodle. Like all other resources offered on Kerboodle, the Kerboodle Book can also be accessed using a range of devices.

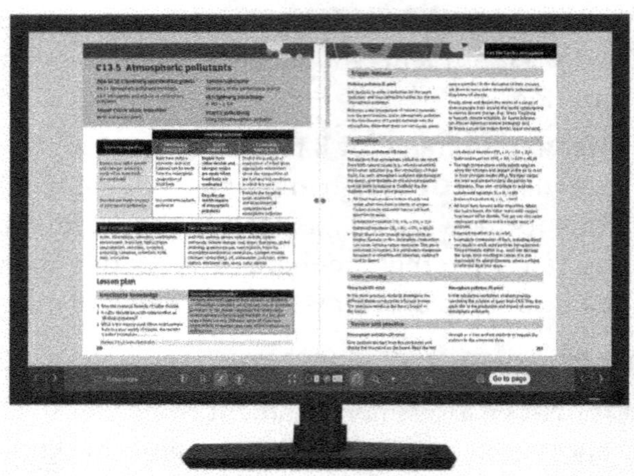

Lessons

Click on the 'Lessons' tab to access the *Oxford Smart* AQA GCSE Sciences Unit planners, Schemes of Work, and Lesson presentations. Each presentation features *resources required*, *Reactivate* question starters, *Something to think about*, lesson objectives, *Trigger Interest* content, activity guidance, key diagrams, questions, extension, plenaries, and homework suggestions. You can further personalise these by adding in your own resources and notes.

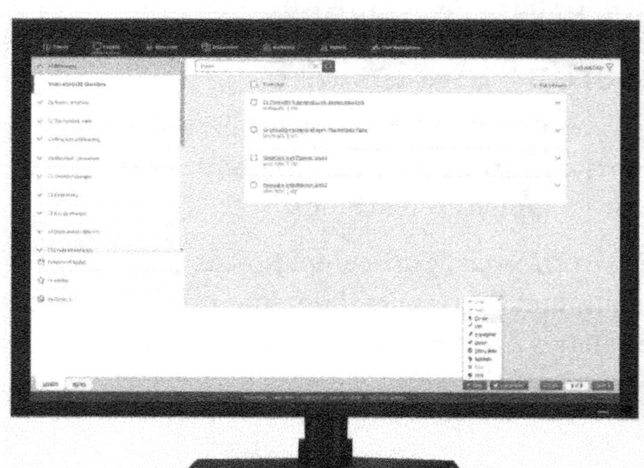

Resources

Click on the 'Resources' tab to access the full list of *Oxford Smart* AQA GCSE Sciences lesson resources.

The resource section includes but are not limited to:

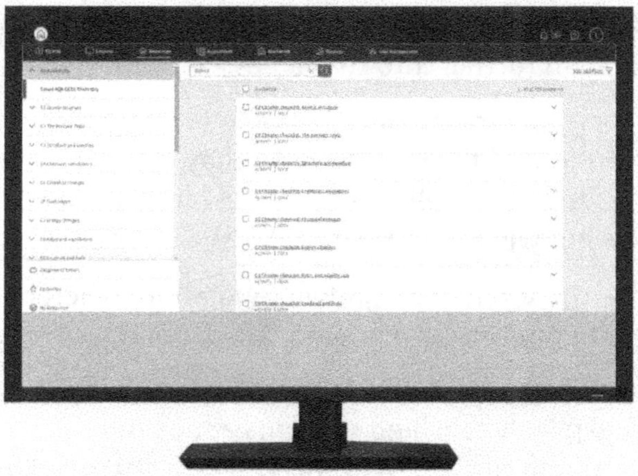

- **Practicals and Activities worksheets** Fully-editable resources to use in class. In addition to an Activity sheet and a Support sheet, which are accessible by pupils, a Teacher and technician sheet provides further ideas on differentiation, answers, and a list of resources required by technicians.
- **Required Practical worksheets** Available in three versions (Aiming for Grades 3/5/8) along with accompanying teacher and technician notes. Every required practical has at least one mini clip to guide students through the methods and data analysis.
- **Maths, Working Scientifically, Literacy, Extension, and Bump up your grade worksheets**
- **Knowledge organisers**
- **Checklists**
- **Glossaries**
- **Metacognition support**

Assessment and Markbook

All the assessment material in Kerboodle has been quality assured by our assessment expert. Click on the Assessment tab to find a range of materials to help you deliver a varied, motivating, and effective assessment programme. It's easy to import class registers and create user accounts for your students. Once your classes are set up, you can assign them assessments to do at home, individually, or as a group. A Markbook with reporting function helps you to keep track of your students' results. This includes both automarked assessments and work that needs to be marked by you. Mini-checkpoint and Checkpoint assessments are available for testing during and after a chapter in addition to end of chapter paper-based assessments. Support has been embedded within assessments to motivate students to develop independent study habits. Personalised assessments combined with immediate next steps ensure students understand their results, core concepts are reinforced, and knowledge and skills are stretched by the follow-up questions.

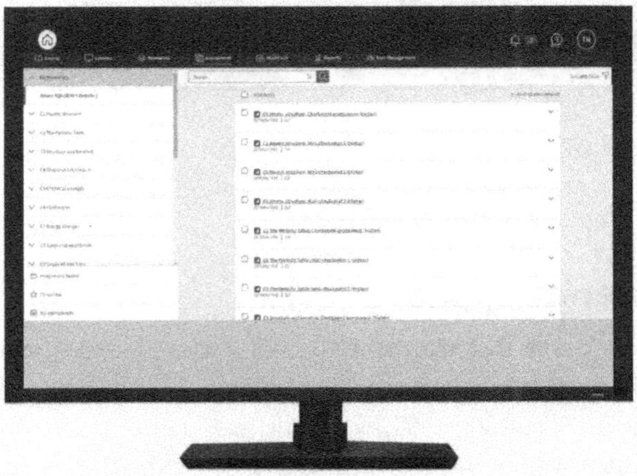

Metacognition

Metacognition is a student's awareness and understanding of their own learning. It is commonly referred to as 'learning about learning' and covers both subject-specific learning as well as wider, independent-study skills. Metacognition can be broken down further into two areas.

Firstly, there is metacognitive knowledge, in which students have a good understanding of the specific tasks they need to complete, as well as the strategies that are available to them with which to carry out the task. Students also have knowledge of themselves as a learner.

Secondly, there is metacognitive regulation, whereby students are able to manage themselves by planning, monitoring, and evaluating their own learning.

An expert learner will naturally follow a cycle like this when they tackle a new task. By explicitly teaching students about this cycle, they will be able to adopt a similar thought process, and this will help them build confidence to successfully complete new and challenging activities.

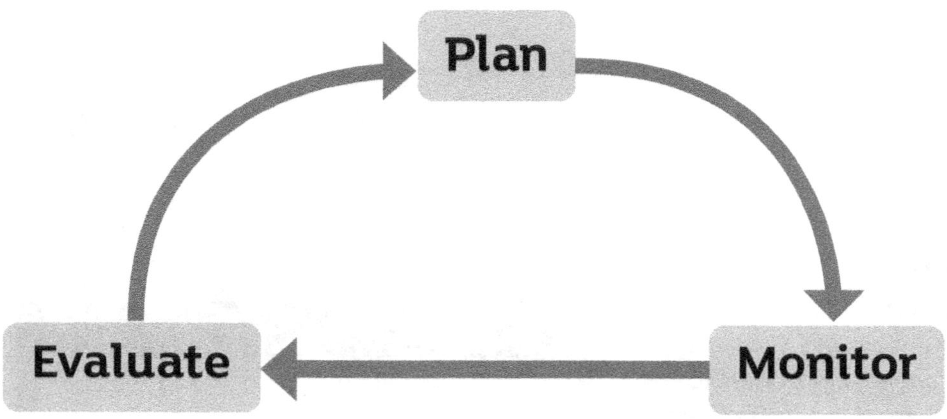

The planning phase

This takes place before the task. An expert learner would plan their approach carefully by thinking about what strategies they could use and what scientific knowledge is relevant. They would also reflect on whether they have completed something similar before and apply this experience to the new task.

The monitoring phase

This takes place during the task. An expert learner would regularly pause and check back to make sure their chosen approach is still right. They would also have the confidence to stop and start again if needed.

The evaluation phase

This takes place at the end of the task. An expert learner would reflect on their performance and carefully consider any areas for improvement. They would plan for how they would use what they have learnt in the future so that they can be successful when doing similar tasks again.

The Seven-Step Approach

The Education Endowment Foundation suggests the following seven-step approach to teaching metacognitive skills.

1 Activate prior knowledge

2 Explicit strategy instruction

3 Modelling of learned strategy

4 Memorisation of strategy

5 Guided practice

6 Independent practice

7 Structured reflection

This approach allows a student to learn about a strategy in relative isolation. When introducing students to a metacognitive strategy, do so in the context of a concept or piece of knowledge that is simple and with which they are already very familiar. This reduces the extra cognitive load that unfamiliar science content would add.

A key aspect of this approach is modelling. Metacognitive modelling is where the focus is on demonstrating how an expert learner would approach a task by verbalising your inner thought processes. For example, you could answer a question 'live' in front of the class. This is helpful for students because they are observing and learning how an expert learner plans, monitors, and evaluates their own learning.

A question tree is a tool that can complement the modelling process. For example, when completing circuit calculations, it is common for values to be placed on a circuit diagram and students then need to choose which values are needed for the calculation. The question tree shown in Figure 1 summarises how students should approach the task.

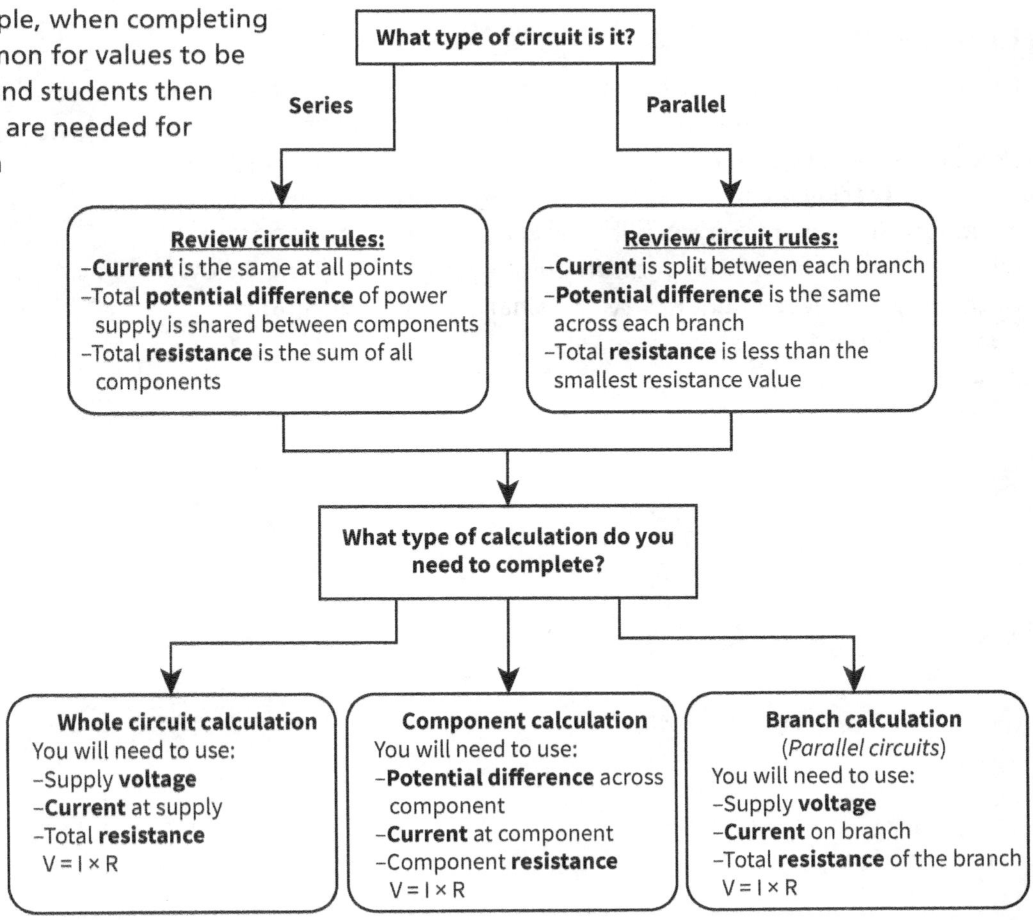

▲ **Figure 1** A question tree for completing circuit calculations

About the authors

Biology

Katie Estruch is an experienced educator with a love of Biology. With 14 years of teaching experience, Katie has held various leadership roles including Head of Science, Head of Biology, and Study Skills Coordinator. In 2019, she took her teaching online too, launching Miss Estruch Biology—an online platform offering biology theory, skills, and revision tips through engaging videos on YouTube, Instagram, and TikTok. Recognising the importance of supporting fellow educators, Katie expanded her reach with the creation of Miss Estruch Teach & Tell. Through this initiative, she shares effective teaching strategies, and lesson plans, and offers support to teachers navigating the ever-evolving educational landscape.

Chemistry

Dr Kristy Turner has a unique role in chemistry education, teaching in both a high school and in the chemistry department at the University of Manchester. A qualified teacher for nearly 20 years, she is a true chemistry specialist with an academic background in synthetic organic chemistry and current research interests including transition from school to higher education and curriculum design. Kristy is a Fellow of the Royal Society of Chemistry and Chair of their education publishing board.

Physics author and Identity Editor

Jonathan Lansley-Gordon obtained his master's degree in theoretical physics from Imperial College London, going on to teach secondary mathematics and science for over 15 years. He is passionate about science communication and believes in the power of physics to inspire, fascinate, and engage all people, regardless of personal characteristics or background. He co-founded The Blackett Lab Family – the first national network of UK based Black physicists – to diversify perceptions of physics and to increase Black representation at all levels in the field. A former Assistant Headteacher and school governor, he now works as a Science Consultant and Director at Physics Forward – an organisation that supports science educators to design and deliver a diverse, inclusive and accessible physics curriculum.

Series Editor

Lawrie Ryan studied Chemistry at University of London before training to be a teacher. He then taught for 20 years, most as Head of Departments, in a variety of large schools in and around Manchester. During this time, he completed a part-time master's degree in science education at Manchester University.

During his career, he has also been a Science Adviser for Norfolk County Council, supporting 50 secondary schools and over 400 primary schools. This involved running many courses for teachers, as well as supporting them in their classrooms with their day-to-day science teaching. During his career he has delivered professional development and worked with teachers in Ireland, Turkey, Barbados, India, Sri Lanka, Lebanon, Jordan, and most of the Gulf states.

He became involved in writing science textbooks whilst teaching in Manchester, and when the opportunity came to pursue this as a full-time career, he became a freelance science education consultant and author/editor. He has been involved in the production of lots of best-selling science resources including Spotlight Science, Chemistry for You (GCSE and A Level editions), the AQA GCSE Sciences series and the Cambridge IGCSE & O Level Essential Science series. As such, he has made science more accessible and enjoyable for students around the world.

1 Cells and organisation

Introduction to unit

This unit allows students to build upon their existing knowledge of cells as the fundamental building blocks of life (in plants, animals, and prokaryotes). Links will be made between the need for molecules to enter organisms, the production of new molecules and cells, and organism growth and differentiation. Students will investigate cellular action through practical work.

Assessment

Suggested minimum guided learning hours: 30 hours

Exam: Paper 1

Prerequisite knowledge from KS3 (chapter references from *Oxford Smart* Activate)

B1.1.1 Cells

- All living organisms are made of cells.
- Microscopes can be used to view cells.

B1.1.5 Unicellular organisms

- Some organisms are made of multiple cells and others are unicellular.

B1.1.3 Specialised cell

- Different cells have different functions so therefore have different structures.

B1.1.2 Plant and animal cells

- Animal and plant cells have similarities and differences.

B3.1.5 Diffusion

- Particles can move by diffusion.

B1.2.1 Levels of organisation

- Cells are organised into tissues, organs, and organ systems.

B2.1.4 Digestive system

- Food is broken down in the digestive system.

AQA GCSE Biology specification links

4.1 Cell biology

- 4.1.1 Cell structure
- 4.1.2 Cell division
- 4.1.3 Transport in cells

4.2 Organisation

- 4.2.1 Principles of organisation
- 4.2.2 Animal tissues, organ, and organ systems
- 4.2.3 Plant tissues, organs, and systems

Identity, awe, and wonder

Life itself can seem miraculous and students have an inherent desire to understand the functioning of their own body and how they fit into the world. This unit allows them to see how developments in technology have furthered our understanding of the human body but there is still much we do not know. Students should be familiar with the idea of the cell as the basic unit of life, but they may never have considered how new cells are formed, where the different types of cells originate, or how they themselves have grown up. The topic of cell division and stem cells allows for discussion of ethical issues in this area. This unit includes plant-based topics, which increase students' awe and wonder of this kingdom. As a species we rely on plants for food, building materials, medicines etc. Plants are an integral part of ecosystems; it is imperative that plant research is carried out to further our knowledge of these vital organisms.

Required practicals

AQA required practicals	Practical skills	Topic
Required practical 1: Microscopy Use a light microscope to observe, draw and label biological specimens.	AT 1 – use appropriate apparatus to record length and area AT 7 – use a microscope to make observations of biological specimens and produce labelled scientific diagrams	B1.2
Required practical 3: Osmosis Investigate the effect of a range of concentrations of salt or sugar solution on the mass of plant tissue.	AT 1 – use appropriate apparatus to record mass and time AT 3 – use appropriate apparatus and techniques to observe and measure the process of osmosis AT 5 – measure the rate of osmosis by water uptake	B1.8
Required practical 4: Food tests Use qualitative reagents to test for a range of carbohydrates, lipids, and proteins. To include: Benedict's test for sugars, iodine test for starch and Biuret reagent for protein.	AT 2 – safe use of a Bunsen burner and a boiling water bath AT 8 – use of qualitative reagents to identify biological molecules	B3.3
Required practical 5: Enzymes Investigate the effect of pH on the rate of reaction of amylase enzyme.	AT 1 – use appropriate apparatus to record the volumes of liquids, time, and pH AT 2 – safe use of a water bath or electric heater AT 5 – measure the rate of reaction by the colour change of iodine indicator AT 8 – use of qualitative iodine reagent to identify starch by continuous sampling	B3.6

Working scientifically

Students explore how scientific methods and theories develop over time. They consider the capabilities and limitations of science and its connection to technological progress. Students plan, make predictions, and carry out experiments to make observations. They use a range of apparatus and procedures, including production and observation of microscope slides, solutions, and aseptic techniques. They also evaluate methods and consider measurement uncertainty. Students discuss the ethical issues of using embryonic stem cells. Discussions around cardiovascular disease provide opportunity for evaluating risks in a wider societal context, including perception of risk in relation to data and consequences. This unit develops key vocabulary needed later in the course.

Maths skills

Students develop their numerical skills (including unit conversions, orders of magnitude, standard form, calculations of rate, surface area, and volume), and algebraic skills (including rearranging equations). They process experimental data, use significant figures, calculate means and percentage change, represent and interpret data in various forms (including graphs), and draw tangents to calculate reaction rates.

Literacy links

Students develop scientific/instructional writing skills by writing step-by-step methods. In B1.4, students assimilate and represent written information about specialised cells. In B1.7, they write extended responses. In B2.4, students read different viewpoints and summarise information to be presented orally, plus develop their persuasive writing skills. In B3.1, they assimilate and represent written information about stomach tissue. In B4.4, students practise summarising and justifying decisions on different treatments for heart conditions.

B1 Cell structure and transport

B1.1 The world of the microscope

AQA GCSE Biology specification points

4.1.1.5 Microscopy

Smart Curriculum Narrative

How organisms work

Theme/subtheme

Cells/Microscope

Disciplinary knowledge

- WS 1.1, 4.4
- MS 1a, 1b, 1d, 2h, 3a

Practical

Observing cells under a microscope

	Learning outcomes		
Learning objectives	**Developing Aiming for 3**	**Secure Aiming for 5**	**Extending Aiming for 8**
Describe the difference between a light microscope and an electron microscope	Describe how microscopy techniques have developed over time	**Describe the difference between a light microscope and an electron microscope**	Discuss the advantages and disadvantages of using a light and an electron microscope
Describe the difference between magnification and resolution	State that microscopes magnify the image of an object and this makes them useful in the study of cell biology	**Describe the difference between magnification and resolution**	Compare and contrast the magnification and resolution obtained by using light and electron microscopes, and use this to explain how electron microscopy has increased understanding of sub-cellular structures
Use the formula: Magnification = $\frac{\text{size of image}}{\text{size of real object}}$	Calculate total magnification	**Use the formula: Magnification =** $\frac{\textbf{size of image}}{\textbf{size of real object}}$	Rearrange the magnification formula and use it to measure magnification, size of image, and the real size of cells

Tier 2 vocabulary	Tier 3 vocabulary
cell, electron, *focus*, image, magnification, millimetre, object, resolution	electron microscope, eyepiece, light microscope, nanometre, objective lens, resolving power

Lesson plan

Reactivate knowledge

1 What are the building blocks of living organisms?
2 Name the main instrument used to see very small objects in biology.
3 What does magnification mean?

Answers: 1 Cells *2* Microscope *3* Making things look bigger than they are

Common misconception

Students can confuse magnification and resolution, thinking that the higher the magnification, the clearer an image is.

Trigger interest

What is it? (5 mins)

Show a range of micrographs of cells (e.g., a red blood cell under a light, scanning, and transmission electron microscope) and ask students to match up each image with options provided of how magnified each image is. Ask students to discuss the advantages of using microscopes in biology. The development of different types of microscopes allows us to understand organelles much better than the past.

Exposition

Light and electron microscopes (15 mins)

Remind students of the different part of a light microscope. Show students where to place the slide on the stage and how to secure it in position with the stage clips. Then show students how to use the coarse and fine focus knobs. If available, you could attach the light microscope to a visualiser so that students can see the image come into focus.

Inform the class that there is a limit on what can be seen using a light microscope and introduce the electron microscope. Give students the definitions of magnification and resolution and clarify any misconceptions here.

Ask the students to use the information from the Student Book to describe the main advantages and disadvantages of using a light and an electron microscope. Alternatively, provide students with facts about the magnification, resolution, and other properties of both types of microscopes and ask them to arrange this into a table showing the pros and cons of each type.

Introduce students to the magnification calculation and ask them to rearrange the formula to make the image size and then the actual size the subject of the equation.

This is a good opportunity to mention careers that incorporate microscope technology such as pathology, nanotechnology, or criminal justice forensics.

Main activity

Observing cells under a microscope (25 mins)

Students observe animal (cheek) cells using a light microscope and draw diagrams of what they see. Ensure that students with visual impairments are paired with sighted students, who should be encouraged to support with descriptions as necessary. Students then answer the questions on the worksheet and then carry out magnification calculations.

Review and practice

Exam style question (10 mins)

Ask students to answer the following summary question from the Student Book: 'Evaluate the use of an electron microscope and a light microscope, giving one example where each type of microscope might be used.' Some students may require scaffolding to help answer the question. They can refer back to their notes here or use the Student Book.

B1.2 Animal and plant cells

AQA GCSE Biology specification points

4.1.1.2 Animal and plant cells

Smart Curriculum Narrative

How organisms work

Theme/subtheme

Cells/Types of cell

Disciplinary knowledge

- WS 1.2
- MS 1b, 1d, 3a

Required Practical 1

Microscopy

	Learning outcomes		
Learning objectives	**Developing** **Aiming for 3**	**Secure** **Aiming for 5**	**Extending** **Aiming for 8**
Describe the structures and functions of the main parts of cells	Name the main parts of cells	**Describe the structures and functions of the main parts of cells**	Explain how the main structures of cells are related to their functions
Compare the structure of plant and animal cells	Identify a plant and an animal cell from a diagram	**Compare the structure of plant and animal cells**	Compare and contrast the structure and functions of animal cells and plant cells
Use a light microscope to observe, draw, and label one example of animal cells and one example of plant cells, including a magnification scale on each	Identify the main parts of a light microscope and use it to observe animal and plant cells	**Use a light microscope to observe, draw, and label one example of animal cells and one example of plant cells, including a magnification scale on each**	Use a light microscope to observe, draw, and label a selection of animal and plant cells, including a magnification scale on each drawing. Compare the sizes of cells using units of length and standard form

Tier 2 vocabulary	Tier 3 vocabulary
animal cell, magnification, microscope, observe, plant cell	algae, cell membrane, cellulose, cell wall, chlorophyll, chloroplast, cytoplasm, mitochondria, nucleus, permanent vacuole, ribosome

Lesson plan

Reactivate knowledge

1 Name some parts of an animal cell.

2 Name the parts of a plant cell only.

3 What does resolution mean?

Common misconception

Students may think that *all* plant cells contain chloroplasts. Clarify that this is not the case. For example, root hair cells do not contain chloroplasts as they are underground, where there is no light to photosynthesise.

Answers: 1 Nucleus, cell membrane, cytoplasm *2* Cell wall, chloroplast, vacuole *3* The minimum distance between two objects when you can see them both clearly

Trigger interest

Label the cells (5 mins)

Show (and describe if required) unlabelled diagrams of plant and animal cells and ask students to work in small groups to name as many of the cell parts as they can. Reveal the answers and highlight the structures that are found only in a plant cell.

Exposition

Cell (5 mins)

Students use the Interactive: *Cells* to match the organelles of a cell with their function. They should be able to recall this information from KS3 but if students require additional support, prompt them and build confidence by providing images of cells. Students then identify if statements on the organelles of cells are true or false.

Address the misconception here to explain why not all cells contain chloroplasts.

Main activity

Microscopy (40 mins)

Students must make their way through the worksheet to complete the required practical. Here they make slides of plant and algal cells.

Students should make scientific drawings of their observations (this should be limited to drawing one to three cells per slide). Ensure that students with visual impairments are paired with sighted students, who should be encouraged to support with descriptions as necessary. Discuss why they cannot see all the structures in the cells, and link back to work done in B1.1 on magnification, resolution, and structures visible in the animal cheek cells.

Review and practice

Which cell? (5 mins)

Show the class micrographs of cells and ask them to identify the cells as plant or animal. Some students may need support with this so ask them to consider whether the cells have a chloroplast, vacuole, or cell wall.

B1.3 Eukaryotic and prokaryotic cells

AQA GCSE Biology specification points

4.1.1.1 Eukaryotes and prokaryotes

Smart Curriculum Narrative

How organisms work

Theme/subtheme

Cells/Types of cell

Disciplinary knowledge

- WS 1.2, 4.4
- MS 1b, 1d, 2a, 2h, 3a

Activity

Orders of magnitude

	Learning outcomes		
Learning objectives	**Developing Aiming for 3**	**Secure Aiming for 5**	**Extending Aiming for 8**
Describe the functions of the parts of a prokaryotic cell	Identify structures in prokaryotic cells	**Describe the functions of the parts of a prokaryotic cell**	Explain how the main structures of prokaryotic cells are related to their functions
Compare prokaryotic and eukaryotic cells	State that prokaryotic cells do not contain a nucleus and eukaryotic cells do	**Compare prokaryotic and eukaryotic cells**	Compare and contrast the structure of prokaryotic cells, animal cells, and plant cells
Use orders of magnitude to compare the sizes of organisms	Use orders of magnitude to correctly order objects according to size	**Use orders of magnitude to compare the sizes of organisms**	Perform calculations to work out orders of magnitude

Tier 2 vocabulary	Tier 3 vocabulary
bacteria, cell, cell membrane, cell wall, cytoplasm, DNA, *scale*	eukaryotic cell, flagellum, genetic material, nucleus, order of magnitude, prokaryotic cell, slime capsule

Lesson plan

Reactivate knowledge

1 Which is bigger, an animal cell or a plant cell?

2 Where is the genetic material stored in plant and animal cells?

3 Name a part of a plant cell that is not found in an animal cell.

Answers: *1* Plant cell *2* Nucleus *3* Chloroplast, vacuole, cell wall

Common misconception

Students may think viruses are prokaryotic organisms. Clarify that viruses are acellular, much smaller, and non-living.

Trigger interest

Eukaryotic and prokaryotic cells (5 mins)

Introduce to the class the idea that eukaryotic cells have a nucleus and organelles, whilst prokaryotic cells are single-celled organisms. Show (and describe if required) students a range of eukaryotic and prokaryotic cells and ask students to sort them accordingly. Students then sort some statements from the slides about eukaryotic and prokaryotic cells.

Exposition

Eukaryotic vs prokaryotic (20 mins)

Show (and describe if required) the class a diagram of a bacterial cell without labels. Introduce this as a bacterial or prokaryotic cell. Ask students how this cell is different from the plant and animal (eukaryotic) cells they studied in B1.2. Show (and describe if required) an image of a virus and an image of a prokaryote and ask them to consider how a virus is different from a prokaryote. Address the misconception that viruses are prokaryotic organisms.

Students copy the diagram from the Student Book and use information to label the parts. Ask them to highlight any features not found in the eukaryotic cells previously studied, and to write in the functions of these parts.

Main activity

Orders of magnitude (20 mins)

Students answer a series of questions using the Maths Skills: *Orders of magnitude* resource as a paired mini-whiteboard exercise. Students then independently complete the activity Working Scientifically: *Prefixes and orders of magnitude.*

Review and practice

Scale and size (10 mins)

Use the Activity: *Scale and size* to show students a series of images of different objects and organisms of different sizes, for example, a buckyball, a grain of sand, a virus, a human hair, a child, and so on. Provide students with a range of sizes, all in metres, and have them match each image to the correct size.

B1.4 Specialisation in animal cells

AQA GCSE Biology specification points

4.1.1.3 Cell specialisation

***Smart Curriculum* Narrative**

How organisms work

Theme/subtheme

Cells/Types of cell

Activity

Specialised cells exam questions

Learning objectives	Learning outcomes: Developing – Aiming for 3	Secure – Aiming for 5	Extending – Aiming for 8
Understand that cell differentiation in animals usually occurs when they are young and developing. As a cell differentiates it becomes specialised to carry out particular functions	State that as organisms develop, cells differentiate and change to form different types of cells	**Understand that cell differentiation in animals usually occurs when they are young and developing. As a cell differentiates it becomes specialised to carry out particular functions**	Describe the effect of differentiation and specialisation in animal cells and explain the importance of differentiation
Compare the structure of a specialised and a generalised animal cell	Identify specialised animal cells from diagrams	**Compare the structure of a specialised and a generalised animal cell**	Suggest the function of a specialised cell, familiar or unfamiliar, based on its structure
Describe the structural adaptations of three types of specialised animal cells	Name three types of specialised animal cells and describe their functions	**Describe the structural adaptations of three types of specialised animal cells**	Discuss how the structures of different examples of specialised animal cells are related to their functions within the organism

Tier 2 vocabulary	Tier 3 vocabulary
cell, *differentiation*, function, nucleus, *organ*, organism, organ system, specialisation, sub-cellular, *tissue*	axon, muscle cell, myelin sheath, nerve cell, sperm cell, palisade cell, root hair cell

Lesson plan

Reactivate knowledge

1 What is a specialised cell?

2 Name two types of specialised cells.

3 Give two adaptations of sperm cells.

Common misconception

Students may think that each specialised cell contains different DNA, causing them to be structurally different. Clarify to students that all body cells in an organism contain the same DNA, but as cells become specialised, each cell will express different parts of the DNA.

Answers: 1 A cell suited to carry out a particular job *2* Sperm cell/red blood cell *3* Long tail, lots of mitochondria

Trigger interest

Which cell am I? (5 mins)

Give students a description of the structure or function of different specialised cells and ask them to identify which specialised cell it is. Students can be challenged here and asked where the cell can actually be found. (Description: *this specialised cell has a biconcave shape*. Answer: *red blood cell, found in the blood of animals*.)

Extend by asking if students can remember the functions or adaptations of these cells.

Exposition

Types of specialised cells (20 mins)

Provide students with information, or use the Student Book, so they can label a specialised animal cell. Students then complete the table in the activity sheet to describe the structure of specialised animal cells and explain how they are adapted to their function.

Main activity

Specialised cells exam questions (20 mins)

Students apply their knowledge of specialised cells to complete the exam style questions activity sheet. Students with prior lower attainment will be able to answer the first few questions but may struggle with the final question. Here, provide some scaffolding or guidance of how to answer compare and contrast questions. Student can peer-mark their answers.

Review and practice

Specialisation in animal cells (5 mins)

Students match specialised animal cells to their structures and functions. Use the Interactive: *Specialisation in animal cells* resource.

True or false? (5 mins)

Ask students to write 'true' on one side of their mini-whiteboard and 'false' on the other. Read aloud the following statements and ask students to hold up their mini-whiteboards to indicate whether they think each statement is true or false.

- Cell differentiation in animals happen when they are young. (True)
- Sperm cells have myelin sheath around the tail. (False)
- Muscle cells have many mitochondria to transfer energy. (True)

B1.5 Specialisation in plant cells

AQA GCSE Biology specification points

4.1.1.3 Cell specialisation

Smart Curriculum Narrative

How organisms work

Theme/subtheme

Cells/Types of cell

Disciplinary knowledge

- WS 1.2, 3.1, 4.4

Practical

Observing root hair cells

	Learning outcomes		
Learning objectives	Developing Aiming for 3	Secure Aiming for 5	Extending Aiming for 8
Understand that many types of plant cells continue to differentiate throughout the life of the plant	State that cells differentiate and form specialised cells in plants as well as animals	**Understand that many types of plant cells continue to differentiate throughout the life of the plant**	Discuss the importance to a plant of cell differentiation continuing throughout life
Compare the structure of a specialised and a generalised plant cell	Identify specialised plant cells from diagrams	**Compare the structure of a specialised and a generalised plant cell**	Suggest the function of a specialised plant cell, familiar or unfamiliar, based on its structure
Describe the structural adaptations of three types of specialised plant cells	Name three types of specialised plant cells and describe their functions	**Describe the structural adaptations of three types of specialised plant cells**	Discuss how the structures seen in different examples of specialised plant cells are related to their functions within the organism
Observe, draw, and label a root hair cell using a light microscope, including a magnification scale	Observe and draw a root hair cell using a light microscope	**Observe, draw, and label a root hair cell using a light microscope, including a magnification scale**	Observe, draw, and label a root hair cell using a light microscope, including a magnification scale, and calculate the length of the root hair

Tier 2 vocabulary	Tier 3 vocabulary
cell, *organ*, organism, organ system, specialised, *tissue*	companion cell, phloem, phloem vessel, photosynthetic cell, root hair cell, sieve plate, xylem

Lesson plan

Reactivate knowledge

1 Name an adaptation of a leaf cell.

2 Name an adaptation of a root hair cell.

3 Which chemical reaction takes place in the mitochondria of cells?

Answers: 1 Lots of chloroplasts *2* Large surface area *3* Respiration

Common misconception

Students can think that *all* plant cells contain chloroplasts. Clarify that not all plant cells do. For example, root hair cells do not contain chloroplasts as they are underground, where there is no light to photosynthesise.

Trigger interest

Plant cell (5 mins)

Show (and describe if required) a diagram of a leaf cross-section and explain what it is showing. Ask students to identify a cell that is similar to the generalised plant cell previously studied (palisade cell). Discuss how the other cells have different structures and ask students to suggest why this might be. This is also an opportunity to show students a cross-section of a xerophyte as an example of how the structure can differ depending on a plant's needs.

Exposition

Plant cell differentiation (5 mins)

Show students an image of a spider plant where budding of 'mini' spider plants can be seen, or an actual example of such a plant, if available. Use this to begin a discussion of how plant cells can continue to become specialised throughout the life of the plant.

Specialised plant cells (20 mins)

Show (and describe if required) students images of a root hair cell, xylem, phloem, and palisade cell. Ask them to label any cell structures they recognise. (They should all be able to identify the cell membrane, cell wall, cytoplasm, nucleus, and vacuole. Some may also be able to identify the chloroplasts and mitochondria if they are visible in the images you provide.)

Inform the students of the function of each of these specialised cells. In pairs, they should discuss how the cell structures and cell shapes would help the cells perform their functions. If students need prompts, suggest they consider the quantity of mitochondria in the cell and the surface area of the cell.

Main activity

Observing root hair cells (20 mins)

Students study prepared slides of the root hair cells of mung beans using a light microscope. They should clearly be able to identify organelles but if students are struggling, they could be shown an example image with labels. They draw scientific drawings based on their observations and answer questions about the adaptations of these cells. If there is more time, students can make their own slides. They place the roots of sprouting mung bean seeds on slides and dye with iodine solution.

Review and practice

What am I? (5 mins)

Write the names of different specialised animal and plant cells on cards. Ask students to work in groups and give each student a card that they must hold in front of them without looking at it. Students then ask questions to determine which cell they are.

For example, students could ask:

- Do I have a cell wall?
- Do I have chloroplasts?
- Do I have a nucleus?
- Am I biconcave?

B1.6 Diffusion

AQA GCSE Biology specification points

4.1.3.1 Diffusion

Smart Curriculum Narrative

How organisms work

Theme/subtheme

Cells/Transport in cells

Disciplinary knowledge

- WS 1.2, 1.5
- MS 1c

Practical

Investigating diffusion

	Learning outcomes		
Learning objectives	**Developing Aiming for 3**	**Secure Aiming for 5**	**Extending Aiming for 8**
Describe diffusion as a net movement of particles in a solution or a gas from an area of higher concentration to an area of lower concentration	State that diffusion is the spreading out of the particles of any substance in solution, or particles of a gas	**Describe diffusion as a net movement of particles in a solution or a gas from an area of higher concentration to an area of lower concentration**	Describe a model of diffusion in terms of the movement of particles in solution or in a gas down a concentration gradient
Describe how concentration gradients and diffusion are involved in gas exchange of oxygen and carbon dioxide across cell membranes	State that substances may move into and out of cells across the cell membranes by diffusion. Oxygen moves into the cells and the waste products carbon dioxide and urea move out of the cells by diffusion	**Describe how concentration gradients and diffusion are involved in gas exchange of oxygen and carbon dioxide across cell membranes**	Explain the role of diffusion in gas exchange and in the removal of the waste product urea from cells into the blood plasma for excretion in the kidney
Explain why surface area affects the rate of diffusion	List the factors that affect the rate of diffusion	**Explain why surface area affects the rate of diffusion**	Explain how temperature and concentration gradient affect the rate of diffusion
Write a hypothesis to explain an example of diffusion using scientific knowledge	Write a simple hypothesis to explain an example of diffusion	**Write a hypothesis to explain an example of diffusion using scientific knowledge**	Write a hypothesis to explain an example of diffusion detailed scientific knowledge, and explain how it could be tested

Tier 2 vocabulary	Tier 3 vocabulary
concentration gradient, particle, surface area, temperature, transport	diffusion, rate of diffusion

Lesson plan

Reactivate knowledge

1 Name two substances that move into cells.
2 Name two substances that move out of cells.
3 Write down the meaning of diffusion.

Common misconception

Students may get confused about the difference between 'a high and low concentration of particles', and 'a concentrated and a dilute solution'. Clarify that the more concentrated a solution is, the more solute is dissolved in the solvent.

Answers: 1 Glucose and oxygen *2* Carbon dioxide and urea *3* The movement of substances from an area of high to low concentration

Trigger interest

Perfume spray (5 mins)

Spray perfume/deodorant in one corner of the lab. Ask students to raise their hands, or say 'yes' when they can smell it.

Get students to suggest why students closest to where the perfume was sprayed smelt it first, and those furthest away smelt it last.

Exposition

Diffusion (15 mins)

Give students the definition of diffusion and get them to work in pairs to consider what factors would make diffusion occur faster. To support, tell them to consider kinetic theory or states of matter and what factors they know that would make particles in a liquid and a gas move more rapidly. Before feeding back answers, address the misconception then discuss with them why temperature, surface area, and concentration affect the rate of diffusion.

Examples of diffusion (10 mins)

Ask students to list any examples in plants or the human body where they think diffusion is occurring.

The students need be aware of gas exchange in plants and animals but are unlikely to know that urea diffuses out of the blood into the kidney, so use the slides to explain the image of the blood containing a large concentration of urea compared to a lower concentration of urea in the filtrate in the kidney.

Main activity

Investigating diffusion (20 mins)

Students complete the practical and answer the questions.

To extend, tell students that smelling food that is being cooked is an example of diffusion. Students can create a comic strip to explain how the smell of food cooking reaches you as you walk past the school kitchens.

Students should include key terms, such as diffusion, particles, concentration gradient, temperature, and kinetic energy, within their comic strip explanations.

Ask students to also explain why we can smell hot food more than cold food.

Review and practice

Intestinal cell (5 mins)

Show students a diagram of a microvillus. Ask them to discuss in pairs how it is adapted for efficient diffusion. Students must mention, large surface area, good blood supply, short diffusion pathway.

B1.7 Osmosis

AQA GCSE Biology specification points

4.1.3.2 Osmosis

Smart Curriculum Narrative

How organisms work

Theme/subtheme

Cells/Transport in cells

Disciplinary knowledge

- WS 1.2

Activity

Keeping things constant

	Learning outcomes		
Learning objectives	**Developing Aiming for 3**	**Secure Aiming for 5**	**Extending Aiming for 8**
State the differences between osmosis and diffusion	State that osmosis is the diffusion of water from a dilute solution to a concentrated solution through a partially permeable membrane	**State the differences between osmosis and diffusion**	Describe a model used to demonstrate osmosis in a cell and explain how it works
Describe the effect of osmosis on animal cells if the external solution is more dilute or more concentrated than the cell contents	State that if animal cells lose or gain too much water by osmosis they can stop working properly	**Describe the effect of osmosis on animal cells if the external solution is more dilute or more concentrated than the cell contents**	Use the terms isotonic, hypotonic, or hypertonic to explain the movement of water across a cell membrane and use this to explain the importance of maintaining relatively constant internal conditions in living organisms
Write a prediction about the effect of a given change in circumstances on the cells of the body using scientific knowledge of osmosis	Write a prediction about the effect of a given change in circumstances on the cells of the body	**Write a prediction about the effect of a given change in circumstances on the cells of the body using scientific knowledge of osmosis**	Using scientific knowledge of osmosis and scientific terms write a prediction about the effect of a given change in circumstances on the cells of the body

Tier 2 vocabulary	Tier 3 vocabulary
concentrated, dilute, solute, *solution*, transport	hypertonic, hypotonic, isotonic, osmosis

Lesson plan

Reactivate knowledge

1 What is diffusion?

2 Give an example of diffusion occurring during breathing.

3 Name two factors that affect the rate of diffusion.

Answers: 1 Movement of substances *2* Oxygen and carbon dioxide *3* Temperature and surface area

Common misconception

Students may think that the solute is moving in osmosis, rather than the water. Emphasise that osmosis is only the movement of water.

Trigger interest

Which way? (5 mins)

Draw a diagram of an animal cell on the board and draw dots inside to show water particles. Draw a higher concentration of dots outside the cell. Ask students to discuss in pairs what they think will happen to the movement of water. Go through their ideas and introduce osmosis as a type of diffusion where water moves across a partially permeable membrane.

Exposition

Investigating osmosis (20 mins)

Set up a demonstration as shown in the Student Book (Figure 2). Explain that Visking tubing is partially permeable, like the cell membrane, and explain what this means. Draw diagrams on the board using different coloured dots to show the concentration of solutions inside and outside the Visking tubing bags. Ask students to write predictions of what they think will happen and why. The demonstration will continue throughout the lesson and re-visited at the end.

Main activity

Keeping things constant (20 mins)

Tell the class that the concentration of water in the blood is kept constant. Ask them to discuss in pairs why this is important and feed back to the rest of the class.

Give students two scenarios:

1) A very hot day and someone who has not drunk much water.

2) A cold day and someone who has drunk a lot of water very quickly.

Ask students to write an explanation of why these scenarios could both be harmful. Give students some prompts to support them with this. What might happen to the volume of water in your blood if you do not drink on a hot day? (Answer: *it would decrease*), What might happen to a cell if you are dehydrated? (Answer: *the cell would shrivel*), What might happen to your blood pressure if you drink too much water? (Answer: *it would increase*), What might happen to a cell if it contains too much water? (Answer: *it might burst*).

Review and practice

Model cells with Visking tubing: results (10 mins)

Return to the model cells set up earlier.

Show the class the results and ask them to compare these results to their predictions. Ask students to state whether the outside solutions were isotonic, hypertonic, or hypotonic, and to explain what happened in terms of osmosis. Address the misconception at this point and clarify that it was the water moving across the membrane of the Visking tubing, and not the solute.

B1.8 Osmosis in plants

AQA GCSE Biology specification points

4.1.3.2 Osmosis

Smart Curriculum Narrative

How organisms work

Theme/subtheme

Cells/Transport in cells

Disciplinary knowledge

- WS 1.2
- MS 1a, 1c, 4a, 4b, 4c, 4d

Required Practical 3

Osmosis

	Learning outcomes		
Learning objectives	**Developing Aiming for 3**	**Secure Aiming for 5**	**Extending Aiming for 8**
Use osmosis to explain the effect on plant cells of changing the concentration of a salt or sugar solution surrounding them	State that plants rely on osmosis to support their stems and leaves by keeping the cells turgid (firm). If a plant loses too much water from its cells, then they become flaccid (soft) and the plant wilts	**Use osmosis to explain the effect on plant cells of changing the concentration of a salt or sugar solution surrounding them**	Explain the mechanisms that lead to turgid or flaccid plant cells and plasmolysis
Write a simple plan to investigate the effect of different concentrations of salt or sugar solutions on the mass of a sample of plant tissue	With support, write a simple method to investigate osmosis in plant tissue	**Write a simple plan to investigate the effect of different concentrations of salt or sugar solutions on the mass of a sample of plant tissue**	Independently write a detailed plan for an investigation into the effect of a range of concentrations of salt or sugar solutions on the mass of plant tissue, and use your scientific knowledge of osmosis to predict the outcome of your investigation
Calculate percentage change and use this to plot a line graph with negative numbers and draw a line of best fit	Use given data to plot a suitable graph with some support	**Calculate percentage change and use this to plot a line graph with negative numbers and draw a line of best fit**	Use a line graph to estimate the concentration of the cell sap solution inside a plant cell

Tier 2 vocabulary	Tier 3 vocabulary
cell, concentration gradient, movement, plant tissue, rigid, solute, solvent, transport, water	flaccid, hypertonic, hypotonic, isotonic, osmosis, partially permeable membrane, plasmolysis, turgid, turgor

Lesson plan

Reactivate knowledge

Common misconception

Students may think that the solute is moving, rather than the water. Emphasise that osmosis is only the movement of water.

1 Name the part of a cell that controls the movement of substances.
2 What part of a plant cell helps to keep its shape?
3 Name the process by which water moves into a root hair cell.

Answers: 1 Cell membrane *2* Vacuole *3* Osmosis

Trigger interest

Wilting plant (5 mins)

Show the class a wilted plant if available, or an image of one. Ask students to discuss in pairs what they think has happened to the plant, why it has happened, and how they could restore the plant. As a prompt, ask students to consider what a plant needs to live and to photosynthesise. Ask students to consider diffusion, osmosis, and water uptake from root hair cells.

Exposition

Investigating osmosis in plant cells (10 mins)

Supply students with a diagram of two plant cells showing a clear permanent vacuole. Explain that the vacuole is full of sap that contains a low concentration of sugars and minerals. Ask students to modify the first diagram to show what would happen to the cell if it was placed in a concentrated sugar solution, and the second diagram to show what would happen if the cell was placed in water.

After checking their answers, ask students to use information from the Student Book to add the terms turgor, turgid, plasmolysis, and flaccid to their diagrams.

Main activity

Osmosis (30 mins)

Students then plan their own investigation to find out how salt or sugar solutions affect plant tissue.

Allow them to do some initial research to find examples of suitable methods (including use of the Student Book or research online). Students need to plan to collect quantitative data. They should write down their independent, dependent, and control variables before writing a simple plan. You may need to remind students of the meanings of independent, dependent, and control variables to help them apply this knowledge to their plan.

A preliminary test could be done so students can check the suitability of their method, however this is not necessary. Students then gather and analyse their results. There should be time at the end for students to calculate percentage change and plot a line graph with a line of best fit. They write their conclusion and evaluate their method. When discussing the conclusion, address the misconception and clarify that it is the water that has moved, not the solute.

Review and practice

Osmosis in plants (5 mins)

Students use the Interactive: *Osmosis in plants* to check their understanding of this topic by completing a paragraph about osmosis in plants.

Water the plants (5 mins)

Ask students to write a short explanation about why house plants need to be watered, using what they have learnt about osmosis in plant cells.

B1.9 Active transport

AQA GCSE Biology specification points

4.1.3.3 Active transport

Smart Curriculum Narrative

How organisms work

Theme/subtheme

Cells/Transport in cells

Disciplinary knowledge

- WS 1.2, 4.4
- MS 1b, 1d, 2a, 2h, 3a

Activity

Active transport models/Exchanging of materials

Learning objectives	Learning outcomes: Developing Aiming for 3	Learning outcomes: Secure Aiming for 5	Learning outcomes: Extending Aiming for 8
Describe active transport and explain why it requires energy from cellular respiration	Define active transport as the movement of a substance against a concentration gradient using energy	**Describe active transport and explain why it requires energy from cellular respiration**	Explain the differences between diffusion, osmosis, and active transport
Explain why active transport is important for living organisms	Identify one site of active transport in plants, and one site of active transport in animals	**Explain why active transport is important for living organisms**	Suggest two structural adaptations often seen in cells carrying out active transport and explain how these adaptations support active transport

Tier 2 vocabulary	Tier 3 vocabulary
concentration gradient, energy, ion, mineral, respiration, transport	active transport

Lesson plan

Reactivate knowledge

1 What substance moves across a membrane in osmosis?

2 What is the role of the cell membrane?

3 Name the plant cell which helps absorb water and minerals from the soil.

Common misconception

Students can get confused about the difference between 'a high and a low concentration of particles' and 'a concentrated and a dilute solution'. Clarify the concept of solutions and solutes, and that the more concentrated a solution is, the more solute is dissolved in the solvent.

Answers: 1 Water *2* To control what moves in and out of the cell *3* Root hair cell

Trigger interest

Absorption of mineral ions (10 mins)

Show (and describe if required) an unlabelled diagram showing active transport in plant roots and ask students why mineral ions cannot move from the soil into the roots via diffusion. Reveal the labels and explain that mineral ions must move from a low to a high concentration (against the concentration gradient). A process called active transport is therefore used, which requires energy.

Exposition

Active transport (10 mins)

Ask the students to explain why they think it is important that sugar molecules are absorbed by active transport in the gut. They can use information from the Student Book to support them if needed.

To support the students further, you could ask them to consider the following questions:

- What is glucose used for in the body? (Answer: *respiration*.)
- What would happen if sugar was only absorbed by diffusion? (Answer: *when equilibrium was reached, no more glucose would be absorbed from the gut into the blood, and excess glucose would be lost in faeces*.)
- Why is it therefore important that all glucose is absorbed? (Answer: *so that respiration can occur to provide energy*.)

Main activity

Active transport models (15 mins)

Students work in small groups to design a model of active transport. Ask for a couple of groups to demonstrate their model, whilst the rest of the students evaluate its effectiveness and comment on possible improvements.

Exchanging of materials (10 mins)

Give students a series of statements on osmosis, diffusion, and active transport. Students can match up the statements and the types of transport.

Review and practice

Compare and contrast (10 mins)

Students answer the exam style question: 'Give an example where each of the different types of transport occur in a plant.'

Answer:

- Osmosis occurs in the root hair cells to absorb water from the soil.
- Active transport occurs in the root hair cells to absorb mineral ions from the soil.
- Carbon dioxide diffuses into plant leaves.
- Oxygen diffuses out of plant leaves.

B1.10 Exchanging materials

AQA GCSE Biology specification points

4.1.3.1 Diffusion

Smart Curriculum Narrative

How organisms work

Theme/subtheme

Cells/Types of cell

Disciplinary knowledge

- WS 1.2
- MS 1c, 5c

Practical

Investigating surface area to volume ratios

Learning objectives	Learning outcomes: Developing Aiming for 3	Learning outcomes: Secure Aiming for 5	Learning outcomes: Extending Aiming for 8
Calculate the surface area to volume ratio of a cube and compare different sized cubes	Calculate the surface area to volume ratio of a cube for given values of surface area and volume	**Calculate the surface area to volume ratio of a cube and compare different sized cubes**	Calculate and compare surface area to volume ratios of different sized cubes and cylinders
Use ideas about surface area to volume ratio to describe why multicellular organisms need exchange surfaces	State that a single-celled organism has a relatively large surface area to volume ratio	**Use ideas about surface area to volume ratio to describe why multicellular organisms need exchange surfaces**	Use ideas about surface area to volume ratios to explain the need for transport systems in some multicellular organisms
Describe how the structure of the exchange surfaces in mammals, fish, and plants are adapted for their functions	Name the following exchange systems: small intestine and lungs in mammals, the gills in fish, and the roots and leaves in plants. State what is exchanged at each surface and why	**Describe how the structure of the exchange surfaces in mammals, fish, and plants are adapted for their functions**	Explain how the adaptations of the exchange surfaces in mammals, fish, plants, and other organisms increase their effectiveness

Tier 2 vocabulary	Tier 3 vocabulary
concentration gradient, diffusion, fish gill, organism, ratio, specialised cell	alveolus, filament, gas exchange, multicellular organism, operculum, rich blood supply, single-celled, stomata, surface area, ventilation, *volume*

Lesson plan

Reactivate knowledge

1 What is a unicellular organism?

2 What is a multicellular organism?

3 Name three substances that move in and out of cells.

Common misconception

Students may think that the larger an organism is, the larger the surface area: volume ratio would be.

Answers: 1 Organism made of one cell *2* Organism made of more than one cell *3* Water/oxygen/waste products

Trigger interest

Exchange of materials (5 mins)

Students match up the mixed statements from the slides about surface area to volume ratio and about factors affecting the effectiveness of an exchange surface.

Exposition

Types of exchange surface (5 mins)

Provide students with an image of each of the structures listed and ask them to write what the three structures have in common and how this is an adaptation to maximise exchange of substances. (Answer: *a large surface area to speed up diffusion/osmosis/active transport*.)

Discuss why these structures are important in multicellular organisms and explain that as an organism gets bigger, the distance between the surface and the centre of the organism increases. This means that simple diffusion is not enough to exchange materials.

Write the following exchange surfaces and answers on the board and make sure they are mixed up. Ask students to match the names of some exchange surfaces in animals and plants with their functions. For example:

- Root hair cell – absorption of water and mineral ions
- Microvilli – absorption of digested food
- Leaves – site of gas exchange

Main activity

Investigating surface area to volume ratios (40 mins)

Students calculate surface area to volume ratios for different 3D shapes (cubes, cylinders, and spheres). Modelling clay could be provided for students to make the shapes. Address the common misconception here, using the maths to show that larger objects have a smaller surface area to volume ratio.

Students decide on the best shape for a single-celled organism and explain why this shape is best.

Review and practice

Surface area to volume ratio (5 mins)

Students calculate the difference in total surface area of one 10 cm × 10 cm × 10 cm cube versus ten 1 cm × 1 cm × 1 cm cubes.

B1 Cell structure and transport

Introduction to checkpoint intervention

The checkpoint intervention provides suggestions for a lesson to follow up the *B1 Cell structure and transport* checkpoint assessment. Depending on the outcomes of the assessment, these suggestions can be used to move a class towards secure, to target specific learning outcomes, or to consolidate knowledge for students achieving secure outcomes. Use students' outcomes from the checkpoint assessment to address general misconceptions from the content covered. Be prepared to re-cover content as required; three key concepts from the chapter are covered in more detail below.

Checkpoint secure learning outcomes

- Compare plant and animal cells.
- Compare prokaryotic and eukaryotic cells.
- Compare the structure of a specialised and a generalised plant cell.
- Explain why surface area affects the rate of diffusion.
- Use ideas about surface area to volume ratio to describe why multicellular organisms need exchange surfaces.
- Write a suitable plan to investigate the effect of salt or sugar solutions on plant tissue.

Student reflection

Encourage students to reflect on whether there were any checkpoint questions they found difficult or straightforward as well as their preparation for the checkpoint assessment, including how they revised, what they revised, the time spent on revision, and what areas they could improve upon. For more strategies, see the Metacognition spread.

Intervention activity: Compare plant and animal cells.

Ask students to draw an animal and a plant cell from memory. Students should then add labels to state the name of each structure and describe the function of each structure. Use this as an opportunity to test the recall of the theory, whilst also emphasising the difference between the command word 'state' and 'describe'. To support students, provide them with the list of structures to draw and label (nucleus, cell membrane, cytoplasm, ribosome, mitochondria, cell wall, chloroplast, vacuole).

Students then use their diagram to answer the exam questions. 'Compare and contrast an animal and a plant cell' (4 marks)

Answer:

Similarities (max 2 marks)

Both have:

- a nucleus
- cytoplasm
- a cell membrane
- ribosomes
- mitochondria

Differences (max 2 marks)

- Plant cells have a chloroplast, whereas animal cells do not.
- Plant cells have a permanent vacuole, whereas animal cells do not.
- Plant cells have a cell wall, whereas animal cells do not.

Intervention activity: Compare the structure of a specialised and a generalised plant cell.

Show (and describe if required) pictures of a general plant cell, a root hair cell, and a palisade cell on the board and discuss the similarities and differences of the structures. They should be able to identify that they contain many of the same organelles, but the quantity may vary, and the shape of the cells will vary.

For each difference, ask students to suggest a reason for this difference in the specialised cell.

Intervention activity: Explain why surface area affects the rate of diffusion.

Use a diagram of the root hair cell and explain the importance of surface area. Ask students to use the Student Book to find the answers to the following questions:

- What do root hair cells absorb from the soil and by which types of transport? (Answer: Root hair cells absorb water by osmosis and minerals by active transport.)
- Explain how the shape of the root hair cell increases the rate of osmosis. (Answer: A larger surface area results in an increased rate of water absorption.)
- List any other specialised cells that have a large surface area to increase the rate of diffusion. (Possible answers: Red blood cells have a biconcave shape and microvilli are long and thin.)

B2 Cell division

B2.1 Cell division

AQA GCSE Biology specification points

4.1.2.1 Chromosomes

4.1.2.2 Mitosis and the cell cycle

***Smart Curriculum* Narrative**

How organisms work

Theme/subtheme

Cells/cell division

Disciplinary knowledge

- WS 1.2

Activity

Observing mitosis

	Learning outcomes		
Learning objectives	**Developing Aiming for 3**	**Secure Aiming for 5**	**Extending Aiming for 8**
Describe the structure of chromosomes as made up of DNA and containing a large number of genes	State that the nucleus of a cell contains chromosomes containing genes	**Describe the structure of chromosomes as made up of DNA and containing a large number of genes**	Explain genes as small sections of DNA molecules
Describe the inheritance of the chromosome pairs from the male and female parents	State that human body cells contain 46 chromosomes arranged in 23 pairs and that the gametes contain only 23 chromosomes	**Describe the inheritance of the chromosome pairs from the male and female parents**	Compare the chromosome number in normal body cells and gametes and explain the difference
Describe the stages of the cell cycle, including the process of mitosis. Describe how this produces two identical cells and identify when cell division occurs in the life of an organism	Give a simple description of the cell cycle, identify mitosis as a stage in cell division, and state the importance of cell division to an organism	**Describe the stages of the cell cycle, including the process of mitosis. Describe how this produces two identical cells and identify when cell division occurs in the life of an organism**	Give a detailed description of the cell cycle, and explain the importance of each stage in the production of two identical cells
Use key scientific terms to describe the process of mitosis	Define the following key scientific terms – mitosis, chromosomes, gene, gametes	**Use key scientific terms to describe the process of mitosis**	Recognise and describe situations in given contexts where mitosis is occurring and explain its importance
Describe each gene as containing the information controlling a characteristic or part of a characteristic	State that genes carry the genetic information needed to make a new cell or organism	**Describe each gene as containing the information controlling a characteristic or part of a characteristic**	Explain how most characteristics are inherited through many different genes

Tier 2 vocabulary	Tier 3 vocabulary
cell division, development, genetic material, growth, identical, nucleus, pair, repair, replicate	cell cycle, chromosome, daughter cell, mitosis

Lesson plan

Reactivate knowledge

1 What are the basic unit of tissues and organs in animals and plants?
2 What is found in the nucleus of a cell?
3 During early pregnancy, what happens to the cells in the embryo?

Answers: 1 Cells *2* Genetic material *3* Divide and specialise

Common misconception

Students may think that a zygote divides by meiosis. Clarify that the zygote divides by mitosis and then the cells differentiate to become specialised.

Trigger interest

Cell key words (10 mins)

Show (and describe if necessary) images to represent a nucleus, gene, DNA, and chromosomes and ask students to match up each image with the correct key term to review their understanding from KS3.

Exposition

Chromosomes (15 mins)

Show an image of a chromosome for students to copy into their notes. Students can label the image as a chromosome made up of DNA. Students can then label sections of this chromosome as genes.

Provide students with micrograph images and drawings of the stages of mitosis and ask them to match them up. To extend, students could use the images to explain why the parent and daughter cells are genetically identical and diploid.

Main activity

Observing mitosis (20 mins)

Show (and describe if necessary) a diagram of the cell cycle. Ask students to estimate the percentage of time involved in each of the three stages, calculate how long the interphase stage would take if the whole cycle took 24 hours, and suggest why the interphase stage is the longest.

Students then complete the practical, where they use light microscopes to observe prepared slides showing mitosis. Ask them to spot cells at different stages and draw examples using Figure 3 in the Student Book to help them.

Review and practice

Mitosis (10 mins)

Students use the Interactive: *Mitosis* to match the key words from this lesson to their meanings and then carry out a cloze exercise about mitosis.

B2.2 Growth and differentiation

AQA GCSE Biology specification points

4.1.1.4 Cell differentiation

4.1.2.3 Stem cells

***Smart Curriculum* Narrative**

How organisms work

Theme/subtheme

Cells/Cell division

Practical

Cauliflower cloning

Learning objectives	Learning outcomes: Developing Aiming for 3	Learning outcomes: Secure Aiming for 5	Learning outcomes: Extending Aiming for 8
Describe the function of stem cells in embryos, in adult animals, and in plants	Define a stem cell	**Describe the function of stem cells in embryos, adult animals, and plants**	Explain the difference in differentiation patterns seen in animals and plants
Describe the production of a plant cell clone	Define the terms cloning and a clone	**Describe the production of a plant cell clone**	Compare and contrast cloning animals and plants

Tier 2 vocabulary	Tier 3 vocabulary
animal cell, development, differentiation, *division*, elongation, plant cell, root, *shoot*	adult stem cell, cloning, embryonic stem cell, mitosis, specialised cell, stem cell

Lesson plan

Reactivate knowledge

1 What is a specialised cell?

2 During early pregnancy, what happens to the cells in the embryo?

3 What is the process for the growth and repair of cells?

Common misconception

Students may think that each specialised cell contains different DNA, structurally different. Clarify that all body cells in an organism contain the same DNA, but as cells become specialised, each cell will express different parts (genes) of the DNA.

Answers: 1 A cell which is suited to carry out a particular job in an animal or plant *2* Divide and specialise *3* Mitosis

Trigger interest

Human life (10 mins)

Show (and describe if necessary) an image of an embryo containing a few cells and an image of a person in their eighties. Ask students to design a flow chart to show what happens to the cells of a person as they grow older, starting from when a sperm and egg fuse and ending with an adult. Students could be provided with the following terms to help them: fertilisation, mitosis, and differentiation.

Discuss students' ideas and explain that in development from an embryo to an adult, cell division is used for growth (increase in size) as well as for repair and replacement. In embryos, cells divide by mitosis to make identical copies. These cells later differentiate into specialised cells. Ask students to name two specialised cells found in an animal. In adults, it is mainly just used for repair and replacement. Introduce the word growth to mean a permanent increase in size.

Exposition

Differentiation (10 mins)

Introduce students to the concept of differentiation by asking them to consider the following question: 'Specialised cells are different in structure, but they all contain the same genetic information. How is it possible for them to look so different?'

Give students the definition of differentiation and let them know that this is how cells become specialised. Only parts of the DNA are expressed (used), and that is how cells that contain the same DNA can look different. This will address the misconception that different specialised cells contain different DNA.

Main activity

Cauliflower cloning (30 mins)

Tell the class that plant cells contain undifferentiated cells in tissues called meristems, and that this makes it possible to make plant clones from adult plants by using cuttings.

Students work in pairs to follow the practical sheet and clone a new cauliflower plant using these cells. They must pay careful attention to working accurately and safely to ensure cloning is successful. Ask students to consider why it might be difficult to produce successful plant clones. Prompt them to consider aseptic technique and the impact pathogens may have.

Students then answer questions about differentiation and cloning in plants. Talk through the answers with the students so that they can self-mark their answers.

Review and practice

Differentiation (5 mins)

Ask students to return to the flow chart they designed at the start of the lesson and add in information about where differentiation occurs and what it is used for. Students might consider differentiation of cells in a foetus to form different specialised cells, which in turn form tissues and organs.

B2.3 Stem cells

AQA GCSE Biology specification points

4.1.2.3 Stem cells

Smart Curriculum Narrative

How organisms work

Theme/subtheme

Cells/Stem cells

Disciplinary knowledge

- WS 1.3

Activity

Exam style questions

	Learning outcomes		
Learning objectives	**Developing** **Aiming for 3**	**Secure** **Aiming for 5**	**Extending** **Aiming for 8**
Describe differences between embryonic and adult stem cells	Define a stem cell	**Describe differences between embryonic and adult stem cells**	Suggest why embryonic stem cells are potentially more useful for treating medical conditions than adult stem cells
Describe how plant clones are used	State that plant stem cells can be used to create plant clones	**Describe how plant clones are used**	Explain the importance of plant clones in conservation, scientific research, and agriculture
Describe how stem cells may be used to treat different medical conditions	State two medical conditions which may be treated using stem cell treatments	**Describe how stem cells may be used to treat different medical conditions**	Explain in detail how stem cells may be used to treat medical conditions

Tier 2 vocabulary	Tier 3 vocabulary
clone, *differentiate*, stem cell, undifferentiated	adult stem cell, embryonic stem cell, meristem, zygote

Lesson plan

Reactivate knowledge

1 What is meant by the term zygote?

2 During early pregnancy, what happens to the cells in the embryo?

3 What is the main difference between differentiation in animals and plants?

Common misconception

Students may think that an embryo divides by meiosis. Clarify that the embryo divides by mitosis and then the cells differentiate to become specialised.

Answers: 1 Fertilised egg cell *2* Divide and specialise *3* Animals: most takes place early in life, Plants: continues throughout life

Trigger interest

Differentiation (15 mins)

Ask students to write down and recall a biological definition of the term differentiation. An image showing a single stem cell differentiating into various specialised cells can be displayed to support students with this.

Gather ideas to come to an agreed meaning and write this on the board.

Then show images of heart, muscle, skin, brain, and bones and ask students what they have in common. Students will likely provide many correct answers (e.g., all contain cells, all are found in humans, etc.). Let students know that they also have in common that scientists have used stem cells to create all of these tissues/organs in the lab. Ask students if they have heard of stem cells on the news or social media. For example, they may have heard of the story of Dolly the sheep.

Exposition

Plant copies (15 mins)

Show students an image of a runner on a spider plant or strawberry plant. Ask students to discuss in pairs why a new clone of the plant can grow off the existing plant, and to suggest reasons why people might want to clone plants. Gather together their ideas as a class.

Students might consider:

- The runner is a clone of the parent plant.
- The clone of the adult plant is made from stem cells.
- Cloning plants is useful as it provides new plants for free.
- Cloning plants enables identical plants to be produced, which is useful if the parent plant had advantageous features.
- Cloning is faster, as you are not starting from seed.

Embryonic vs adult stem cells (15 mins)

Ask students to write out a comparison table of the properties of embryonic stem cells, adult stem cells, and meristems.

Main activity

Exam style questions (20 mins)

To check the students' understanding of the key terms and concepts, they should complete the activity.

The exam questions should take eight - ten minutes to complete, and students can then self-mark their answers. Some students may need support with this so prompts or use of the Student Book can be provided.

Review and practice

Stem cells (5 mins)

Use the Interactive: *Stem cells*, where students match the correct words to statements and then select words to fill in the gaps in a paragraph about stem cells.

This will be a good opportunity for students to ask any questions and remember definitions.

B2.4 Stem cell dilemmas

AQA GCSE Biology specification points

4.1.2.3 Stem cells

***Smart Curriculum* Narrative**

How organisms work

Theme/subtheme

Cells/Stem cells

Disciplinary knowledge

- WS 1.3

Activity

Viewpoints

	Learning outcomes		
Learning objectives	**Developing Aiming for 3**	**Secure Aiming for 5**	**Extending Aiming for 8**
Describe what therapeutic cloning may be used for	Define therapeutic cloning	**Describe what therapeutic cloning may be used for**	Explain the process of therapeutic cloning
Describe one ethical, one religious, and one medical issue with the use of stem cells in medicine	List some arguments for and against the use of stem cells	**Describe one ethical, one religious, and one medical issue with the use of stem cells in medicine**	Evaluate the practical risks and benefits, as well as social and ethical issues, in the use of stem cells in medical research and treatments
Verbally communicate well-constructed arguments	Verbally communicate simple ideas during a group discussion	**Verbally communicate well-constructed arguments**	Clearly communicate strong, well-researched arguments in a persuasive manner

Tier 2 vocabulary	Tier 3 vocabulary
development, embryo, ethics, growth, rejection, religion, repair, risk	adult stem cell, embryonic stem cell, immunosuppressant drug, stem cell, therapeutic cloning

Lesson plan

Reactivate knowledge

1 What is an embryonic stem cell?

2 What is an adult stem cell?

3 How do people use plant stem cells?

Common misconception

Students may think that each specialised cell contains different DNA, making them structurally different. Clarify to students that all body cells in an organism contain the same DNA, but as cells become specialised, each cell will express different parts (genes) of the DNA.

Answers: 1 Stem cells in the early embryo that differentiate to form different cells *2* Stem cell in adults which differentiate to form a particular type of cell *3* To make identical clones or research

Trigger interest

Stem cell decision (5 mins)

Patients with Type 1 diabetes inject themselves with insulin every day. Ask students to imagine they are a doctor, and a patient has been diagnosed with diabetes. There is a new treatment available using stem cells grown from embryos. Would they advise their patient to use the new treatment?

Ask the class to discuss in small groups what information they would need before they could make a decision. Students may want to consider the source of the stem cells, whether consent was given to use the stem cells, and how effective the stem cell treatment is.

Listen to ideas from the class and discuss how there are possible risks of the treatment, and these must be weighed up alongside possible benefits. Students should also discuss religious and ethical questions raised by the use of embryonic stem cells. Encourage students to value and consider different contributions here. This will provide a positive behaviour template for similar activities later in this lesson.

Exposition

Ethical dilemmas (5 mins)

Read out the following list of situations:

- Organ transplants using donated organs from people who have died.
- Organ transplants from living donors.
- Taking the morning-after pill.
- Using contraception.
- Using unwanted embryos for medical research (they will be destroyed after use).
- Testing new medical treatments on healthy volunteers.
- Testing new medical treatments on animals.

After each item in the list, ask any student who feels comfortable doing so to raise their hand and share their viewpoints about it with the class. It is important to acknowledge that there is no right or wrong opinion, and that these are very complex ethical situations with no correct answers. Encourage students to be respectful of others' opinions, and to not speak over each other.

Main activity

Viewpoints (35 mins)

Assign students to the different roles shown on the activity sheet and ask them to read through their character's thoughts on the use of stem cells. They should then prepare a 60-second speech arguing their character's viewpoint. Ask students to form groups containing all the roles and to each present their viewpoint. Ask each group for the main points they found out during their presentations and record these on the board.

As a class, discuss all the points mentioned, and then have students individually write two arguments in support of using the stem cells, two arguments against the use of stem cells, and their own conclusions on the ethics of stem cell research. Students should focus on writing a persuasive argument including religious and ethical arguments and using evidence or data that has been discussed within the lesson. Reassure students that their conclusions will not be shared with the class.

Review and practice

Stem cell research (10 mins)

Tell students about the story of Henrietta Lacks, HeLa cells, and their importance to stem cell research. Discuss with the class how HeLa cells have lead to medical breakthroughs but also the ethical and moral dilemmas surrounding this.

B2 Cell division

Introduction to checkpoint intervention

The checkpoint intervention provides suggestions for a lesson to follow up the *B2 Cell division* checkpoint assessment. Depending on the outcomes of the assessment, these suggestions can be used to move a class towards secure, to target specific learning outcomes, or to consolidate knowledge for students achieving secure outcomes. Use students' outcomes from the checkpoint assessment to address general misconceptions from the content covered. Be prepared to re-cover content as required; three key concepts from the chapter are covered in more detail below.

Checkpoint secure learning outcomes

- Describe the stages of the cell cycle, including the process of mitosis. Describe how this produces two identical cells and identify when cell division occurs in the life of an organism.
- Use the key words to describe the process of mitosis.
- Describe the production of a plant cell clone.
- Describe differences between embryonic and adult stem cells.
- Describe how stem cells can be used to treat medical conditions.
- Describe one ethical, one religious, and one medical issue with the use of stem cells in medicine.

Student reflection

Encourage students to reflect on whether there were any checkpoint questions they found difficult or straightforward as well as their preparation for the checkpoint assessment, including how they revised, what they revised, the time spent on revision, and what areas they could improve upon. For more strategies, see the Metacognition spread.

Intervention activity: Describe the stages of the cell cycle, including the process of mitosis. Describe how this produces two identical cells and identify when cell division occurs in the life of an organism.

Students perform a collective memory blurt. Provide students with a plain sheet of A4 paper and ask them to write their names on the sheet.

Give students two minutes to draw the first stage of the cell cycle, including key terms to describe what is happening.

When the time is up, students pass their A4 sheet on to the student next to them. Students then have another two minutes to add the second stage of the cell cycle to the new sheet, including key terms to describe what is happening. Repeat this so students can write the third stage of the cell cycle on a new sheet.

Return the A4 sheet to the original student and give them two minutes to add any missing details, or correct anything they think is incorrect.

Finally, project an image of the cell cycle and discuss the key terms they should include so that students can assess their collective diagram.

Answer:

Stage 1: Replication of DNA to form two copies of each chromosome and synthesis of new sub-cellular structures. This is the longest stage of the cell cycle.

Stage 2: Mitosis and nuclear division occurs.

Stage 3: Cell divides in two forming two identical daughter cells.

Intervention activity: Use the key words to describe the process of mitosis.

For this activity, find a video about the process of mitosis online. Students can complete the B2.1 Go Further: *Mitosis* activity sheet after the video has played. Project the teacher answer sheet for students to self-mark.

Intervention activity: Describe one ethical, one religious, and one medical issue with the use of stem cells in medicine.

Present students with the six-mark question, 'Evaluate the use of stem cells from embryos instead of from adult bone marrow for treating human diseases'.

Students can split this six-mark question into a fish-bone scaffold as shown below. They should add two reasons why the use of embryonic stem cells would be an advantage over the use of adult bone marrow, and two disadvantages. The conclusion box is where they come to a judgment, but let students know their conclusions won't be shared with the class.

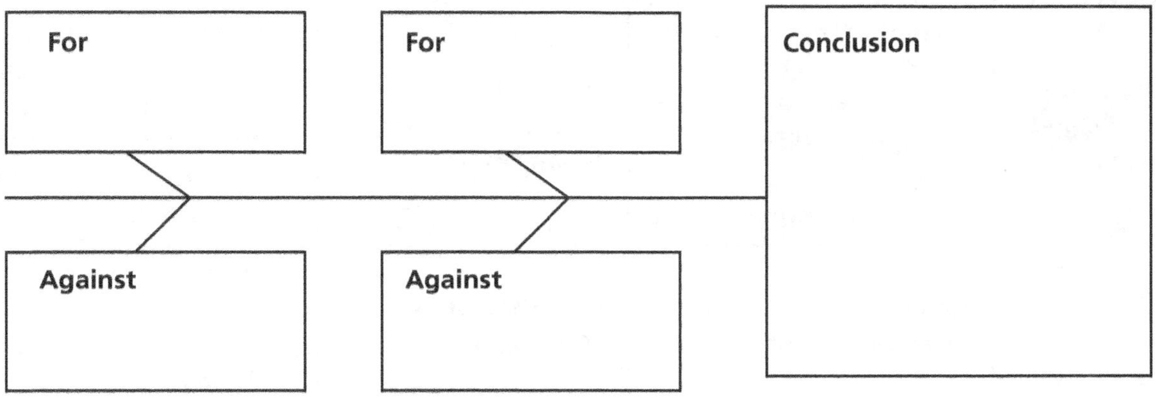

Potential answers students could give:

For the use of embryonic stem cells over adult bone marrow:

- can treat a wide variety / lots of diseases / problems
- many available / plentiful
- using them better than wasting them
- painless to the embryo
- collecting adult bone marrow requires an operation
- collecting adult bone marrow is painful so may deter donors.

Against the use of embryonic stell cells compared to adult bone marrow:

- there are no ethical issues in collection as adults can give permission
- there is a quick recovery after a bone marrow transplant
- collecting adult bone marrow is (relatively) safe
- using embryonic stem cells causes harm / death to embryo
- embryo can't be 'asked' / 'embryo rights' idea
- religious and ethical beliefs about right to life beginning at fertilisation.

B3 The human digestive system

B3.1 Principles of organisation

AQA GCSE Biology specification points
4.2.1 Principles of organisation

***Smart Curriculum* Narrative**
How organisms work

Theme/subtheme
Organisation / organ systems

Activity
Tissue research

	Learning outcomes		
Learning objectives	**Developing Aiming for 3**	**Secure Aiming for 5**	**Extending Aiming for 8**
Define the terms tissue, organ, and organ system	State that cells are the building blocks of all living organisms and give examples of different cells, tissues, organs, and organ systems	**Define the terms tissue, organ, and organ system**	Relate levels of organisation to familiar organ systems and demonstrate understanding of the different size and scale of cells, tissues, organs, and organ systems
Identify tissues that make up organs	Name organs found in a given organ system	**Identify tissues that make up organs**	Explain why the cells of multicellular organisms are organised into tissues, organs, and organ systems
Describe the function of named organs and organ systems	Order cells, tissues, organs, organ systems, and organisms according to their relative sizes	**Describe the function of named organs and organ systems**	Suggest the function of different tissues, e.g., glandular, epithelial, and muscular tissue in different organs

Tier 2 vocabulary	Tier 3 vocabulary
cell, *organ*, organisation, organ system, *tissue*	epithelial tissue, glandular tissue, multicellular organism, muscular tissue

Lesson plan

Reactivate knowledge

1 What is a tissue?
2 What is an organ?
3 Name the levels of organisation in a large multicellular organism from smallest to largest.

Common misconception
Students may think that each specialised cell contains different DNA, making them structurally different. Clarify to students that all body cells in an organism contain the same DNA, but as cells become specialised, each cell will express different parts (genes) of the DNA.

Answers: 1 A group of specialised cells working together to carry out a particular function *2* A collection of tissues all working together to carry out a particular function *3* Cells, tissues, organs, organ systems, organism.

Trigger interest

Organising an organism (5 mins)

Ask students to rearrange the key words organ, tissue, organism, cell, and organ system from smallest to largest. Then remind students of the hierarchy of organisation, using these key words.

Provide students with the following example to demonstrate this:

Cell (epithelial cell), tissue (epithelium tissue), organ (stomach), organ system (digestive system), organism (human).

Address the misconception that different specialised cells have different DNA.

Exposition

How organisms are made (5 mins)

Students use the Interactive: *How organisms are made* to match key words from the lesson with their definitions. They then sort terms according to whether they are tissues, organs, or organ systems.

Happy organ systems (20 mins)

Supply each group of four students with a set of cards, each card showing a different organ from the Activity: *Happy organ systems*. Students shuffle all the cards together, then deal them out evenly between the members of the group. Each student must collect the three organ cards needed to make up an organ system. Students take it in turns to ask another student for an organ. If that student has the organ in their hand, they must give it to the asker, and the asker then gets to ask for another organ. If they don't have it in their hand, then that student gets to ask for an organ. Once a student has collected an entire organ system, they put that group to one side.

Once all the organ systems have been collected, students then play for whole organ systems. Whichever student collects all the organ systems wins.

The organ systems to include are nervous (brain, eyes, spinal cord), digestive (stomach, pancreas, oesophagus), excretory (kidneys, bladder, liver), respiratory (lungs, trachea, bronchi), and circulatory (heart, arteries, veins).

Main activity

Tissue research (20 mins)

Ask students to work in groups of four. Students use the Student Book or the internet to research which tissues are present in the stomach. They also write down the function of each tissue in the organ.

Students should collate the following information:

Tissue	Function
Muscular	Muscles contract to churn the food and digestive juices of the stomach together
Glandular	Produces the digestive juices that break down food
Epithelial	Covers the inside and outside of the organ

Review and practice

Organisation match up (5 mins)

Students match images of various cells, tissues, organs, and organ systems to each level of organisation (e.g., show students an image of the eye and have students state it is an organ, or show the phloem and they state it is a tissue.)

Students could display their answers on mini-whiteboards.

B3.2 The human digestive system

AQA GCSE Biology specification points

4.2.2.1 The human digestive system

***Smart Curriculum* Narrative**

How organisms work

Theme/subtheme

Organisation/digestive system

Activity

The digestive system

	Learning outcomes		
Learning objectives	**Developing** **Aiming for 3**	**Secure** **Aiming for 5**	**Extending** **Aiming for 8**
Identify all of the main organs of the human digestive system	Identify some of the organs of the human digestive system	**Identify all of the main organs of the human digestive system**	Explain the need for a digestive system in the human body
Describe the functions of the organs of the human digestive system	State the function of some of the organs of the human digestive system	**Describe the functions of the organs of the human digestive system**	Explain in detail how the small intestine is adapted to its functions
Summarise the process of digestion	State simply what happens to food during digestion	**Summarise the process of digestion**	Explain in detail the main processes taking place along the human digestive system

Tier 2 vocabulary	Tier 3 vocabulary
cell, large intestine, small intestine, stomach, surface area	digestive system, enzyme, villi

Lesson plan

Reactivate knowledge

1 Where is food broken down?
2 What is the first organ in the digestive system?
3 Which are the main organs where food is digested?

Answers: 1 Digestive system *2* Mouth *3* Stomach, small intestine

Common misconception

When discussing the digestive system, students sometimes say 'digestion' when they mean 'absorption'. Make sure students have the exact definitions of each to avoid confusion.

Trigger interest

What happens to my food? (10 mins)

Ask students to draw a flow chart to show what they know already about what happens to food once it has been eaten. Take time to discuss prior knowledge and correct any possible misconceptions at this stage.

Exposition

The small intestine (20 mins)

Introduce the small intestine and explain its adaptations. Show (and describe if necessary) images of the small intestine, villi, and microvilli and ask them to recall adaptations for efficient absorption of digested food. To support students, ask them to think back to factors that speed up the rate of diffusion.

Make sure to cover that the microvilli have many mitochondria for respiration to provide energy for absorption from respiration.

A network of capillaries in each villus provides a short diffusion distance and maintains concentration gradients as it carries absorbed digested food molecules away in the blood.

Students could draw these three structures, or be provided with images, and then annotate the structural adaptations and write an explanation of how the structures increase the efficiency of absorption.

Main activity

The digestive system (15 mins)

Provide students with an unlabelled diagram of the human digestive system from the activity resource.

Students label the diagram and state the function of each organ. The names of each organ could be provided as a start point if needed or they can use Figure 1 in the Student Book to help them.

Review and practice

Digestive system colour coding (10 mins)

Ask students to use a different colour or pattern and a key to label their diagrams, identifying the organs in the digestive system where the following occur: absorption of food, absorption of water, chemical digestion, and physical digestion.

B3.3 The chemistry of food

AQA GCSE Biology specification points

4.2.2.1 The human digestive system

Smart Curriculum Narrative

How organisms work

Theme/subtheme

Organisation/digestive system

Disciplinary knowledge

- WS 1.2, 2.4

Required Practical 4

Food tests

	Learning outcomes		
Learning objectives	**Developing Aiming for 3**	**Secure Aiming for 5**	**Extending Aiming for 8**
Describe the structure of simple sugars, complex carbohydrates, lipids, and proteins	Recall that food contains carbohydrates, lipids, and proteins, which are all large molecules	**Describe the structure of simple sugars, complex carbohydrates, lipids, and proteins**	Explain how the structure of protein molecules influences their shape and function
Describe how carbohydrates, proteins, and lipids are used in the body	State one function of each of the three main types of food molecules in the diet	**Describe how carbohydrates, proteins, and lipids are used in the body**	Apply knowledge of the function of food molecules in the body to explain the need for a balanced diet
Describe the food tests for starch, sugars, protein, and lipids	State that the main food groups are identified using standard food tests	**Describe the food tests for starch, sugars, protein, and lipids**	Describe how to carry out the food tests for starch, sugars, protein, and lipids
Suggest what a food contains using results from food tests, evaluating the observed data collected	Carry out a food test on a known compound and record results in a table	**Suggest what a food contains using results from food tests, evaluating the observed data collected**	Carry out and evaluate multiple food tests in an organised manner

Tier 2 vocabulary	Tier 3 vocabulary
chemistry, *diet*, energy, food, molecule, structure, sugar, *unit*	amino acid, Benedict's solution, Biuret reagent, carbohydrate, denature, enzyme, ethanol, fatty acid, glucose, glycerol, iodine, lipid, protein, starch, sucrose, *tissue*

Lesson plan

Reactivate knowledge

1 What are the three main types of food molecules?
2 Why is protein needed in a balanced diet?
3 Why do we need a digestive system?

Common misconception

Students often confuse Benedict's reagent and Biuret reagent as they are both blue. Make sure students have written in their notes what each reagent tests for, so they can refer back to this to check.

Answers: *1* Carbohydrates, proteins, lipids *2* Growth and repair *3* Break down large molecules to smaller ones

Trigger interest

Nutrient names (10 mins)

Display (and describe if necessary) an image of a meal and ask students to write down the different types of nutrients found in this meal. They then compare lists in small groups, before a teacher-led group discussion on the function of each nutrient. Alternatively, show images of a collection of national dishes from various countries to reflect the different cultures in the class. Students can write down the food groups of each dish.

Support students here by asking if they can identify food items which contain carbohydrates, fats, and proteins.

If time permits, higher attaining students can be challenged to study the nutritional information on a ready-meal package and calculate the percentage of each nutrient in the food.

Exposition

Nutrients (10 mins)

Introduce students to the structure of carbohydrates, proteins, and lipids, and explain why each nutrient is needed in the diet. Link this to how each nutrient functions within the body.

Main activity

Food tests (30 mins)

Students conduct food tests on samples of food and record their results using the required practical worksheet.

Students analyse their results to identify different nutrients present in each food source. Address the misconception of confusing Benedict's reagent and Biuret reagent at this stage.

Review and practice

Food molecules (5 mins)

Students use the Interactive: *Food molecules* to match nutrients to their functions. They then sort the uses of nutrients into the correct categories.

Students can be stretched by asking them to consider how the structure of protein molecules influences their shape and function.

Students should consider that the sequence of amino acids in the protein determines how the protein folds and its final 3D shape and function.

B3.4 Catalysts and enzymes

AQA GCSE Biology specification points

4.2.2.1 The human digestive system

4.2.3 Plant tissues, organs, and systems

Smart Curriculum Narrative

How organisms work

Theme/subtheme

Organisation / digestive system

Disciplinary knowledge

- WS 1.2
- MS 1a, 1c

Practical

Breaking down hydrogen peroxide

	Learning outcomes		
Learning objectives	**Developing** **Aiming for 3**	**Secure** **Aiming for 5**	**Extending** **Aiming for 8**
Define metabolism and describe the role of enzymes in three different types of metabolic reactions	Recall that enzymes are proteins that are biological catalysts	**Define metabolism and describe the role of enzymes in three different types of metabolic reactions**	Explain the importance of enzymes in controlling cell metabolism
Use the lock and key theory to explain why the shape of an enzyme is vital for it to function as a catalyst	State that the shape of the active site of an enzyme is vital for the enzyme to work as a catalyst	**Use the lock and key theory to explain why the shape of an enzyme is vital for it to function as a catalyst**	Produce an annotated diagram as a simplified model to explain the specificity of enzyme action
Identify the key variables in a given investigation	State the independent variable in a given investigation	**Identify the key variables in a given investigation**	Plan an experiment to investigate how different catalysts affect the rate of a reaction

Tier 2 vocabulary	Tier 3 vocabulary
chemical reaction, metabolism, product, protein	active site, biological catalyst, enzyme, 'lock and key' theory, substrate

Lesson plan

Reactivate knowledge

1 What is a catalyst?

2 What is an enzyme?

3 Name the three main digestive enzymes.

Common misconception

Students may think that enzymes are alive and therefore can be killed by high temperatures. Clarify that enzymes are molecules and are not living. They can be broken down, but they cannot die.

Answers: 1 A substance that speeds up a reaction but is not used up/affected by it *2* A biological catalyst *3* Carbohydrase, protease, lipase

Trigger interest

Food in the fridge discussion (5 mins)

Ask students the following question and discuss their ideas: 'Why does putting food in the fridge slow down the reactions that cause food to go off?'

Students should consider what temperature does to the rate of chemical reactions in terms of particle movement and therefore the rate of successful collisions between substrates.

Exposition

Enzyme action (15 mins)

Show students the Animation: *Enzyme action* to explain the concept of enzymes as catalysts and the lock and key theory. Students should write down the key points and draw and label a diagram of the lock and key model. They can also explain why amylase cannot break down proteins.

Metabolic reactions

Discuss the metabolic reactions that enzymes are involved in and make students aware that the role of enzymes in digestion will be covered in a later lesson.

Main activity

Breaking down hydrogen peroxide (25 mins)

Students complete the Practical: *Breaking down hydrogen peroxide*. This is a short enzyme practical with follow-up questions to test students' understanding of enzyme theory and the necessary disciplinary knowledge. This will help students to understand the role enzymes play in speeding up reactions.

Review and practice

Spot the errors activity (10 mins)

For the first five minutes, discuss and explore the roles that enzymes play in wider industry such as paper recycling, laundry detergents, and textiles and fashion. Students can do their own research online to find more examples.

Then, read aloud or show the students the following five statements and ask them to correct the error in each statement:

- Catalysts decrease the rate of chemical reactions without changing chemically themselves. (correct to increase)
- Enzymes are biological catalysts and catalyse specific reactions in living organisms due to the shape of their cell site. This is the lock and key theory of enzyme action. (correct to active)
- Enzymes are lipids. (correct to proteins)
- The substrate binds to the active site and the reaction is slowed by the enzyme. (correct to catalysed)
- Metabolism is the sum of all the enzymes in a cell or the body. (correct to reactions)

B3.5 Factors affecting enzyme action

AQA GCSE Biology specification points

4.2.2.1 The human digestive system

***Smart Curriculum* Narrative**

How organisms work

Theme/subtheme

Organisation / enzymes

Practical

The effect of temperature on enzymes
Calculating the rate of reaction

	Learning outcomes		
Learning objectives	**Developing Aiming for 3**	**Secure Aiming for 5**	**Extending Aiming for 8**
Describe how changes in temperature and pH affect the way enzymes catalyse reactions and explain why high temperatures stop enzymes working	State that changes in temperature and pH affect how well an enzyme works and that very high temperatures stop enzymes working completely	**Describe how changes in temperature and pH affect the way enzymes catalyse reactions and explain why high temperatures stop enzymes working**	Explain how a change in temperature or pH affects the rate of an enzyme-catalysed reaction
Determine the gradient of a graph and calculate the rate of a reaction with guidance	Plot a line graph	**Determine the gradient of a graph and calculate the rate of a reaction with guidance**	Determine the gradient of a graph in order to calculate the rate of a reaction
Plot a line graph and use it to draw conclusions about how temperature and pH affect the rate of an enzyme-catalysed reaction	State simply what a line graph shows about how temperature or pH affects the rate of an enzyme-catalysed reaction	**Plot a line graph and use it to draw conclusions about how temperature and pH affect the rate of an enzyme-catalysed reaction**	Carry out rate calculations for given enzyme reactions and explain how changes in temperature and pH cause the observed rate changes

Tier 2 vocabulary	Tier 3 vocabulary
pH, rate of reaction, temperature	active site, denature, enzyme, optimum temperature

Lesson plan

Reactivate knowledge

Common misconception

Enzymes are alive and therefore can be killed by high temperatures. Clarify that enzymes are molecules and are not living. They can be broken down, but they cannot die.

1 What is an advantage of an enzyme?

2 What is the name of the part of the enzyme that binds to a specific substrate?

3 Each enzyme has an optimum temperature and pH. What does this mean?

Answers: 1 It doesn't get used up *2* Active site *3* Each enzyme has a pH and temperature at which it works best

Trigger interest

Lock and key model diagram (5 mins)

Provide a diagram of the lock and key model and ask students to annotate labels from what they can remember from KS3. To support students, give them the following labels to help them: enzyme, substrate, enzyme-substrate complex, product.

Exposition

The effect of temperature on enzymes (30 mins)

Introduce the factors that affect enzyme reactions – temperature and pH. Students plot a line graph using the data on the worksheet. They should explain what the graph shows using ideas about collision theory, active sites, and denaturation in their answer.

Tell students that the gradient of the line is the rate of reaction, but we have to work out the gradient from a straight line. Explain to the students that a tangent is a straight line drawn at a point on a curve to be able to work out the gradient. Model how to draw a tangent on the graph. Next show students how to calculate the gradient of a straight line.

Main activity

Calculating the rate of reaction (15 mins)

Students use the line graphs they plotted earlier and draw tangents to at least three different places on the line. They use these tangents to calculate the rate of the reaction at different places.

Review and practice

Reaction rate (5 mins)

In this interactive, students study a graph to select an enzyme's optimum temperature. They then study a graph and select statements about it that are true.

B3.6 How the digestive system works

AQA GCSE Biology specification points

4.2.2.1 The human digestive system

***Smart Curriculum* Narrative**

How organisms work

Theme/subtheme

Organisations/enzymes

Disciplinary knowledge

- WS 2.1, 2.4, 2.5, 2.6, 3.1, 3.2
- MS 2b, 4a, 4c

Required Practical 5

Enzymes

Learning objectives	Learning outcomes Developing Aiming for 3	Secure Aiming for 5	Extending Aiming for 8
Explain why enzymes are needed for digestion	State that enzymes are used in digestion to break down food molecules	**Explain why enzymes are needed for digestion**	Explain why the digestive system produces many different enzymes
For each food molecule, name the enzyme that acts on it, where it is produced, and which products are formed	Identify that carbohydrases break down carbohydrates, proteases break down proteins, and lipases break down lipids	**For each food molecule, name the enzyme that acts on it, where it is produced, and which products are formed**	Describe how to set up a model gut and test for substrates and products
Plan and carry out an investigation in order to gather accurate results	Plan a simple method to carry out an investigation	**Plan and carry out an investigation in order to gather accurate results**	Analyse results in order to evaluate a method, explaining suggestions for possible improvements

Tier 2 vocabulary	Tier 3 vocabulary
absorption, carbohydrate, digestion, insoluble, lipid, pancreas, pH, protein, small intestine, soluble, temperature	amino acid, amylase, carbohydrase, digestive enzyme, fatty acid, glycerol, lipase, protease

Lesson plan

Reactivate knowledge

1 If an enzyme gets too hot, it is denatured. What does this mean?

2 Name the three main digestive enzymes.

3 How does a change in pH affect the way an enzyme works?

Common misconception

Students may think that when the blue/black colour of iodine changes to orange, it indicates starch is present. Clarify that there is starch present initially, which is why there is a blue/black colour with iodine. When the starch is digested, it is no longer present (as sugar has formed) and this is why the iodine turns orange.

Answers: 1 The protein structure breaks down so the shape of the active site is lost. The enzyme doesn't work *2* Carbohydrase, protease, lipase *3* The active site is denatured and the enzyme does not work

Trigger interest

Question creation (5 mins)

Present students with the following keywords: enzyme, catalyst, active site, substrate, lock and key model, denature, digestion, carbohydrate, lipid, and protein. Ask them to write a question that has one of the keywords as the answer. Ask volunteers to read out their questions so the class can answer them.

Exposition

Required practical: Planning (15 mins)

Students plan an experiment to investigate how pH affects the rate of amylase activity.

Students should be guided to choose a suitable range for the independent variable and to use control variables. Use the worksheet Required practical: *The effect of pH* on the rate of reaction of amylase.

Main activity

Enzymes (25 mins)

Before starting the required practical, mention that pH is parts of hydrogen which changes from acidic to alkaline. Ask students to name organs where pH is important.

Address the misconception around the cause of the colour change of iodine in this experiment to ensure students understand the results they are expected to observe.

Students conduct the practical and use their data to plot a line graph. They answer questions to help them draw conclusions from the data and evaluate the method. More time can be devoted to this investigation next lesson or as a homework activity.

Review and practice

Digestive enzymes (10 mins)

Students use the Student Book to find out the names of the main digestive enzymes, where in the digestive system they are produced, their site of action, what food molecules they break down, and the names of the products. They should summarise this information as a table.

This could also be done as a homework activity.

B3.7 Making digestion efficient

AQA GCSE Biology specification points

4.2.2.1 The human digestive system

Smart Curriculum Narrative

How organisms work

Theme/subtheme

Organisation/digestive system

Disciplinary knowledge

- WS 1.2

Practical

Breaking down protein

	Learning outcomes		
Learning objectives	**Developing Aiming for 3**	**Secure Aiming for 5**	**Extending Aiming for 8**
State that the hydrochloric acid in the stomach results in a low pH	State that the stomach contains acid	**State that the hydrochloric acid in the stomach results in a low pH**	Explain the impact of a low pH in the stomach on the efficiency of pepsin
Describe the functions of bile	State that the liver produces bile	**Describe the functions of bile**	Explain how bile increases the efficiency of fat digestion
Carry out an investigation into the effect of pH on the rate of the reaction of the enzyme amylase on starch using a continuous sampling technique	Follow instructions to carry out a simple investigation into the effect of pH on the activity of amylase on starch	**Carry out an investigation into the effect of pH on the rate of the reaction of the enzyme amylase on starch using a continuous sampling technique**	Plan, carry out, and evaluate an investigation into the effect of pH on the enzyme amylase using a continuous sampling technique

Tier 2 vocabulary	Tier 3 vocabulary
acidic, alkaline, digestive system, gall bladder, pancreas, pH, small intestine, stomach	bile, emulsification, neutralisation, pepsin, protease, secretion, surface area

Lesson plan

Reactivate knowledge

1 What is the function of bile?

2 What do lipids break down into?

3 Which enzyme breaks down proteins into amino acids?

Answers: 1 Neutralises acids and emulsifies fats *2* Fatty acids and glycerol *3* Protease

Common misconception

Students may think bile is an enzyme.

Trigger interest

The gall bladder (10 mins)

Ask students to recall the organs in the digestive system and its functions. Then show students a diagram of the gall bladder and tell them that it releases an alkaline substance called bile. Ask students to suggest why bile needs to be released into the small intestine.

Exposition

Functions of bile (20 mins)

Discuss with the class why washing-up liquid is used to help clean dirty dishes. Explain that it breaks large fat globules in food down into smaller droplets that are mixed (emulsified) into the water and so are washed away more easily.

Carry out the demonstration as outlined on the Working scientifically: *Function of bile* sheet. Ask students to use information in the sheet to come up with a hypothesis for the experiment. Show them the results – the washing-up liquid emulsifies the lipids, so this reaction should be quicker.

Main activity

Breaking down protein (20 mins)

Set up the practical as detailed in the worksheet. Ask students to write a hypothesis for what they think will happen to the meat chunks.

Provide students with some data showing the time taken for a piece of egg white to break down in different pH solutions. Inform students that continuous sampling would be used to collect this data, as the breakdown of the egg would be continuously monitored every set amount of time (e.g., 30 seconds) to find the end point when all the egg white has been broken down.

Students calculate the mean rate of the reaction at each pH value. Then ask students to review their hypothesis and make any changes if necessary.

The experiment will need to be revisited in a future lesson to confirm whether students' hypotheses were correct.

To support students, a Maths skills activity is available for students to further practise calculating mean rates. Remind students how to calculate a mean.

To extend, students can analyse rates data in the form of results tables and graphs to draw conclusions on how pH affects enzyme activity.

Review and practice

Bile and enzymes (5 mins)

Ask students to list all the enzymes involved in digestion and what they break down (lipase breaks down lipids, carbohydrase/amylase breaks down carbohydrates, protease/pepsin breaks down proteins). Use this as an opportunity to address the misconception that bile is an enzyme, and correct any students who list it.

B3 The human digestive system

Introduction to checkpoint intervention

The checkpoint intervention provides suggestions for a lesson to follow up the *B3 The human digestive system* checkpoint assessment. Depending on the outcomes of the assessment, these suggestions can be used to move a class towards secure, to target specific learning outcomes, or to consolidate knowledge for students achieving secure outcomes. Use students' outcomes from the checkpoint assessment to address general misconceptions from the content covered. Be prepared to re-cover content as required; three key concepts from the chapter are covered in more detail below.

Checkpoint secure learning outcomes

- Describe the food tests for starch, sugars, protein, and lipids.
- Define the term tissue, organ, and organ system.
- Describe the functions of the organs of the digestive system.
- Use the lock and key theory to explain why the shape of an enzyme is vital for it to function.
- Summarise the process of digestion.
- Plan and carry out an investigation in order to gather accurate results.

Student reflection

Encourage students to reflect on whether there were any checkpoint questions they found difficult or straightforward as well as their preparation for the checkpoint assessment, including how they revised, what they revised, the time spent on revision, and what areas they could improve upon. For more strategies, see the Metacognition spread.

Intervention activity: Describe the food tests for starch, sugars, protein, and lipids.

Using the table of observations below, students determine what each unknown solution must contain.

Solution	Benedict's Test	Adding Biuret	Adding Starch	Adding ethanol and then water
A	Blue	Purple	Orange	Colourless
B	Red	Blue	Orange	Colourless
C	Blue	Blue	Blue/black	Colourless
D	Blue	Blue	Orange	White emulsion
E	Yellow	Blue	Orange	Colourless

Intervention activity: Define the terms tissue, organ, and organ system.

Students work in pairs to arrange the cards according to different instructions.

Check the students' understanding of the terms tissues, organ, and organ systems on the mini-whiteboards. Then, using the suggested cards below, provide each pair with a set of the cards.

Ask students to arrange the cards to show which are examples of cells, tissues, organs, and organ systems. Students usually struggle with the tissues and might not realise that the pacemaker, phloem, and blood are tissues. Use this to emphasise the definition of a tissue.

Blood	Phloem	Pacemaker	Muscle cells
Epithelium	Heart	Lungs	Digestive system
Nervous system	Neurone	Red blood cell	Egg
Sperm	Leaf	Mesophyll	Glandular layer
Liver	Small intestines	Large intestines	Stomach

Intervention activity: Describe the functions of the organs of the digestive system.

Ask students to select the cells, tissues, and organs found in the digestive system from the table below.

Blood	Phloem	Pacemaker	Muscle cells
Epithelium	Heart	Lungs	Digestive system
Nervous system	Neurone	Red blood cell	Egg
Sperm	Leaf	Mesophyll	Glandular layer
Liver	Small intestines	Large intestines	Stomach

Students should be able to select the following:

Cells: muscle cells, Tissues: glandular layer, epithelium, Organs: liver, small intestines, large intestines, stomach.

Then ask students to describe the function of each of organs in the digestive system.

To extend, ask students to explain how the glandular, epithelium, and muscular tissue layers help the stomach perform its function.

To support, students could match up the organ to the function using the table below:

A	Liver	1	Breaking down large insoluble molecules into smaller soluble molecules and absorption
B	Stomach	2	Absorbing water from undigested food
C	Small intestines	3	Producing bile
D	Large intestines	4	Breaking down large insoluble molecules into smaller soluble molecules

B4 Organisation in animals and plants

B4.1 The blood

AQA GCSE Biology specification points

4.2.2.3 Blood

Smart Curriculum Narrative

How organisms work

Theme/subtheme

Organisation/Circulatory system

Disciplinary knowledge

- WS 1.4, 1.5, 3.5

Practical

Blood components

	Learning outcomes		
Learning objectives	**Developing Aiming for 3**	**Secure Aiming for 5**	**Extending Aiming for 8**
Describe the functions of each of the main components of blood	State the main components in blood	**Describe the functions of each of the main components of blood**	Summarise the role of different blood components in the process of blood clotting
View blood under a light microscope and recognise components	Recognise the components of blood from photomicrographs and diagrams	**View blood under a light microscope and recognise components**	Estimate the diameter of red blood cells and white blood cells and comment on the variation seen
Explain how white blood cells are adapted to their functions	Explain how red blood cells are adapted to their function	**Explain how white blood cells are adapted to their functions**	Explain how all of the main components of the blood are adapted to their functions

Tier 2 vocabulary	Tier 3 vocabulary
blood, *transport system*, urine	biconcave, circulatory system, fibrin, haemoglobin, *plasma*, platelet, red blood cell, urea, white blood cell

Lesson plan

Reactivate knowledge

1 What is a multicellular organism?
2 Which organ moves blood around the body?
3 Name the process by which substances move in and out of cells.

Common misconception

Students may assume that tissues are solid structures, and so may not realise that the blood is an example of a tissue made up of different cells.

Answers: 1 An organism made up of lots of cells *2* The heart *3* Diffusion

Trigger interest

Blood cells (5 mins)

Show students a photomicrograph of a sample of blood. Ask them to write down any facts they know about blood on sticky notes. This could be something they've studied previously in Key Stage 3 or knowledge from outside the classroom. Invite contributions from the class to discuss facts or misconceptions.

Exposition

Components of blood table (10 mins)

Students use the Student Book or another relevant source to complete the table in the worksheet.

Main activity

Blood components (30 mins)

Allow students to work in pairs to study prepared microscope slides of blood smears under a light microscope.

Students should be able to use Figure 4 in the Student Book, or a similar image to identify and name red blood cells and white blood cells.

Extend students' knowledge by asking them to estimate the diameter of a red blood cell and comment on the uncertainty. As red blood cells are small, students will have to measure the length of a group of cells and then divide by how many there are to find a mean diameter.

Review and practice

Blood model (5 mins)

Build a model of blood by adding sunflower oil (plasma), small red counters (red blood cells), white marshmallows (white blood cells), and lentils (platelets) to a large beaker. Ask students what each part represents, what its function is, and how much of each they would need to put in to make an accurate representation of the composition of blood.

Blood and blood cells (5 mins)

Use the Interactive: *Blood and blood cells* where students summarise how different components of the blood are adapted for their function. They then decide whether statements are true or false.

B4.2 The blood vessels

AQA GCSE Biology specification points

4.2.2.2 The heart and blood vessels

Smart Curriculum Narrative

How organisms work

Theme/subtheme

Organisation/Circulatory system

Disciplinary knowledge

- MS 1a, 1c

Practical

Measuring the pulse

	Learning outcomes		
Learning objectives	**Developing Aiming for 3**	**Secure Aiming for 5**	**Extending Aiming for 8**
Describe the structure of arteries, capillaries, and veins	State the three main types of blood vessel and identify them from diagrams	**Describe the structure of arteries, capillaries, and veins**	Explain how the structure of arteries, capillaries, and veins are related to their functions
Describe how the valves function in veins to prevent backflow of blood	State that arteries have a pulse and veins contain valves	**Describe how the valves function in veins to prevent backflow of blood**	Explain why valves are present in veins but not in arteries or capillaries
Describe the two main elements of a double circulatory system	Define a double circulatory system	**Describe the two main elements of a double circulatory system**	Explain why a double circulatory system is necessary in active animals such as humans
Comment on how accurate estimations are	Estimate your own heart rate	**Comment on how accurate estimations are**	Explain how to make estimates more accurate in terms of precision of data

Tier 2 vocabulary	Tier 3 vocabulary
blood, heart	artery, blood vessel, capillary, double circulatory system, valve, vein

Lesson plan

Reactivate knowledge

1 What are the four main components of the blood?

2 What colour is plasma?

3 What is the role in the body of red blood cells, white blood cells, and platelets?

Common misconception

Students may think that blood in the veins is blue. Inform students that oxygenated blood is bright red and deoxygenated blood is dark red. The blue colour of veins is an optical illusion caused by the wavelengths of light that the skin filters.

Answers: 1 Plasma, red blood cells, white blood cells, and platelets *2* Yellow *3* RBC – carrying oxygen to the cells; WBC – defence against invading microorganisms; Platelets – clotting of the blood

Trigger interest

Circulatory systems (5 mins)

Show students an unlabelled image of a human double circulatory system. Ask them to identify any structures they recognise – prompt students to consider the names of any organs or blood vessels they can see.

Ask students to determine what they think the term 'double circulatory system' means from looking at the diagram.

Exposition

Blood flow (30 mins)

Ask students to write the name of all three blood vessels on mini-whiteboards. Ask students the following questions to recap their functions:

1 Which blood vessels carry blood away from the heart? (artery)
2 Which blood vessels carry blood back towards the heart? (vein)
3 Which blood vessels have a wall once cell thick to enable exchange of substances? (capillary)

Using a diagram of the circulatory system, discuss with students the path the blood takes through the body. Provide students with a copy of this diagram so that they can draw arrows on it to show the route that the blood takes through the circulatory system. They could also code the blood vessels to indicate whether they carry oxygenated (shaded) and deoxygenated (unshaded) blood. At this point, clarify the misconception that deoxygenated blood is blue.

Explain that humans have a double circulatory system. It is called this because blood passes through the heart twice per circuit. The right pump sends deoxygenated blood to the lungs where it becomes oxygenated and returns back to the heart. The left pump sends the newly oxygenated blood around the body. By the time this blood returns to the heart, it has returned to a deoxygenated state.

Link the explanations to blood pressure and the need for blood to reach all respiring cells in the body, but also the need to have a lower pressure in the lungs to prevent damage to the capillaries and to allow time for diffusion to occur.

Point out the locations of arteries, veins, and capillaries on the diagram of the circulatory system. Ask students to use information from the Student Book to draw sketches of each type of blood vessel (or provide students with an image), and to add labels to describe the structure and function of each.

Main activity

Measuring the pulse (15 mins)

Allow students to work in pairs to measure each other's pulse. First, they should count how many beats they can feel in five seconds, then ten seconds, 30 seconds, and finally 60 seconds.

Students then use their measurements for each timeframe to estimate the beats per minute.

Discuss why estimating is useful, but not as accurate as measuring. Link this to any variation they observe in the estimates for the different time periods they measured their pulse for.

Review and practice

Blood donation (5 mins)

Discuss the importance of blood, the need for blood donors, and an equitable supply of blood for different groups of people. Students could read and reflect on articles online about the ethnic minority blood donor crisis.

B4.3 The heart

AQA GCSE Biology specification points

4.2.2.2 The heart and blood vessels

Smart Curriculum Narrative

How organisms work

Theme/subtheme

Organisation/Circulatory system

Disciplinary knowledge

- WS 1.4, 1.3
- MS 1a, 1c

Practical

Sheep heart dissection

	Learning outcomes		
Learning objectives	**Developing Aiming for 3**	**Secure Aiming for 5**	**Extending Aiming for 8**
Describe the main structures of the human heart	Describe the function of the heart and identify the main structures of a heart on a diagram	**Describe the main structures of the human heart**	Explain in detail how the structure of the different parts of the human heart is related to their function
Describe the flow of blood around the heart	State that the right side of the heart pumps blood to the lungs and the left side pumps blood around the body	**Describe the flow of blood around the heart**	Explain the cause of the sounds of the heartbeat and the differential muscle thickness of the right and left ventricle walls
Describe how coronary heart disease develops	State one common cause of coronary artery disease in humans	**Describe how coronary heart disease develops**	Explain how common treatments for coronary artery disease prevent heart problems
Suggest advantages and disadvantages of using stents and statins	State that two common treatments for coronary heart disease are stents and statins	**Suggest advantages and disadvantages of using stents and statins**	Evaluate the use of stents, bypass surgery, and statins in treating problems with blood vessels

Tier 2 vocabulary	Tier 3 vocabulary
blood, heart, lung	aorta, *atrium*, cholesterol, coronary artery, coronary heart disease, pulmonary artery, pulmonary vein, statin, stent, valve, vena cava, ventricle

Lesson plan

Reactivate knowledge

1 Which type of blood vessel usually carries oxygenated blood?

2 Which type of blood vessel has valves to stop the blood flowing backwards?

3 What happens in the capillaries?

Common misconception

Students sometimes confuse the roles of the pulmonary and coronary artery. Clarify that the pulmonary artery carries deoxygenated blood from the heart to the lungs, whereas the coronary artery supplies the cardiac muscle with oxygenated blood.

Answers: 1 Arteries *2* Veins *3* Substances are exchanged between the blood and the cells of the body

Trigger interest

The heart (10 mins)

Display an image of the heart and a list of its structures. Ask students to match up the structures listed to where they are found in the heart. This could be done using mini-whiteboards or by making paper copies of the image. Review this with the students and provide them with a correctly labelled image of the heart for their notes.

Exposition

The structure of the heart (10 mins)

Discuss the structure of the heart with students, including:

- why the left ventricular wall is thicker than the right ventricular wall
- the role of the valves in the heart
- the direction that blood flows through the heart
- the role of the coronary arteries (address the misconception here).

Main activity

Sheep heart dissection (30 mins)

Demonstrate a sheep heart dissection, ideally displayed on the screen using a visualiser. Show students the following during the dissection:

- The coronary arteries.
- The aorta, pulmonary artery, pulmonary vein, and vena cava (if all are present).
- The difference in the size of the atria and ventricles and the difference in the thickness of their muscular walls.
- The difference in the thickness of the left and right muscular walls.
- The valves.

Then, position the heart so that some of the above features are visible and freeze the visualiser image in this position. Ask students to draw a scientific drawing of the heart. Remind students that scientific drawings must be in pencil, with no colouring, shading, or overlapping lines, and that the key focus is to show size, proportion, and structures. Using the labelled diagram from earlier, students should label the structures on their scientific drawing. As an extension, students could also add descriptions of the structures and functions of each.

Students who do not carry out the practical can instead complete the Student Book questions. They can then complete the Bump up your grade worksheet to help them understand the role of valves in controlling the direction of blood flow in the heart.

Review and practice

Blocked coronary arteries (5 mins)

Project an image of a blocked artery and ask students to predict what the effect of a blockage in one of the coronary arteries would be. Discuss how stents and statins could be used as treatments.

B4.4 Helping the heart

AQA GCSE Biology specification points

4.2.2.2 The heart and blood vessels

4.2.2.4 Coronary heart disease: a non-communicable disease

Smart Curriculum Narrative

Health and disease

Theme/subtheme

Health and disease/Non-communicable diseases

Disciplinary knowledge

- WS 1.3, 1.4
- MS 1a, 1c

Activity

Which one?

	Learning outcomes		
Learning objectives	**Developing Aiming for 3**	**Secure Aiming for 5**	**Extending Aiming for 8**
Explain why an irregular heartbeat causes health problems	State that the heartbeat is maintained by a group of cells that acts as a pacemaker	**Explain why an irregular heartbeat causes health problems**	Suggest how an artificial pacemaker regulates an irregular heartbeat
Describe the two main types of replacement heart valves	State two ways in which heart valves may fail to function healthily	**Describe the two main types of replacement heart valves**	Compare and contrast mechanical and biological replacement heart valves
Describe an artificial heart and give two reasons why they may be used	State that if the heart becomes seriously damaged, a heart or heart and lung transplant may be needed	**Describe an artificial heart and give two reasons why they may be used**	Compare and contrast mechanical and biological replacement hearts
Summarise the advantages and disadvantages of the three main methods of treating cardiovascular diseases	List the three main ways of treating cardiovascular disease as drugs, mechanical devices, or by transplants	**Summarise the advantages and disadvantages of the three main methods of treating cardiovascular diseases**	Evaluate in detail the benefits and the risks of the different methods used in the treatment of heart problems

Tier 2 vocabulary	Tier 3 vocabulary
blood, heart, heartbeat, muscle	artificial heart, artificial pacemaker, *atrium*, biological valve, heart transplant, mechanical valve, natural pacemaker, valve

Lesson plan

Reactivate knowledge

1 Which chamber of the heart pumps blood around the body?
2 Which blood vessels supply oxygenated blood to the heart muscle?
3 What is a stent?

Common misconception

Students may think blood goes into one side of the heart and then the other, rather than both atria filling simultaneously.

Answers: *1* Left ventricle *2* Coronary arteries *3* A metal mesh placed in an artery to hold it open if it is blocked

Trigger interest

Heart problems (10 mins)

Show the students a diagram of the outside of the heart. Ask them to discuss in small groups what can go wrong with the heart. Listen to their ideas as a class. Be aware that students may have friends or relatives affected by some form of heart disease and so sensitivity is required. Heart disease is one of the leading causes of death so this is a useful learning opportunity to link the lesson to everyday experience.

Exposition

Controlling the heartbeat (15 mins)

Ask students to consider what causes the heart to pump. Students could write their ideas on mini-whiteboards, or select students to offer their ideas. Reveal that it works from an electric current that the human body produces.

Fill a water balloon with water. Use a pen to draw the four heart chambers on the balloon. Explain that an electrical current produced by a group of cells called the pacemaker in the right atrium causes the top of the heart to contract. Squeeze the balloon at the top so all the water goes to the bottom.

Explain that the electrical current then moves down the heart. Squeeze the balloon at the bottom, so all the water goes back up to the top. Emphasise that although the pacemaker is in the right atrium, it causes both atria to squeeze and pump blood simultaneously, and clarify any misconceptions around this idea.

Link this movement to the cardiac cycle studied in B4.3. Then use this model to show an irregular heartbeat where the top and bottom do not pump in a coordinated manner. Be considerate of students who may know a person with heart conditions.

Main activity

Which one? (25 mins)

Discuss, or provide students with the information from the Student Book about, the advantages, disadvantages, and ethics of:

artificial pacemakers, biological valves, mechanical valves, artificial hearts, heart transplants.

Have students write whether they would choose an artificial heart or heart transplant. Ask them to include the reasons for their choice (including why they opted for the treatment they chose, and why they didn't choose the other option). Be considerate of students who may disagree with certain medical procedures on religious grounds.

Review and practice

How does it work? (5 mins)

Use the Interactive: *How does it work?*, which lists many different treatments for cardiovascular disease. Students link each treatment to the explanation of how it works.

B4.5 Breathing and gas exchange

AQA GCSE Biology specification points

4.2.2.2 The heart and blood vessels

Smart Curriculum Narrative

How organisms work

Theme/subtheme

Organisation/Respiratory system

Disciplinary knowledge

- MS 1c

Activity

Ventilation

	Learning outcomes		
Learning objectives	Developing Aiming for 3	Secure Aiming for 5	Extending Aiming for 8
Describe the function of the main structures of the human gas exchange system	Identify the main structures of the human gas exchange system	**Describe the function of the main structures of the human gas exchange system**	Evaluate in detail a model of the lungs
Describe the structure of an alveolus and how gas exchange takes place	State that gas exchange happens in the alveoli	**Describe the structure of an alveolus and how gas exchange takes place**	Explain in detail how the adaptations of alveoli result in efficient gas exchange
Describe and differentiate between the processes of ventilation and gas exchange	Use data in the form of percentages to describe the differences between the composition of inhaled and exhaled air	**Describe and differentiate between the processes of ventilation and gas exchange**	Explain the differences between the composition of inhaled and exhaled air in terms of events in the gas exchange system

Tier 2 vocabulary	Tier 3 vocabulary
carbon dioxide, concentration gradient, diffusion, lung, oxygen	alveolus, bronchiole, bronchus, capillary, diaphragm, gas exchange, intercostal muscle, rib, thorax, trachea, ventilation

Lesson plan

Reactivate knowledge

1 Why are people with faulty heart valves often breathless?

2 What is the name of the area of the heart that controls the heart rate?

3 What are the two types of artificial heart valves?

Common misconception

Students might think that inhalation involves the forcing/pushing of air into the lungs. Clarify that air enters the lungs due to the relatively lower pressure in the thorax compared to the atmosphere, causing it to move down the pressure gradient.

Answers: 1 The heart is less efficient and not enough oxygen reaches the tissues *2* Pacemaker *3* Mechanical valves (metal and plastics) and biological valves (based on animal valves)

Trigger interest

How do we breathe? (5 mins)

Show students an image of the thorax with the ribs and diaphragm visible. Ask students to use their knowledge from KS3 to write how they think air enters and exits and lungs. They may use mini-whiteboards for this.

Exposition

Lung model (15 mins)

Present to the class the bell jar model of the lungs. A large model may already be available or one can easily be made using a plastic bottle and balloons.

Ask students how they would use this model to show inhalation and exhalation.

Many students will think that air is blown into the neck of the bottle. Clarify the misconception here and demonstrate how it is in fact the contraction and relaxation of the diaphragm that changes the volume and hence the pressure of the inside of the chest cavity. Show students how this results in inflation and deflation of the lungs.

Main activity

Ventilation (25 mins)

Explain to students the mechanisms of ventilation and the exchange of gases in the alveoli.

Then, ask students to label a diagram of the parts of the gas exchange system. They should describe the function of each part, give any adaptations of each structure, and explain what happens during gaseous exchange.

Support students with prior lower attainment by providing labels, adaptations, and definitions of keywords. They can then try to match up the labels and adaptations of each structure.

Review and practice

How we breathe in and out (10 mins)

Use the worksheet Bump up your grade: *How we breathe in and out*, where students use a model of the lungs to understand how volume changes result in pressure changes and cause inhalation and exhalation. This can be started in the lesson and then completed in a follow-up lesson or for homework.

B4.6 Tissues and organs in plants

AQA GCSE Biology specification points

4.2.3.1 Plant tissues

4.2.3.2 Plant organ systems

Smart Curriculum Narrative

How organisms work

Theme/subtheme

Organisation/Plant organ systems

Disciplinary knowledge

- MS 1a,1c, 2a, 2c, 2d, 4a, 4c, 5c

Practical

Looking at leaves

	Learning outcomes		
Learning objectives	**Developing** **Aiming for 3**	**Secure** **Aiming for 5**	**Extending** **Aiming for 8**
Explain how the structures of different plant tissues are related to their functions	Name six different plant tissues and state their functions in a plant	**Explain how the structures of different plant tissues are related to their functions**	Suggest reasons why some tissues are present in all plant organs and some are not
Describe the arrangement of tissues in the cross-section of a plant leaf and the arrangement of transport tissues in the roots, stems, and leaves	Recognise examples of plant organs and state their functions	**Describe the arrangement of tissues in the cross-section of a plant leaf and the arrangement of transport tissues in the roots, stems, and leaves**	Explain how the structure of a leaf is adapted to its function
Use a microscope to draw and identify the different tissues in a cross-section of a leaf, with a magnification scale	Use a light microscope to view a cross-section of a leaf	**Use a microscope to draw and identify the different tissues in a cross-section of a leaf, with a magnification scale**	Use observations made with a light microscope to draw a cross-section of a leaf and calculate the length of palisade cells and the diameter of the main vascular bundle

Tier 2 vocabulary	Tier 3 vocabulary
chloroplast, leaf, *organ*, organ system, photosynthesis, root, stem, *tissue*, *transport*	epidermal tissue, guard cell, meristem tissue, palisade mesophyll tissue, phloem, spongy mesophyll tissue, stomata, vascular bundle, xylem

Lesson plan

Reactivate knowledge

Common misconception

Students sometimes think that a leaf is a tissue. Clarify the definitions of tissue and organ, and use this to explain why a leaf is an organ.

1 What is a tissue?
2 What is an organ?
3 Name a human tissue and a human organ?

Answers: 1 Group of cells with similar structure and function working together *2* Collection of different tissues working together tc perform a specific function *3* Tissue – muscle, glandular, nervous; Organ – stomach, heart, lungs, kidney

Trigger interest

Plant organs (5 mins)

Use the Interactive: *Plant organs*, where students match plant organs with their functions and then label a diagram to show the main structures of a plant stem and a root cross-section.

Exposition

Name the organ (5 mins)

Display a range of plant organs from different plants (roots, stems, leaves, and flowers) and ask students to name what type of organ each is. Recap the definition of a tissue and an organ and address the misconception here.

Tissue adaptations (5 mins)

List the different tissues in a leaf and ask students to state one adaptation of each. This could be done using mini-whiteboards or in their notes.

Main activity

Looking at leaves (35 mins)

Supply groups of students with labelled diagrams of the plant tissues xylem, phloem, meristem, and epidermal. Ask them to discuss the organs where each tissue is found, their functions, and how each is adapted for its function.

Students then use light microscopes to study prepared slides of leaf cross-sections.

Students should draw and label a diagram from their observations using the light microscope. They should identify the cells and tissues they can see and describe their functions. Ask students to explain how the functions of these cells and tissues are related to their structures.

To extend, students should use the image size and magnification to calculate the length of a palisade cell and the diameter of the main vascular bundle.

Review and practice

Maths skills: Calculating leaf surface area activity (5 mins)

Students complete the maths questions in the interactive activity, providing answers on their mini-whiteboards.

B4.7 Transport systems in plants

AQA GCSE Biology specification points

4.2.3.2 Plant organ systems

Smart Curriculum Narrative

How organisms work

Theme/subtheme

Cells/Types of cell

Disciplinary knowledge

- MS 1a, 1c, 2a, 2c, 2d, 4a, 4c, 5c

Practical

Adaptations of xylem and phloem

Learning objectives	Learning outcomes: Developing Aiming for 3	Learning outcomes: Secure Aiming for 5	Learning outcomes: Extending Aiming for 8
Describe the structure of xylem and phloem tissue	Describe the function of xylem and phloem tissue	**Describe the structure of xylem and phloem tissue**	Explain how xylem and phloem tissues are adapted for their different functions
Explain why transport in plants is important	State what substances are moved around plants from roots to leaves and from leaves around the plant body	**Explain why transport in plants is important**	Describe evidence for movement of water through xylem

Tier 2 vocabulary	Tier 3 vocabulary
glucose, leaf, mineral ion, photosynthesis, root, stem, *transport system*, water	phloem tissue, storage organ, translocation, vascular bundle, xylem tissue

Lesson plan

Reactivate knowledge

1 Name a plant tissue.

2 What is the function of the tissue you have named?

3 What are the main plant organs?

Common misconception

Students might think that the phloem and xylem are made up of one type of cell, e.g., a 'phloem cell'. Clarify that both are tissues made up of different types of cells.

Answers: 1 Any from: Epidermis, palisade mesophyll, spongy mesophyll, xylem, phloem *2* Any from: Epidermis – Protect the leaf cells; palisade mesophyll – photosynthesis; spongy mesophyll – photosynthesis and gas exchange; xylem – transport of water and minerals; phloem – transport of dissolved food *3* Roots, stems, leaves

Trigger interest

Substance journeys (10 mins)

State the names of some substances that are transported in plants (water, mineral ions, sugars) and ask students to describe their journey – where did they start, where are they going, and how will they get there?

If students are struggling, show them the different journeys and ask students to choose the one that best describes each substance. For example,

soil → roots → stem → leaf

The answer would be water.

Challenge students by asking them to explain what will happen to the substance once it has reached its destination.

Exposition

Label the diagram (10 mins)

Remind students of the specialised cells in phloem and xylem from B1.5. This could be through showing them a labelled diagram of the cells.

Then, ask students to write 'phloem' on one side of their mini-whiteboard and 'xylem' on the other side.

Display a series of images showing transverse and cross-sections of roots, stems, and leaves to show the vascular systems. Ask students to hold up their white-boards to indicate whether each image is showing xylem or phloem.

Main activity

Adaptations of xylem and phloem (30 mins)

Supply students with celery sticks that have been left overnight in ink. They cut the sticks to see how the dye has stained the xylem walls. Students then look back at B1.5 in the Student Book and explain how the structure of xylem cells is adapted for their function.

Then introduce how sap is tapped from trees in order to make maple syrup. If possible, show photos or a video from the Internet. Discuss with the class where the sugars in the tree trunk (stem) come from and how they are transported, or translocated, in the phloem. Ask students to look back at B1.5 in the Student Book and explain how phloem cells are adapted for their function. At this point, clarify the misconception around phloem being a single type of cell, rather than a tissue.

Review and practice

Xylem and phloem (5 mins)

Use the Interactive: *Xylem and phloem*, where students complete sentences about xylem and phloem. They then choose whether statements describe xylem or phloem.

B4.8 Evaporation and transpiration

AQA GCSE Biology specification points

4.2.3.2 Plant organ systems

Smart Curriculum Narrative

How organisms work

Theme/subtheme

Organisation/Plant organ systems

Disciplinary knowledge

- MS 2b, 2d

Practical

Investigating stomata

	Learning outcomes		
Learning objectives	**Developing** **Aiming for 3**	**Secure** **Aiming for 5**	**Extending** **Aiming for 8**
Describe how transpiration maintains the movement of water from root hair cells in the roots to leaves	State that transpiration is the evaporation of water vapour from the leaves	**Describe how transpiration maintains the movement of water from root hair cells in the roots to leaves**	Evaluate drinking from a straw as a model for transpiration
Describe how the opening and closing of stomata is controlled by guard cells	State the function of stomata	**Describe how the opening and closing of stomata is controlled by guard cells**	Explain in detail the role of stomata in gas exchange and water loss in a plant
Use sampling to estimate the number of stomata on a leaf	Calculate the mean number of stomata on a given area of leaf	**Use sampling to estimate the number of stomata on a leaf**	Describe the differences observed in the number and distribution of stomata on different areas of a leaf and explain the importance of these adaptations to the plant

Tier 2 vocabulary	Tier 3 vocabulary
carbon dioxide, diffusion, evaporation, gas *exchange*, leaf, osmosis, oxygen, water vapour	guard cell, stomata, transpiration, transpiration stream, xylem

Lesson plan

Reactivate knowledge

1 How are leaves adapted for photosynthesis?

2 Which plant transport tissue is made up of living cells?

3 Which plant transport tissue only moves things in one direction?

Answers: 1 Broad – light capture; thin – short diffusion distances; green – contain chlorophyll to capture light; stomata for gas exchange; xylem – water and mineral ions transport; phloem – dissolved food transport *2* Phloem *3* Xylem

Common misconception

Students may think that a stoma is a cell on the epidermis. Clarify that the stoma is the hole created by the bending of two guard cells.

Trigger interest

Stomata (5 mins)

Show the class a magnified image of a stoma and ask them to discuss what it could be and its function.

Students may recall that the stomata allows for gas exchange to occur and they are usually found on a leaf.

Exposition

Introducing transpiration (10 mins)

Explain that water vapour can move out of a leaf through the stomata. Introduce this as transpiration.

Ask students to define the words transpiration, transpiration stream, stomata, and guard cells.

Main activity

Investigating stomata (30 mins)

Demonstrate how you can carry out an epidermal peel to view stomata under the microscope. Clarify the misconception at this point, explaining that the stoma is the hole and not a cell.

Ask students to discuss in small groups what they could investigate about the stomata using this method and how they would go about doing so. Examples include comparing the number of stomata on different types of plant, on different leaves on one plant, and in different areas of a leaf.

Discuss the use of sampling to study different leaves and different areas on a leaf.

Students then carry out the practical in small groups. They count the number of stomata in several fields of view before calculating a mean and writing a conclusion.

Challenge students by asking them to make predictions about the distribution of stomata in floating and submerged water plants.

Review and practice

Plant adaptations (10 mins)

Show students images of the leaves of marram grass and cacti. Ask them to suggest how the stomata on these plants may be specially adapted, and why.

Students should be able to identify the locations of these plants, hence how the plant can be adapted to the weather.

B4.9 Factors affecting transpiration

AQA GCSE Biology specification points

4.2.3.2 Plant organ systems

Smart Curriculum Narrative

How organisms work

Theme/subtheme

Organisation/Plant organ systems

Disciplinary knowledge

- MS 1a, 1c, 2a, 2c, 2d, 4a, 4c, 5c

Activity

Investigating transpiration

	Learning outcomes		
Learning objectives	**Developing Aiming for 3**	**Secure Aiming for 5**	**Extending Aiming for 8**
Explain why temperature, humidity, light intensity, and air movement affect the rate of transpiration	List the factors that affect transpiration	**Explain why temperature, humidity, light intensity, and air movement affect the rate of transpiration**	Apply the particle model to explain in detail why temperature, humidity, light intensity, and air movement affect the rate of transpiration
Describe the differences between a moving bubble potometer and a mass potometer	Label a diagram of a potometer and state that it is used to measure the uptake of water by a plant	**Describe the differences between a moving bubble potometer and a mass potometer**	Explain the difference between potometer measurements and transpiration
Predict, using scientific knowledge, the impact of changing an environmental factor on the rate of transpiration and use data from a practical investigation to test your prediction	Identify control variables when investigating rate of transpiration	**Predict, using scientific knowledge, the impact of changing an environmental factor on the rate of transpiration and use data from a practical investigation to test your prediction**	Give scientific explanations for observed measurements when investigating the impact of different environmental factors on the rate of transpiration

Tier 2 vocabulary	Tier 3 vocabulary
air movement, concentration gradient, diffusion, environmental factor, evaporation, gas exchange, humidity, light intensity, photosynthesis, temperature, water vapour	*cuticle*, potometer, stomata, transpiration, wilting

Lesson plan

Reactivate knowledge

1 What controls the opening and closing of the stomata?
2 What is transpiration?
3 What is the transpiration stream?

Common misconception

Students may think a higher light intensity causes evaporation to happen faster. Clarify that a higher light intensity increases the rate of photosynthesis, and therefore more stomata open. The increase in open stomata causes more evaporation.

Answers: 1 Guard cells *2* The loss of water vapour from the leaves of a plant by evaporation (and diffusion) *3* The constant movement of water molecules through the xylem from the roots to the leaves and then out into the air by evaporation

Trigger interest

Drying clothes (10 mins)

Display an image of clothes drying outside on a washing line. Use the Interactive: *Drying clothes,* where students select the factors that would affect how quickly washing dries and then sort descriptions into whether they would increase or decrease the rate of drying.

Then, discuss the fact that clothes dry because of evaporation, the rate of which is increased by high temperature, low humidity, and degree of air flow (wind).

Exposition

Factors affecting transpiration (10 mins)

Explain to students that the rate of evaporation increases as temperature increases, humidity decreases, and air flow increases. Explain that higher light intensities also increase the rate of transpiration, and clarify the misconception here.

Main activity

Investigating transpiration (30 mins)

Demonstrate how a mass potometer and a moving bubble potometer can be used to measure the uptake of water by a plant. Discuss how leaving the equipment in different conditions would affect the rate of water uptake, and why. Mention why light intensity increases the rate of transpiration (it does not affect evaporation). Ask students to draw diagrams of the equipment and describe briefly how it works. Use Activity: *Factors affecting transpiration.*

Students then make a prediction and plan a fair test to investigate one of the factors that affect the movement of water through a plant by transpiration. A second lesson will be required for students to carry out their planned practical and to analyse their data.

Review and practice

Hand up, hand down (5 mins)

Read out a series of conditions and, for each one, ask students to put their hand up if they think it will increase the rate of transpiration or put their hand down if they think it will decrease the rate of transpiration.

B4 Organisation in animals and plants

Introduction to checkpoint intervention

The checkpoint intervention provides suggestions for a lesson to follow up the *B4 Organisation in animals and plants* checkpoint assessment. Depending on the outcomes of the assessment, these suggestions can be used to move a class towards secure, to target specific learning outcomes, or to consolidate knowledge for students achieving secure outcomes. Use students' outcomes from the checkpoint assessment to address general misconceptions from the content covered. Be prepared to re-cover content as required; three key concepts from the chapter are covered in more detail below.

Checkpoint secure learning outcomes

- Explain how red blood cells are adapted to their function.
- Describe how coronary heart disease develops.
- Describe the structure of an alveolus and how gas exchange takes place.
- Describe and differentiate between the processes of ventilation and gas exchange.
- Describe the arrangement of tissues in the cross-section of a plant leaf and the arrangement of transport tissues in the roots, stems, and leaves.
- Describe how transpiration maintains the movement of water from roots to leaves.

Student reflection

Encourage students to reflect on whether there were any checkpoint questions they found difficult or straightforward, as well as their preparation for the checkpoint assessment, including how they revised, what they revised, the time spent on revision, and what areas they could improve upon. For more strategies, see the Metacognition spread.

Intervention activity: Explain how red blood cells are adapted to their function.

Ask students to draw a red blood cell to fill half of a page. Then ask students to annotate their drawings with descriptions of the unique structures of this specialised cell (e.g., biconcave shape of cell).

Finally, under each structure they label, ask students to explain the advantage of that structure to the function of the cell.

Use this activity to recall their knowledge of how a red blood cell is adapted to its function, while also emphasising the difference between the command words 'describe' and 'explain'.

Answers students could provide include:

- They have a biconcave shape to increase the surface area for diffusion of oxygen into the cell.
- They contain haemoglobin which oxygen binds to.
- They have no nucleus – this provides more space for haemoglobin, so more oxygen can be transported.
- They are small and flexible so that they can fit through narrow capillaries.

Intervention activity: Describe and differentiate between the processes of ventilation and gas exchange.

Ask students to create a flow diagram of what happens during inhalation and exhalation.

To support students, you could prompt them to consider whether:

- the diaphragm is contracting/relaxing
- the intercostal muscles are contracting/relaxing
- the ribcage is moving up and out/in and down
- the volume of the thorax has increased/decreased
- the pressure inside the thorax has increased/decreased.

Students' flow diagrams could include this:

Inhalation

- The diaphragm contracts and lowers and intercostal muscles contract, and the rib cage moves up and out.
- The volume of the thorax increases.
- The pressure inside the thorax decreases.
- Air flows in to the lungs.

Exhalation

- The diaphragm relaxes and domes upwards and intercostal muscles relax, and the rib cage moves down and in.
- The volume of the thorax decreases.
- The pressure inside the thorax increases.
- Air is forced out of the lungs.

Intervention activity: Describe how transpiration maintains the movement of water from roots to leaves.

Students complete the B4.7 *Movement through a xylem* practical and questions. Students can self-mark their answers to these questions.

2 Disease and bioenergetics

Introduction to unit

This unit extends students' knowledge of digestive, respiratory, and circulatory systems studied at Key Stage 3. They will study the difference between communicable and non-communicable diseases, plant and human defences against disease, and the development of medicines and vaccination. Misconceptions may still exist from Key Stage 3, and these should be checked before proceeding.

Prerequisite knowledge from KS3 (chapter references from *Oxford Smart* Activate)

B3.1 Cells

- Living organisms are composed of cells.
- Some organisms are multicellular, and others are unicellular.
- A disease is a disfunction of an organism.
- Some diseases are communicable, and others are not.
- Cells replicate by cell division.

B2.2 Biological processes

- Plants respire and photosynthesise.
- Plants obtain water through roots and carbon dioxide through leaves.
- Photosynthesis requires light.
- Animals respire but do not photosynthesise.
- Respiration is a chemical process carried out inside cells. It is not another name for breathing.

Assessment

Suggested minimum guided learning hours: 30 hours

Exam: Paper 1

AQA GCSE Biology specification links

4.2 Organisation

- 4.2.2 Animal tissues, organs, and organ systems

4.3 Infection and response

- 4.3.1 Communicable diseases
- (H) 4.3.2 Monoclonal antibodies (biology only)
- 4.3.3 Plant disease (biology only)

4.4 Bioenergetics

- 4.4.1 Photosynthesis
- 4.4.2 Respiration

Identity, awe, and wonder

Students can easily make links between health and real-life experiences such as the events of the COVID-19 pandemic. Reflection can be made as to how organisms unseen to the naked eye causes so much disruption to normal life. Discuss the leaps made in vaccine technology, the testing of medical drugs, and the unprecedented levels of cooperation within the scientific community. Lateral flow tests, used to diagnose COVID-19, use monoclonal antibodies. Furthermore, plant disease may not be at the forefront of a student's mind, but it is a critical area of research for our future. For our food security, scientists need to understand more about plant pathogens to enable development of resistant strains of crops to improve the yield as human population continues to grow. Plant pathogens can devastate a crop and possibly a whole region, costing livelihoods of farmers. The Xylella infection of olive trees and grape vines would be an excellent example.

Required practicals

AQA required practicals	Practical skills	Topic
Required practical 2: Microbiology Investigate the effect of antiseptics or antibiotics on bacterial growth using agar plates and measuring zones of inhibition.	AT 1 – use appropriate apparatus to record length and area. AT 3 – use appropriate apparatus and techniques to observe and measure the process of bacterial growth. AT 4 – safe and ethical use of bacteria to measure physiological function and response to antibiotics and antiseptics in the environment. AT 8 – the use of appropriate techniques and qualitative reagents in problem-solving contexts to find the best antibiotic to use or the best concentration of antiseptic to use.	B5.4
Required practical 6: Photosynthesis Investigate the effect of light intensity on the rate of photosynthesis using an aquatic organism such as pondweed.	AT 1 – use appropriate apparatus to record the rate of production of oxygen gas produced; and to measure and control the temperature of water in a large beaker that acts as a 'heat shield'. AT 2 – use a thermometer to measure and control the temperature of a water bath. AT 3 – use appropriate apparatus and techniques to observe and measure the process of oxygen gas production. AT 4 – safe and ethical use and disposal of living pondweed to measure physiological functions and responses to light. AT 5 – measuring rate of reaction by oxygen gas production.	B8.2

Working scientifically

Lessons covering the use of vaccination and medical drugs provides opportunity for explanations of everyday technological applications of science. Discussions on the use of antibiotics and developing resistance will allow students to appreciate the power and limitations of science, evaluate implications, and make decisions based on the evaluation of evidence and arguments. Required practical 2 provides an opportunity for the evaluation of practical risks. The analysis of research and data on aspects of health such as smoking, drinking alcohol, and becoming infected with a pathogen allow the evaluation of risk in a wider societal context. This can include recognition of the importance of peer review of results and of communicating results to a range of audiences.

Maths skills

Students develop their numerical and algebraic skills when handling results and exponential growth in lesson B5.4. They will process data from experiments; calculate means and use significant figures. There are opportunities in B7.1, B8.2, B8.4, and B9.2 to develop graph plotting skills from primary and secondary data including higher tier students justifying their choice of graph.

Literacy links

Any of the practical investigations could be used to develop scientific/instructional writing skills such as writing a step-by-step method. The topics of communicable and non-communicable disease provide opportunities for students to develop their skills of oracy, persuasive writing, and extended writing. B5.5 and B7.2 provide students with the opportunity to both assimilate and represent written information about different diseases. B9.4 also includes a comprehension task.

B5 Communicable diseases

B5.1 Health and disease

AQA GCSE Biology specification points

4.2.2.5 Health issues

Smart Curriculum Narrative

Health and disease

Theme/subtheme

Health and Disease/Communicable diseases

Disciplinary knowledge

- MS 2c, 2d, 2g, 4a

Activity

Interpreting health data

	Learning outcomes		
Learning objectives	**Developing Aiming for 3**	**Secure Aiming for 5**	**Extending Aiming for 8**
Describe the difference between communicable and non-communicable diseases	Describe health as a state of physical and mental wellbeing	**Describe the difference between communicable and non-communicable diseases**	Suggest examples of communicable and non-communicable diseases which may be affected by the environment of an individual
Describe the cause of communicable diseases, and some causes of non-communicable diseases	State some causes of ill health	**Describe the cause of communicable diseases, and some causes of non-communicable diseases**	Suggest how different types of health problems may interact
Use a scatter diagram to identify a correlation between two variables	Recognise a scatter diagram and a bar graph	**Use a scatter diagram to identify a correlation between two variables**	Suggest links between lifestyle and health
Construct and interpret bar charts, frequency tables, frequency diagrams, and histograms	Draw a simple conclusion from data on health	**Construct and interpret bar charts, frequency tables, frequency diagrams, and histograms**	Discuss the validity of a statement based on evidence in the form of data

Tier 2 vocabulary	Tier 3 vocabulary
cancer, *cause*, *diet*, disease, health, lifestyle, microorganism, stress, well-being	communicable diseases, immune system, *interaction*, non-communicable diseases, pathogen, *virus*

Lesson plan

Reactivate knowledge

1 Name a disease which is not caused by an infection.

2 Name two recreational substances or behaviours that affect human health.

3 Give seven components of a healthy diet.

Answers: 1 Any from: heart disease, cancer, COPD *2* Alcohol, smoking cigarettes, any recreational drugs *3* Carbohydrates, lipids, proteins, vitamins, minerals, dietary fibre, and water

Common misconception

Students can think that 'disease' means communicable diseases only. Clarify that diseases can be communicable and non-communicable.

Trigger interest

Health vs disease (10 mins)

Display the words 'measles' and 'asthma', or an image to represent the two conditions. Ask students to list the differences between the diseases, and prompt them to not just consider symptoms, but also causes.

From this ask students to decide on a definition for the term 'health' and list factors that affect it.

Exposition

Believing the headlines (20 mins)

After discussing the term 'health' ask students if they know of any possible risk factors for the following conditions (define a risk factor if they are unfamiliar with this term first):

- Liver disease (alcohol consumption)
- Lung Cancer (smoking)
- Diabetes (obesity)
- Brain function (cannabis and alcohol)
- Under-developed/small babies at birth (smoking during pregnancy)

Provide students with examples of health-related news headlines in the past. For example, some headlines in the past have stated 'coffee can cure cancer' and 'coffee can cause cancer'. Ask students how they would know which to believe or if any are true. Prompt students to consider data from experiments, who the experiments were conducted on, how large the sample size was, whether the study was peer reviewed, etc.

Explain that headlines are usually based on data from research but that sometimes the data is interpreted incorrectly.

Discuss the need to back up a correlation with a plausible cause before it can be validated and accepted as meaningful. A correlation might be just an association between factors, if no causal link can be established.

Main activity

Interpreting health data (20 mins)

Demonstrate how to interpret a scatter graph to work out the correlation using the Activity: *Interpreting Health Data*. Review the answers as a class, getting students to self-mark.

Review and practice

What does it show? (5 mins)

Students use this interactive to study a graph that shows the interaction between different factors and health. Students answer questions about what the graph is showing on their mini-whiteboards or have a whole class discussion.

B5.2 Pathogens and disease

AQA GCSE Biology specification points

4.3.1.1 Communicable (infectious) diseases

Smart Curriculum Narrative

Health and disease

Theme/subtheme

Health and disease/Communicable diseases

Disciplinary knowledge

- WS 1.4

Activity

Pathogens causing disease

	Learning outcomes		
Learning objectives	**Developing** **Aiming for 3**	**Secure** **Aiming for 5**	**Extending** **Aiming for 8**
Describe how bacteria and viruses cause disease	State that pathogens are microorganisms that cause disease and list the four main types of pathogens	**Describe how bacteria and viruses cause disease**	Explain the cause of common symptoms of infectious diseases
Explain why human communicable diseases often spread rapidly following a natural disaster	Describe three ways in which pathogens are spread	**Explain why human communicable diseases often spread rapidly following a natural disaster**	Compare and contrast the spread of pathogens by air, direct contact, and by water

Tier 2 vocabulary	Tier 3 vocabulary
bacteria, *conditions*, *contact*, contaminated, disease, fungi, health, infect, infection, lifestyle, organism, reproduce, risks, species, *spread*, symptoms	binary fission, communicable, droplet infection, fungal spores, infectious, pathogens, protists, *vectors*, viruses

Lesson plan

Reactivate knowledge

1 What do we mean by health?

2 What is the difference between communicable and non-communicable diseases?

3 Name three factors which affect health.

Common misconception

Students often don't consider that plants can contract communicable diseases too.

Answers: 1 A state of physical and mental well-being (not simply absence of physical disease) *2* A communicable disease can be transferred from person to person but a non-communicable disease cannot be transferred, only inherited *3* Any from: diet, stress levels, life situations (e.g., where you live, gender, financial status, ethnic group, etc.)

Trigger interest

Communicable diseases (5 mins)

Tell the class that communicable diseases are caused by bacteria, fungi, viruses, and protists. Ask the class to list the different communicable (infectious) diseases they have heard of, and, as an extension, which pathogen causes each disease.

Exposition

Pathogens and the spread of diseases (15 mins)

Introduce the term pathogen as a microorganism that causes infectious diseases in living things, including plants. Address the misconception at this point.

Students then create a spider diagram to summarise the ways in which pathogens are spread. Ask them to use their diagram to explain why diseases spread more quickly following a natural disaster like an earthquake.

Main activity

Pathogens causing disease (30 mins)

Students then use the Student Book, or research on a computer, to design an animation about how pathogens, including viruses, protists, and bacteria cause disease. They can design them as a drawn storyboard or, if time allows, use an application on a tablet or computer to make an animation. The animations should include a voiceover script. Students with higher prior attainment can be challenged here and asked to include the common symptoms of infectious diseases as well as the different ways it can be spread.

Review and practice

Bacteria or virus? (5 mins)

Use this interactive where students select the correct definition of a pathogen. They then choose whether statements are describing bacteria or viruses.

B5.3 Growing bacteria in the lab

AQA GCSE Biology specification points

4.1.1.6 Culturing microorganisms

Smart Curriculum Narrative

Growth and disease

Theme/subtheme

Health and disease/Disease treatment and drug development

Disciplinary knowledge

- WS 2.2, 2.4
- MS 1a, 1b, 1d, 2a, 2h, 5c

Practical

Growing useful organisms

	Learning outcomes		
Learning objectives	**Developing Aiming for 3**	**Secure Aiming for 5**	**Extending Aiming for 8**
Describe the importance of providing a suitable culture medium when growing bacteria in the laboratory	State that bacteria reproduce by simple cell division in a process called binary fission	**Describe the importance of providing a suitable culture medium when growing bacteria in the laboratory**	Suggest how the balance of nutrients in a culture medium might affect the growth of bacteria
Explain why it is important to use an uncontaminated culture to investigate bacterial growth	Describe how to prepare a bacterial culture on agar gel	**Explain why it is important to use an uncontaminated culture to investigate bacterial growth**	Explain the importance of controlling the temperature at which a bacterial culture is grown
Describe and explain why each safety rule is needed in order to safely prepare, incubate, and dispose of a culture	State the safety rules to prepare an uncontaminated bacterial culture	**Describe and explain why each safety rule is needed in order to safely prepare, incubate, and dispose of a culture**	Explain how factors such as food and oxygen availability or a build-up of toxins, may affect the growth of a bacterial culture

Tier 2 vocabulary	Tier 3 vocabulary
bacteria, cell division, *colony*, *condition*, contamination, growth, microorganisms, mutation, sterilise, uncontaminated	agar gel, antibiotics, autoclave, binary fission, *culture*, culture medium, disinfectant, incubate, inoculate, pathogen, Petri dish

Lesson plan

Reactivate knowledge

1 What do we call microorganisms that cause disease?
2 How do viruses cause disease?
3 What causes the symptoms of disease?

Common misconception

Students sometimes state that the clear inhibition zone is due to the bacteria not growing. They should be more specific and state that this is where the bacteria have been killed.

Answers: 1 Pathogens *2* The virus takes over the cells of the body and uses it to create new viruses which bursts out and destroys the cells, therefore causing symptoms of the disease *3* Response of body to toxins produced by bacteria and cell damage caused by viruses

Trigger interest

Growing Bacteria (5 mins)

Show students an image of bacterial colonies grown on an agar plate. Ask students to consider how we are able to see the bacteria if they are microscopic.

Students should answer that it is because there are so many bacteria present that they are visible.

From this, remind students that bacteria reproduce by binary fission approximately every 20 minutes.

Exposition

Aseptic Technique (10 mins)

Discuss why it is important to use sterile conditions and produce an uncontaminated culture.

Support students by demonstrating how to sterilise metal equipment and the neck of a McCartney bottle in the roaring flame of a Bunsen burner.

Students then create a list of rules for preparing uncontaminated cultures in the lab.

To challenge students, ask them to consider why it would be harmful to grow the bacteria in a school at temperatures hotter than 20°C.

Main activity

Growing useful organisms (30 mins)

Students investigate the effect of antiseptics or antibiotics on bacterial growth.

Students prepare an uncontaminated culture of bacteria.

Supply students with small circles of filter paper and a range of different antiseptics. They use tweezers to soak the pieces of paper in different antiseptics and place them on different labelled sections of their dish. Leave the dishes to incubate until the next lesson.

Use Practical: *Growing useful organisms* activity sheet.

Review and practice

How to grow bacteria (10 mins)

In this interactive, students put the steps of the method used to prepare a culture of bacteria in the correct order.

B5.4 Preventing bacterial growth

AQA GCSE Biology specification points

4.1.1.6 Culturing microorganisms

Smart Curriculum Narrative

Health and disease

Theme/subtheme

Health and disease/Disease treatment and drug development

Disciplinary knowledge

- WS 2.2, 2.4
- MS 1a, 1b, 1d, 2a, 2h, 5c

Required Practical 2

Microbiology

	Learning outcomes		
Learning objectives	**Developing Aiming for 3**	**Secure Aiming for 5**	**Extending Aiming for 8**
Describe the difference between antiseptics, disinfectants, and antibiotics	State that disinfectants, antiseptics, and antibiotics are all used to kill bacteria or control bacterial growth	**Describe the difference between antiseptics, disinfectants, and antibiotics**	Explain when an antiseptic, a disinfectant, and an antibiotic would be used
Plan an investigation to investigate the impact of a disinfectant, antiseptic, or antibiotic on the growth of a bacterial culture by measuring zones of inhibition	Predict the impact of a disinfectant, antiseptic, or antibiotic on the growth of a bacterial culture in an investigation in terms of zone of inhibition	**Plan an investigation to investigate the impact of a disinfectant, antiseptic, or antibiotic on the growth of a bacterial culture by measuring zones of inhibition**	Use detailed scientific knowledge to explain the predicted impact of a disinfectant, antiseptic, or antibiotic on the growth of a bacterial culture in a planned investigation using zones of inhibition
Calculate the number of bacteria in a population after a certain time if given the mean division time	Using a provided graph, explain the exponential growth of a bacterial colony	**Calculate the number of bacteria in a population after a certain time if given the mean division time**	Calculate the number of bacteria in a sample when using a counting chamber
Calculate the zone of inhibition around antimicrobial discs or wells on bacterial agar plates	Measure the diameter of clear areas (zone of inhibition) on bacterial plates around antimicrobial discs or wells	**Calculate the zone of inhibition around antimicrobial discs or wells on bacterial agar plates**	Apply knowledge of sampling techniques to ensure samples are representative

Tier 2 vocabulary	Tier 3 vocabulary
bacteria, *concentration*, *conditions*, disease, *dividing*, population, factors, growth, microorganisms, reproduce	agar gel, antibiotic, antiseptic, binary fission, *colony*, *culture*, disinfectant

Lesson plan

Reactivate knowledge

1 How do bacteria reproduce?
2 What is needed to grow bacteria in the lab?
3 Why are microorganisms never kept above 25°C in a school laboratory?

Common misconception

Students sometimes state that the clear inhibition zone is due to the bacteria not growing. They should be more specific and state that this is where the bacteria have been killed.

Answers: 1 Binary fission *2* Culture medium, right temperature, and usually oxygen *3* To minimise the risk of harmful organisms growing

Trigger interest

Binary fission (5 mins)

Remind students that bacteria reproduce by dividing in two. Ask students to calculate how many bacteria are present after four hours if a bacterium divides once every hour.

To support students with this, give them the formula 2^n (where n is the number of divisions that occur).

Exposition

Calculating bacterial growth (20 mins)

Work through the maths box in the Student Book to show students how to calculate the number of bacteria in a population. Give students further examples of calculations for them to solve.

Remind students of the plates they set up last lesson and ask them to use their knowledge of the uses of different antiseptics and information from the Student Book to predict what they will see on their plates. Next, ask students to write on their mini-whiteboards what each of the following are:

- Antiseptic
- Antibiotic
- Disinfectant

Discuss their answers and students can write the definitions in their notes using the Student Book, or from the discussion.

Main activity

Microbiology (20 mins)

Allow students to study their plates from the last lesson, but remind them not to open them.

They should measure the clear area around each circle and calculate the area using πr^2. Address the misconception at this point.

They then write a conclusion explaining whether their prediction was correct, and evaluating their investigation.

Support students by asking them to measure the diameter of the clear area.

As an extension, students should pool results, applying sampling techniques to choose clear areas that are representative and calculating the mean area.

Review and practice

Antiseptic, antibiotic, or disinfectant? (10 mins)

Use this interactive and have students match the biological term with its correct meaning in their books. Students then decide what would be best to use in different scenarios.

B5.5 Preventing infections

AQA GCSE Biology specification points
4.3.1.1 Communicable (infectious) diseases

Theme/subtheme
Health and disease/Communicable diseases

***Smart Curriculum* Narrative**
Health and disease

Activity
Public health campaign

	Learning outcomes		
Learning objectives	**Developing Aiming for 3**	**Secure Aiming for 5**	**Extending Aiming for 8**
Describe how the spread of diseases can be reduced or prevented	List some ways in which communicable diseases are spread	**Describe how the spread of diseases can be reduced or prevented**	Use scientific knowledge to explain in detail how different methods reduce or prevent the spread of disease
Produce a communication to inform the general public about how to stop the spread of a disease	Take a role in designing a form of communication to inform the public about how to prevent the spread of a disease	**Produce a communication to inform the general public about how to stop the spread of a disease**	Use examples to explain how the scientific method has been applied to help prevent the spread of disease

Tier 2 vocabulary	Tier 3 vocabulary
cause, cure, disease, hygiene, immune system, isolation, microorganisms, preventative, *spread*, symptoms	antiseptic, communicable, disinfectants, infectious, pathogen, vaccination, *vectors*

Lesson plan

Reactivate knowledge

1 What is exponential growth?

2 What is the difference between a disinfectant and an antiseptic?

3 What is an antibiotic?

Common misconception
Students might think viruses are living. Clarify viruses as non-living whenever referring to a disease causes by a virus.

Answers: 1 When bacteria divide repeatedly so the population initially grows slowly but then increases fast to give an exponential curve *2* A disinfectant kills bacteria in an environment but is toxic to human cells whereas an antiseptic is safe to use on human skin *3* A compound safe to use inside the human body which kills bacteria or stops them growing

Trigger interest

The plague (10 mins)

Tell the class that the Great Plague was spread by fleas. Tell them some methods that people in the past used to try to stop the spread of the plague (isolating infected families, carrying sweet-smelling flowers, covering their faces, praying). Ask them to discuss how useful they think each method was, and what they would recommend instead.

Exposition

How would you control the spread? (5 mins)

Supply students with the name of a communicable disease (e.g., common cold) and how it is spread. Ask them to suggest which control methods would work to stop the spread of the disease. Suggestions could include isolation, covering the nose and mouth, sneezing or coughing into elbows.

Main activity

Public health campaign (35 mins)

Ask students to read the information in the Student Book on preventing the spread of communicable diseases Make students aware that communicable diseases are among the leading causes of death and disability, particularly in low income countries. Divide the class into groups and give details about recent pandemics or epidemics, for example, COVID-19, Ebola in Sierra Leone, MERS in South Korea, cholera in Haiti, Spanish flu in parts of Europe, and so on.

Groups research a disease and how it is spread. They then create a campaign to educate people about how to keep healthy and reduce the spread of the disease. This could include a poster, leaflet, or advertisement, or all three if time allows. Students will need access to a computer and the Internet for this. If this is not possible provide printed information for them.

Ask groups to present their campaigns and ask students to peer assess their work based on scientific accuracy and how useful it would be.

Review and practice

The work of Semmelweis (5 mins)

Using the Interactive: *The work of Semmelweis.* Here, students match the stages of the different scientific methods in his work (hypothesis, predict, test, observe, conclude).

B5.6 Viral diseases

AQA GCSE Biology specification points

4.3.1.2 Viral disease

Smart Curriculum Narrative

Health and disease

Theme/subtheme

Health and disease/ Communicable diseases

Activity

Measles vaccination

	Learning outcomes		
Learning objectives	**Developing** **Aiming for 3**	**Secure** **Aiming for 5**	**Extending** **Aiming for 8**
Describe the main problem with treating viral diseases	Name measles and HIV as examples of human diseases caused by viruses, and tobacco mosaic virus as an example of a viral plant disease	**Describe the main problem with treating viral diseases**	Explain why the prevention of viral infections is key to controlling them in animals and plants
Describe how measles, HIV, and tobacco mosaic virus affect the infected organism, including the progression of HIV infection to AIDS if untreated, and the economic and social impact of these diseases	Describe how measles, HIV, and tobacco mosaic virus are spread and list the main symptoms of each disease	**Describe how measles, HIV, and tobacco mosaic virus affect the infected organism, including the progression of HIV infection to AIDS if untreated, and the economic and social impact of these diseases**	Discuss the difficulties of controlling viral diseases in populations and explain the advances made with antiretroviral drugs in treating HIV
Interpret data to describe how the number of people infected with specific viral diseases has changed over time	Summarise overall trends from graphical representations of data on viral diseases	**Interpret data to describe how the number of people infected with specific viral diseases has changed over time**	Use reliable data to write a persuasive letter to parents urging them to vaccinate their children against measles

Tier 2 vocabulary	Tier 3 vocabulary
contact, control, disease, *exchange*, hygiene, immune system, infect, fever, photosynthesis, *rash*, symptoms, transfusion	antiretroviral drugs, HIV/AIDS, measles, pathogen, tobacco mosaic virus (TMV), vaccination, virus, *vector*

Lesson plan

Reactivate knowledge

Common misconception

Students often don't consider that plants can contract communicable diseases too.

1 Name four ways of reducing the spread of infectious diseases.
2 Give one hygiene measure and explain how it works.
3 What does TMV stand for and how does it affect plants?

Answers: 1 Good hygiene, isolation, controlling or destroying vectors, vaccination *2* Any from: hand washing – stops contact spread; disinfectant – reduces number of pathogens; separate raw meat from other food – prevents spread of pathogens; cough into tissue or elbow – prevents airborne spread *3* Tobacco mosaic virus – discolours the leaves so they can't photosynthesise

Trigger interest

What is the disease (10 mins)

Students complete the interactive to select true or false for a series of statements about viral diseases. They then match the statements to the correct disease in their books.

Exposition

Viral infections (20 mins)

Ask students to read through the information on measles, HIV, and the tobacco mosaic virus in the Student Book, or give them access to a computer room to conduct their own research. They should then summarise the information in a table showing disease, pathogen, ways to reduce spread, symptoms, and treatment.

Main activity

Measles vaccination (20 mins)

Show students a graph that shows the number of cases of measles in the UK since the start of the twentieth century. Ask pairs to discuss what it shows and possible explanations for this.

Discuss their thoughts. Tell them that in the late 1990s many parents stopped giving their children the MMR vaccine because of claims that it was linked to bowel disorders and autism. Tell them that, even though these claims were later shown to be false, many children are still not being vaccinated. Ask students to write an open letter telling parents why they should vaccinate their children against measles.

Review and practice

Treatment (5 mins)

Ask students to write down reasons why viral infections are often more difficult to prevent and treat than bacterial infections.

As an extension, ask students to consider the difficulties of controlling viral diseases in populations and explain the advances made with antiretroviral drugs in treating HIV. Include some case studies of the experiences of a diverse range of people living with HIV in the UK.

Some students may also need a reminder of how viruses replicate.

B5.7 Bacterial diseases

AQA GCSE Biology specification points

4.3.1.3 Bacterial diseases

Theme/subtheme

Health and disease/ Communicable diseases

Smart Curriculum Narrative

Health and disease

Activity

Advice for patients

	Learning outcomes		
Learning objectives	**Developing** **Aiming for 3**	**Secure** **Aiming for 5**	**Extending** **Aiming for 8**
Describe similarities and differences between *Salmonella* and gonorrhoea	Name *Salmonella* food poisoning and gonorrhoea as examples of human diseases caused by bacteria	**Describe similarities and differences between *Salmonella* and gonorrhoea**	Suggest why viral diseases are more likely to have higher mortality rates than bacterial diseases
Describe how the spread of *Salmonella* and gonorrhoea is controlled	Describe how *Salmonella* and gonorrhoea are spread	**Describe how the spread of *Salmonella* and gonorrhoea is controlled**	Explain in detail how methods to control the spread of *Salmonella* and gonorrhoea work

Tier 2 vocabulary	Tier 3 vocabulary
bacteria, diarrhoea, dehydration, *discharge*, diseases, poisoning, resistant, symptoms, *viral*	antibiotics, antibiotic resistance, communicable, gonorrhoea, *incidence*, infectious, *Salmonella*, sexually transmitted disease (STD), sexually transmitted infection (STI), vaccination

Lesson plan

Reactivate knowledge

1 Name two human viral diseases.

2 What are the main symptoms of measles and why is it so dangerous?

3 How does bacteria divide?

Answers: 1 Any from: measles, HIV/AIDS, COVID-19 *2* Fever and red rash: can cause blindness, brain damage, and death *3* Binary fission

Common misconception

Students may think that viruses can be killed by antibiotics. Clarify that antibiotics work against bacteria because they break down the cell wall found in bacteria.

Trigger interest

Antibiotic resistance (5 mins)

Share with the class a headline or news video about antibiotic resistance. Discuss why this is a problem, and what scientists are trying to do about it. Clarify the misconception at this point.

Exposition

***Salmonella* and gonorrhoea (10 mins)**

Introduce *Salmonella* and gonorrhoea as bacterial infections in humans. Tell students the similarities and differences between them (including how they are spread and can be controlled) and ask students to produce a summary table which can then be used later in the lesson.

Main activity

Advice for patients (30 mins)

Students then write advice on *Salmonella* or gonorrhoea suitable for patients at a doctor's surgery. This can be a poster warning people how to avoid getting *Salmonella* from a summer barbeque, or a leaflet on how to tell if you have gonorrhoea and the treatments available.

Share with the class a news story on the fact that gonorrhoea may soon be untreatable due to antibiotic resistance.

Ask students to work in small groups to summarise the story in one sentence. After listening to these, discuss as a class why antibiotic resistance is a problem.

Review and practice

Bacteria true or false (10 mins)

Students use this interactive to identify if a series of statements on bacterial infections are true or false. They then match the statements to the correct disease in their books.

B5.8 Diseases caused by fungi and protists

AQA GCSE Biology specification points

4.3.1.4 Fungal diseases

4.3.1.5 Protist diseases

Smart Curriculum Narrative

Health and disease

Theme/subtheme

Health and disease/Plant diseases, deficiencies, and defence responses

Activity

Stop malaria

	Learning outcomes		
Learning objectives	**Developing Aiming for 3**	**Secure Aiming for 5**	**Extending Aiming for 8**
Describe how rose black spot affects a plant and how it is treated	State that rose black spot is a plant disease caused by fungi and malaria is a human disease caused by protists and spread by mosquitoes	**Describe how rose black spot affects a plant and how it is treated**	Discuss the problems in treating a fungal disease such as rose black spot in plants
Describe how the main ways of controlling malaria have their effects	State how the malaria protist is passed into the human host by mosquitoes and list some ways in which malaria is controlled	**Describe how the main ways of controlling malaria have their effects**	Explain in detail how different methods of controlling and treating malaria are effective at reducing the risk of the disease

Tier 2 vocabulary	Tier 3 vocabulary
disease, fungi, infection, parasite, resistant, *spread*	anti-fungal, fungicide, *host*, insecticide, malaria, pathogen, protist, rose black spot, spores, vaccine, *vector*

Lesson plan

Reactivate knowledge

1 Name two human bacterial diseases.

2 Where do *Salmonella* bacteria live?

3 How can the spread of gonorrhoea be prevented?

Answers: 1 Any from: *Salmonella* food poisoning, gonorrhoea, tonsilitis *2* In the guts of many different animals *3* Barrier methods of contraception; treating affected person and their sexual partners with antibiotics

Common misconception

Students often don't consider that plants can contract communicable diseases too.

Trigger interest

Fungal infections (5 mins)

Show images of a range of fungal infections, for example, athlete's foot, rose black spot, potato blight, and ringworm. Ask students what they think they have in common, and reveal that they are all fungal infections.

Exposition

Rose black spot (15 mins)

Show images of rose black spot. Address the misconception at this point and tell students there are lots of different types of plant diseases.

Tell the class that a rose grower has sent these in and asked for advice on what it is. Students should use the Student Book to research the disease and reply to the rose grower.

They should consider including an explanation of:

- what rose black spot is
- if rose black spot is harmful
- how to stop it from spreading to other plants.

Main activity

Stop malaria (30 mins)

Introduce malaria to students, including how it kills, how many people it affects, and where this is likely to happen in the world.

Explain that malaria is caused by a protist, and discuss the life cycle of the malaria protist.

Ask students to produce a plan to eradicate malaria. They should refer to the life cycle of the protist that causes the disease, and annotate each stage with methods that can be used to prevent the spread.

Review and practice

Fungi and protists (5 mins)

Students complete the interactive and decide whether diseases are cause by fungi or protists. They then type in answers to complete a passage about the spread and control of malaria.

B5.9 Human defence responses

AQA GCSE Biology specification points

4.3.1.6 Human defence system

Smart Curriculum Narrative

Health and disease

Theme/subtheme

Health and disease/Immune system and vaccines

Activity

Phagocytosis

	Learning outcomes		
Learning objectives	**Developing Aiming for 3**	**Secure Aiming for 5**	**Extending Aiming for 8**
Describe how the skin, nose, trachea and bronchi, and the stomach act as non-specific defence systems against pathogens	Name the skin, nose, trachea and bronchi, and the stomach as non-specific defence systems which help protect the human body against pathogens	**Describe how the skin, nose, trachea and bronchi, and the stomach act as non-specific defence systems against pathogens**	Explain why the skin, nose, trachea and bronchi, and stomach are known as part of the non-specific defence system of the body
Describe three ways in which the white blood cells help protect the body against disease-causing pathogens	State that white blood cells are part of the immune system and help defend the body against pathogens	**Describe three ways in which the white blood cells help protect the body against disease-causing pathogens**	Explain in detail how the production of antibodies and antitoxins by the immune system prevents infection by a viral or bacterial pathogen, and how a reduced or overactive immune system may cause disease

Tier 2 vocabulary	Tier 3 vocabulary
bacteria, barrier, clot, defence, disease, immune system, microorganisms, mucus, platelets, virus	antibodies, antigen, antimicrobial, antitoxins, cilia, non-specific defence system, pathogens, white blood cells

Lesson plan

Reactivate knowledge

1 Name a common mild human fungal disease.

2 Which organism is the vector for malaria?

3 Give four ways of preventing the spread of malaria.

Common misconception

Students can think that viruses are living. Clarify that viruses are non-living and can only reproduce inside a cell, which makes it harder for the white blood cells to detect and destroy a virus.

Answers: 1 Athlete's foot *2 Anopheles* mosquito *3* Insecticide-impregnated mosquito nets – insecticides to kill mosquitoes in homes; preventing mosquitoes breeding (no standing water, spraying water with insecticide); vaccination

Trigger interest

Pathogens invading the body (10 mins)

Show images of a pathogen with antigens shown. Explain that it is due to antigens having different shapes which allow the white blood cells to identify them. Ask students to list all the entry points where pathogens can enter the body to cause an infection.

Show images of the skin, eyes, ears, nose, and mouth as prompts to support if needed.

Then ask students to suggest any defences our bodies have to prevent entry at these points.

Exposition

Human defence systems (15 mins)

Discuss that humans have many mechanisms to stop pathogens from entering the body.

Ask students to complete task one of the Activity: *Human Defence Systems*. For this, students label the defence system. After recording their initial thoughts, allow groups to use information from the Student Book to add further information.

Main activity

Phagocytosis (25 mins)

Show the class an animation or video of white blood cells ingesting pathogens. Question students on what it is showing and why it is important for our health.

Students then complete the questions from the human defence systems activity. Review the answers with the class for students to self-mark.

Review and practice

Where is it found? (5 mins)

Students use this interactive to select the correct defence mechanisms in the human body. They then match the defence mechanism to where it is found in the human body.

B5.10 More about plant diseases

AQA GCSE Biology specification points

(H) 4.3.3.1 Detection and identification of plant diseases

Smart Curriculum Narrative

Health and disease

Theme/subtheme

Health and disease/ Plant diseases, deficiencies, and defence responses

Disciplinary knowledge

- WS 1.4

Activity

Plant diseases

	Learning outcomes		
Learning objectives	**Developing Aiming for 3**	**Secure Aiming for 5**	**Extending Aiming for 8**
Explain that plant diseases have a major impact on global food security and economies and describe seven symptoms used to detect disease in plants	Name three types of pathogens that cause disease in plants and list two specific plant diseases	**Explain that plant diseases have a major impact on global food security and economies and describe seven symptoms used to detect disease in plants**	Describe how plant diseases are identified, giving a scientific explanation for each
Describe how aphids attack plants and how they may be controlled	Identify aphids as common plant pests	**Describe how aphids attack plants and how they may be controlled**	Explain two ways in which aphids cause plant disease and evaluate the different methods used to control them
Match signs of plant disease to ion deficiency and explain how nitrate and iron concentrations affect plant growth	State that mineral deficiencies cause non-communicable plant diseases	**Match signs of plant disease to ion deficiency and explain how nitrate and iron concentrations affect plant growth**	Analyse data showing the effect of mineral ion concentrations on plant growth and explain how understanding ion deficiencies allows horticulturists to provide optimal conditions for plants

Tier 2 vocabulary	Tier 3 vocabulary
aphid, bacteria, deficiency, disease, fungi, ions, magnesium, mineral, nitrate, pest, virus	anti-fungal, chlorosis, monoclonal antibodies, pathogen, pesticide, *vector*

Lesson plan

Reactivate knowledge

Common misconception

Students often don't consider that plants can contract communicable diseases.

1 How does the skin protect humans against infections?
2 What are the defences of the respiratory system again pathogens getting into the lungs?
3 What are the three different ways in which white blood cells defend the body against infection?

Answers: 1 Acts as a barrier; produces substances that kill pathogens *2* Mucus/hairs to trap pathogens; cilia in trachea and bronchi that move mucus and pathogens away from the lungs *3* Engulfs pathogens; produce antibodies against specific pathogens; produce antitoxins against bacterial toxins

Trigger interest

True or false? (5 min)

Tell the class that only animals (including humans) get ill. Ask them to discuss this in pairs. Listen to the feedback and make a list of all the ways in which plants can be unhealthy. Use this to address the misconception.

Exposition

Plant doctors (10 mins)

Either collect a variety of unhealthy plant specimens, or show images of unhealthy plants. Ask the students to study the plants and describe their symptoms.

Show students an image of a plant covered in aphids. Inform students that aphids have a sharp mouthpart that can penetrate the phloem so they can feed on the sugar solution in the phloem. Ask students to suggest how this may cause harm to the plant.

Main activity

Plant diseases (35 mins)

Ask students to suggest what causes ill health in plants. They then use the Student Book or computers to conduct their own research to make a list of the causes of plant disease, and how each damages a plant using the Activity: *Plant diseases*. Review the tables using the teacher notes for this activity.

Students then analyse the results of an experiment about the effects of mineral deficiencies on plant growth (these can be photos sourced from the Internet), commenting on changes in the appearance of the plants, differences in the height of the seedlings, and changes in dry mass.

Students then complete the questions from the Activity: *Plant diseases*. Students can self-mark their answers.

Review and practice

What's the disease? (5 mins)

Use the Interactive: What's the disease? Show the class a series of images showing plant diseases, and ask them to name the problem.

B5.11 Plant defence responses

AQA GCSE Biology specification points

4.3.3.2 Plant defence responses

Smart Curriculum Narrative

Health and disease

Theme/subtheme

Health and disease/ Plant diseases, deficiencies, and defence responses

Activity

Plant defences

	Learning outcomes		
Learning objectives	**Developing Aiming for 3**	**Secure Aiming for 5**	**Extending Aiming for 8**
Classify plant defences as physical, chemical, or mechanical	Describe why plants need to defend themselves	**Classify plant defences as physical, chemical, or mechanical**	Explain in detail the difference between physical, chemical, and mechanical plant defences
Describe four physical defences against pathogens and explain how each one works	Give one example of a physical plant defence against attacks by pathogens and insects, and one example of a chemical defence system	**Describe four physical defences against pathogens and explain how each one works**	Discuss the importance of plant defences to human well-being
Carry out research using secondary resources of personal choice to present examples of plant defence responses	Using secondary resources provided, find out more about plant defence systems	**Carry out research using secondary resources of personal choice to present examples of plant defence responses**	Interpret information from a scientific article to explain how plant-to-plant communication can be used as a form of defence

Tier 2 vocabulary	Tier 3 vocabulary
aphids, barrier, chemical, damage, defence, herbivores, mechanical, mimicry, microorganisms, poison, stomata, thorns	antibacterial, pathogen

Lesson plan

Reactivate knowledge

1 How do aphids increase the risk of a plant becoming infected by pathogens?

2 What is the main symptom of magnesium deficiency in plants?

3 Why does nitrate deficiency matter in plants?

Answers: 1 Sharp mouthparts penetrate the phloem and let pathogens in *2* Yellow leaves *3* Nitrates are needed to make proteins for plant growth

Common misconception

Students often don't realise plants have a defence system because they do not have blood and white blood cells.

Trigger interest

What do they have in common? (10 mins)

Show students a bottle of tea tree oil or a product that contains it, and the nettles on a leaf. Ask students to suggest what these have in common.

Explain to students that tea tree oil is used because it has antibacterial properties. Ask them why plants would produce a chemical like this, and what the advantage is for them.

Ask students to consider the advantage of stinging nettles to a plant.

Exposition

Types of defences (10 mins)

Discuss why plants need to defend themselves, and from what.

Provide students with some unusual and interesting methods of plant defences (e.g., mint, witch hazel, hypericum, foxgloves, lab's ear, *Mimosa pudica*, and plant mimicry). Explain the differences between mechanical, physical, and chemical defences and ask students to match up them up to the plants.

Main activity

Plant defences (30 min)

Ask students to draw a picture of a plant and use information from the Student Book to label all the methods plants use to defend themselves. They can use a key to classify them as physical, mechanical, and chemical methods. Use the Activity: *Plant defences* for this.

To challenge students, provide them with a scientific article about how acacia plants communicate with each other as a form of defence. They should then explain what the article is about in their own words.

Review and practice

Physical, mechanical, or chemical? (5 mins)

Use this interactive where students select true or false in a series of statements about plant defence responses. They then sort defence mechanisms into whether they are physical barriers, chemical barriers, or mechanical adaptations.

B5 Communicable diseases

Introduction to checkpoint intervention

The checkpoint intervention provides suggestions for a lesson to follow up the *B5 Communicable diseases* checkpoint assessment. Depending on the outcomes of the assessment, these suggestions can be used to move a class towards secure, to target specific learning outcomes, or to consolidate knowledge for students achieving secure outcomes. Use students' outcomes from the checkpoint assessment to address general misconceptions from the content covered. Be prepared to re-cover content as required; three key concepts from the chapter are covered in more detail below.

Checkpoint secure learning outcomes

- Describe the difference between communicable and non-communicable diseases.
- Calculate the area of clearance around antimicrobial discs or wells on bacterial agar plates.
- Describe how measles, HIV, and tobacco mosaic virus affect the infected organism, including the progression of HIV infection to AIDS if untreated and the economic and social impact of these diseases.
- Describe how the spread of *Salmonella* and gonorrhoea is controlled.
- Explain that plant diseases have a major impact of global food security and economies and describe seven symptoms used to detect disease in plants.
- Describe how the skin, nose, trachea and bronchi, and the stomach act as non-specific defence systems against pathogens.

Student reflection

Encourage students to reflect on whether there were any checkpoint questions they found difficult or straightforward as well as their preparation for the checkpoint assessment, including how they revised, what they revised, the time spent on revision, and what areas they could improve upon. For more strategies, see the Metacognition spread.

Intervention activity: Describe how measles, HIV, and tobacco mosaic virus affect the infected organism, including the progression of HIV infection to AIDS if untreated and the economic and social impact of these diseases.

Students should use the Student Book and conduct research online about measles, HIV, and tobacco mosaic virus.

Then create a table with the following headings:

- How organism is affected?
- Progression of disease
- How does it spread?
- Social impact
- Economic impact

Students fill out their tables with the information they have researched. Then, as a class, plan a 6 mark answer to the following exam question:

Describe and explain the progression and impact of one communicable disease. Use named examples.

Intervention activity: Explain that plant diseases have a major impact of global food security and economies and describe seven symptoms used to detect disease in plants.

Students should create a poster to inform gardeners about the disease rose black spot.

The poster should include:

- which pathogen causes the disease
- what the symptoms are and how this affects plant growth (to support students, tell them to link this to photosynthesis)
- what they should do to prevent the spread of the disease if they discover they have an infected plant.

Intervention activity: Describe how the skin, nose, trachea and bronchi, and the stomach act as non-specific defence systems against pathogens.

Inform students that the reason most people get a cold every year is because the common cold is caused by around 200 different viruses, so each time we have a cold we are being infected by a new virus that we have no immunity to.

Ask students to list all of the defences in the body that protect against infection from viruses that cause the common cold.

To support students, inform them that the virus is transmitted in the air, typically by droplet infection, so they should consider this when identifying the defence mechanisms.

As an extension, challenge students by asking them to explain how each defence mechanism prevents entry of the virus or destroys the virus.

B6 Preventing and treating disease

B6.1 Vaccination

AQA GCSE Biology specification points
4.3.1.7 Vaccination

Smart Curriculum Narrative
Health and disease

Theme/subtheme
Health and disease/Immune system and vaccines

Disciplinary knowledge
- WS 1.4c

Activity
How do vaccinations work?

	Learning outcomes		
Learning objectives	**Developing Aiming for 3**	**Secure Aiming for 5**	**Extending Aiming for 8**
Describe, with reference to the human immune system, how vaccination works	Give two reasons why people are vaccinated	**Describe, with reference to the human immune system, how vaccination works**	Explain the principle of vaccinating whole populations to produce herd immunity and reduce the spread of specific pathogens
Describe effective immunity to a disease in terms of antibodies and white blood cell memory	State that vaccines contain dead or inactive forms of a pathogen	**Describe effective immunity to a disease in terms of antibodies and white blood cell memory**	Apply ideas about the specificity of antibodies to explain the need to have different vaccines for different diseases
Interpret data on behaviour and vaccination patterns on the incidence of disease	Describe the impact of specific vaccinations from graphical data provided	**Interpret data on behaviour and vaccination patterns on the incidence of disease**	Evaluate the effectiveness of vaccination programmes from given data

Tier 2 vocabulary	Tier 3 vocabulary
bacteria, disease, inactive, pathogen, population	antigen, herd immunity, vaccination, white blood cell

Lesson plan

Reactivate knowledge

Common misconception

Students might think that vaccines cure diseases, rather than prevent them.

1 Name four ways to prevent the spread of infectious diseases.
2 Which type of blood cells are involved in defending against invading microorganisms?
3 Give three ways these blood cells protect against infection.

Answers: 1 Good hygiene/hand washing; isolation of infected people; controlling disease vectors; vaccination *2* White blood cells *3* Engulf and digest pathogens; make antibodies; make antitoxins

Trigger interest

Antigens and antibodies (5 mins)

Draw a diagram of a pathogen and ask students to list, and describe three ways that white blood cells will work to destroy the pathogen.

Use this as a link to inform students that the production of antibodies by the white blood cells is fundamental in how vaccines work.

Exposition

Immunity (15 mins)

Ask students to write down what they think the term 'immune' means. Then ask them if they know of any ways that humans can become immune to a disease. Some may suggest having immunity due to previously having had a disease or from vaccinations. Some may suggest being naturally immune from birth or having temporary immunity from breast milk antibodies.

Show students a graph of antibody concentration against time and talk them through the primary and secondary response to infection. Students may question why you can have a cold numerous times. Inform students that there are approximately 200 different viruses that cause the common cold, so when they do get a cold, they are most likely being infected with a new virus each time.

Link this initially to natural immunity. Inform students that the reason they are often ill from the common cold for about a week – feeling most ill in the first three days – is because they are being infected with a new virus and it takes time for the white blood cells to create the correct complementary-shaped antibody. If they are reinfected with the same pathogen, their white blood cells will create the correct shaped antibody rapidly and in large enough quantities that the pathogen should be destroyed before they develop symptoms, and therefore they would be immune. Address the misconception here.

Let students know that scientists have used this knowledge to develop vaccines, so that people could artificially develop immunity without having to endure the disease.

Vaccination – talk students through the concept of herd immunity and use this as a link to talk about people who are too vulnerable to have a vaccine.

Discuss the reasons why some people choose to be vaccinated whilst others don't.

Main activity

How do vaccinations work? (30 mins)

Ask students to use information in the Student Book to create a storyboard explaining how vaccinations work.

Their storyboards should include the following stages:

- Vaccine contains dead or inactive forms of a pathogen.
- Vaccine injected into the body.
- White blood cells produce antibodies against the pathogen.
- (Some time later) live pathogen infects the body.
- White blood cells make antibodies quickly, and the pathogen is destroyed.

Higher tier students should include information on the specificity of antibodies.

Review and practice

Vaccination opinions (5 mins)

Get students to reflect on the public discourse around vaccines, trials, and the history of use and misuse of trialling new treatments. Present students with various headlines and opinions on why people might be wary of vaccines, and that this is nothing new (e.g., whooping cough controversy in 1970s UK).

B6.2 Antibiotics and painkillers

AQA GCSE Biology specification points

4.3.1.8 Antibiotics and painkillers

Smart Curriculum Narrative

Health and disease

Theme/subtheme

Health and disease/Disease treatment and drug development

Disciplinary knowledge

- WS 1.4

Activity

Antibiotic resistance

Learning objectives	Learning outcomes: Developing Aiming for 3	Learning outcomes: Secure Aiming for 5	Learning outcomes: Extending Aiming for 8
Describe how antibiotics work	Define antibiotics	**Describe how antibiotics work**	Explain the importance of using the right specific antibiotic to treat specific types of bacteria
Describe what is meant by antibiotic-resistance in bacteria and explain why it matters	State that some bacteria have become antibiotic-resistant	**Describe what is meant by antibiotic-resistance in bacteria and explain why it matters**	Plan an investigation into antibiotic resistance in a strain of bacteria and suggest why scientists are constantly trying to develop new antibiotics
Describe how painkillers and other over-the-counter drugs are used	Describe a painkiller	**Describe how painkillers and other over-the-counter drugs are used**	Explain the limitations of medicines used to treat symptoms compared to antibiotics which kill or inhibit growth of bacteria
Explain why it is difficult to develop drugs to treat viral infections	State that viral infections cannot be treated with antibiotics	**Explain why it is difficult to develop drugs to treat viral infections**	Discuss the limitations of medicines used to treat symptoms compared with antibiotics and antiviral medicines

Tier 2 vocabulary	Tier 3 vocabulary
bacteria, inactivate, infection, resistant	antibiotics, painkillers, penicillin, virus

Lesson plan

Reactivate knowledge

1 What is an antigen?

2 What is an antibody?

3 Which type of disease is cured by antibiotics?

Answers: 1 Unique protein markers on the surface of cells *2* Proteins made by white blood cells which match up and attach to specific antigens *3* Bacterial infections

Common misconception

Students may think that antibiotics can destroy viruses as well as bacteria. Students may also think that bacteria become 'immune' to antibiotics instead of 'resistant'.

Trigger interest

Antibiotics and painkillers (5 mins)

Ask students to write down the difference between antibiotics and painkillers, when you would take an antibiotic, and when you would take a painkiller.

Address this misconception here – clearly state that antibiotics *only* kill bacteria.

Exposition

Antibiotics (15 mins)

State that antibiotics are medicines that destroy bacteria and explain how they work. This is a good opportunity to show a video of bacteria being destroyed by antibiotics. Students could use the Internet to further research the efficacy of antibiotics.

Ask students to firstly recall how viruses replicate, and then to consider the following:

- Why won't your doctor give you antibiotics for flu?
- Why is it difficult to treat viral infections?

The following will address any misconceptions.

Explain that viruses are very hard to destroy using medicine (e.g., antibiotics) because it would result in damaging the body cells. Therefore, the main treatment/advice is to rest and take painkillers to relieve any symptoms.
Discuss that painkillers will relieve symptoms but will *not* destroy pathogens.

Main activity

Antibiotic resistance (30 mins)

Discuss with students that antibiotics were deemed the 'wonder' drug in the 1940s and they played an important role in World War II. This resulted in the overuse of antibiotics.

Ask students if they know about the issues surrounding antibiotic overuse. At this point, address this misconception – some students may say 'bacteria became immune'. Remind them that immune means you don't develop the symptoms of a disease, therefore it is not the correct term to use.

Link the students' answers of 'antibiotic resistance' to the Required Practical 2 and show them an image of an agar plate with antibiotic discs and inhibition zones.

Discuss the development of methicillin-resistant *Staphylococcus aureus* (MRSA) and how antibiotic resistance arises.

Discuss the impact of antibiotic overuse and how dangerous this is. Explain this is why doctors are reluctant to prescribe antibiotics unless they are essential.

Students complete the Activity: *Doctor Doctor.*

Review and practice

Career paths (5 mins)

Highlight a diverse range of people working in the pharmaceuticals industry? For example, research and development (lab based, clinical trials), manufacturing (engineers), distribution and marketing? Discuss the skills students have that could align well in these areas.

B6.3 Discovering drugs

AQA GCSE Biology specification points

4.3.1.9 Discovery and development of drugs

Smart Curriculum Narrative

Health and disease

Theme/subtheme

Health and disease/Disease treatment and drug development

Disciplinary knowledge

- WS 1.6

Activity

Drug development

	Learning outcomes		
Learning objectives	**Developing Aiming for 3**	**Secure Aiming for 5**	**Extending Aiming for 8**
Describe the origin and uses of the heart drug digitalis and the painkiller and anti-coagulant aspirin	State that traditionally many drugs were extracted from plants and microorganisms	**Describe the origin and uses of the heart drug digitalis and the painkiller and anti-coagulant aspirin**	Suggest the advantages of synthesising aspirin over extracting the original chemical compound from plants
Describe the sequence of events in the discovery of penicillin from discovery to industrial production and explain why the latter step was so important	State that penicillin, the first antibiotic, was discovered by Alexander Fleming from the *Penicillium* mould	**Describe the sequence of events in the discovery of penicillin from discovery to industrial production and explain why the latter step was so important**	Describe the advantages and disadvantages of looking for new drugs from living organisms
State that most new drugs are now synthesised by the pharmaceutical industry and explain that this enables patients to take carefully measured doses of pure active compounds	State why scientists continue to look for new medicines	**State that most new drugs are synthesised in the pharmaceutical industry and explain that this enables patients to take carefully measured doses of pure active compounds**	Analyse data to evaluate the effectiveness of new antibiotics and make a reasoned decision on which one to develop further

Tier 2 vocabulary	Tier 3 vocabulary
anti-coagulant, infection, medicine, microorganisms, painkiller, *synthesised*	anti-inflammatory, aspirin, *cultures*, penicillin

Lesson plan

Reactivate knowledge

1 What are the two main categories of drugs?
2 Give two problems with the use of antibiotics.
3 Why are antiviral drugs so difficult to develop?

Answers: 1 Drugs which relieve symptoms, drugs which cure diseases *2* They don't affect viral infections, some types of bacteria are becoming resistant to antibiotics *3* Viruses go inside the human cells so it is hard to damage them without harming human cells

Common misconception

Students may think that antibiotics, such as penicillin, can destroy viruses.

Trigger interest

Making connections (10 mins)

Show the class images of a willow tree, a foxglove, mould, and a beaver tail. Ask students what these have in common. After a quick discussion, ask them to read the Student Book, then ask for their ideas again. Discuss that these images all represent the sources of substances that modern drugs are based on.

Exposition

The discovery of penicillin (10 mins)

In pairs, provide students with cards from Activity: *The discovery of penicillin* and get them to put the cards into the correct order.

Present students with this statement:
'If Fleming had been a tidier scientist, we would not have any antibiotics.'

In pairs, ask students to write down one argument that supports this statement, and one that refutes it. Indicate to students that if they take GCSE history they will have come across this before, as this drug was tested on soldiers during World War II.

Main activity

Drug development (30 mins)

Students analyse data on the effectiveness of plant and microorganism extracts on the growth of bacteria. They then use this data to decide which extract is the best option to progress to the next stage of drug development. Use the activity Working scientifically: *Drug Development*.

Review and practice

Matching drugs (5 mins)

Call out plants and/or microorganisms for students to then name the drug that was developed from it.

B6.4 Developing drugs

AQA GCSE Biology specification points

4.3.1.9 Discovery and development of drugs

Smart Curriculum Narrative

Health and disease

Theme/subtheme

Health and disease/Disease treatment and drug development

Disciplinary knowledge

- WS 1.6

Activity

Carrying out a double-blind trial

	Learning outcomes		
Learning objectives	**Developing Aiming for 3**	**Secure Aiming for 5**	**Extending Aiming for 8**
Describe the characteristics which a new drug is tested for	State that new medical drugs must be tested and trialled before they are used to check that they are safe and effective	**Describe the characteristics which a new drug is tested for**	Describe and explain the four characteristics of a good medicine
Explain why each procedure in drugs testing and trialling is used	Give the procedures used to trial a new drug in the correct order	**Explain why each procedure in drugs testing and trialling is used**	Describe in some detail how new medical drugs are tested and trialled for safety, effectiveness, toxicity, efficacy, and dose
Describe how a double-blind trial is carried out	State that the final stage of clinical trials of a new drug is known as a double-blind trial	**Describe how a double-blind trial is carried out**	Explain why the results of drug trials are published in journals
Explain why a placebo is used during drug trialling	Describe what is meant by a placebo	**Explain why a placebo is used during drug trialling**	Explain why the placebo in a drug trial is often an active drug rather than a sugar pill

Tier 2 vocabulary	Tier 3 vocabulary
clinical trials, *compounds*, disease, dosage, efficacy, placebo, toxicity	double blind trials, optimum dose, peer review, preclinical testing

Lesson plan

Reactivate knowledge

1 State two drugs which are based on compounds from plants.

2 What was the first antibiotic discovered?

3 Why are synthetic drugs safer than simple plant extracts?

Common misconception

Students may think that drugs do not need to be tested and move from lab to the public.

Answers: 1 Any from: aspirin, digitalis, digoxin *2* Penicillin *3* They are purer and can give reliable doses

Trigger interest

Drug development (10 mins)

Ask students to estimate how long it takes for a new drug to be developed to the stage it is being prescribed. Also ask them to estimate (roughly) how much they think it would cost.

Tell the class it would take approximately 12 years and would cost approximately £1.7 billion to bring a new medicine from first tests to readiness to be prescribed to the public.

Exposition

Drug development (15 mins)

Following on from the starter, talk students through why it takes approximately 12 years and costs approximately £1.7 billion to make a new medicine available to the public. Address the misconception here.

Inform them that new medical drugs must be tested and trialled for safety, effectiveness, toxicity, efficacy, and dose. Discuss the different stages of this testing, linked to clinical trials.

Ask students to use information from the Student Book to create a flow chart outlining:

1. the stages involved in a drug trial
2. what happens at each stage
3. the time and money spent at each stage.

Ask students if they have heard of the 'placebo effect' and if so, what their understanding of it is. Discuss their answers and use this to go through the importance of using double-blind trials.

Main activity

Carrying out a double-blind trial (25 mins)

In pairs, students to carry out a double-blind trial to test if a type of exercise can improve memory. Collate and discuss all the class results.

Use the class results to discuss:

- what is meant by a double-blind trial
- what placebos are
- why placebos are used in drug trials.

Review and practice

What happens during a drug trial? (5 mins)

Students complete the interactive to decide on the correct order for the stages in developing a new drug. They then suggest words to fill in the gaps to complete a paragraph outlining what happens during a double-blind trial.

If there is time, link back to the career paths discussion in B6.2. Mention careers in pharmacology and drug development, discussing skills that students have developed that are suited to this.

Ⓗ B6.5 Making monoclonal antibodies

AQA GCSE Biology specification points

Ⓗ 4.3.2.1 Producing monoclonal antibodies

Smart Curriculum Narrative

Health and disease

Theme/subtheme

Health and disease/Immune system and vaccines

Disciplinary knowledge

- WS 1.3

Activity

Pregnancy test

Learning objectives	Learning outcomes: Developing – Aiming for 3	Secure – Aiming for 5	Extending – Aiming for 8
Describe a monoclonal antibody	Define an antibody	**Describe a monoclonal antibody**	Explain why monoclonal antibodies have such potential for targeting specific cells and tissues in the body
Outline the procedure used to produce monoclonal antibodies	Recognise the main stages in a diagram of the production monoclonal antibodies	**Outline the procedure used to produce monoclonal antibodies**	Explain the similarities and differences between a hybridoma cell and the two different cell types it is made from
Describe some ways in which monoclonal antibodies are used	List five uses of monoclonal antibodies	**Describe some ways in which monoclonal antibodies are used**	Explain in detail how monoclonal antibodies are used in pregnancy tests and to diagnose diseases

Tier 2 vocabulary	Tier 3 vocabulary
antibody, cancer cells, clone, disease, lymphocytes, medicine, pregnancy test	hybridomas, monoclonal antibodies

Lesson plan

Reactivate knowledge

1 What are the four characteristics of a good medicine?
2 What is a placebo?
3 What is a double-blind trial?

Common misconception

Students may think it doesn't take very long to develop a drug.

Answers: 1 Effective, safe, stable, successfully taken into and broken down in the body *2* A medicine given in a drug trial which does not contain the new drug *3* A drug trial where neither doctor nor patient knows if they are being given a placebo or the real drug

Trigger interest

Types of antibodies (5 mins)

Show the class an image of many different antibodies with different antigen-binding sites and label this image 'polyclonal antibodies'. Ask the class to discuss (in pairs) what they think monoclonal antibodies are.

Exposition

Monoclonal antibodies (15 mins)

Discuss with the class what monoclonal antibodies are and explain how they are made.

Ask students to use information in the Student Book to summarise this in the activity 'Monoclonal antibodies'.

Main activity

Pregnancy test (30 mins)

Explain how pregnancy tests use monoclonal antibodies. Ask students to create a leaflet to accompany a pregnancy test, and to explain how it works using simple diagrams.

Students could use the Internet to further research the use of monoclonal antibodies for their leaflets.

Review and practice

ELISA tests (5 mins)

Supply students with information about ELISA tests, including how they work, and their applications. Ask students to write down three key bullet points about them.

Ⓗ B6.6 Uses of monoclonal antibodies

AQA GCSE Biology specification points

Ⓗ 4.3.2.2 Uses of monoclonal antibodies

Smart Curriculum Narrative

Health and disease

Theme/subtheme

Health and disease/Monoclonal antibodies

Disciplinary knowledge

- WS 1.3, 1.5

Activity

Treating cancer

	Learning outcomes		
Learning objectives	**Developing Aiming for 3**	**Secure Aiming for 5**	**Extending Aiming for 8**
Describe the three main ways in which monoclonal antibodies are used to treat cancer	State that there are currently three ways in which monoclonal antibodies are being used to treat cancer	**Describe the three main ways in which monoclonal antibodies are used to treat cancer**	Explain in detail how the three main methods of using monoclonal antibodies to treat cancer work
Outline the advantages and disadvantages of using monoclonal antibodies	State that monoclonal antibodies have many potential uses but that they have created more side effects than expected	**Outline the advantages and disadvantages of using monoclonal antibodies**	Evaluate the use of monoclonal antibodies in treating cancer compared to other treatments

Tier 2 vocabulary	Tier 3 vocabulary
cancer, disease	monoclonal antibodies

Lesson plan

Reactivate knowledge

1 What are monoclonal antibodies?

2 What is a hybridoma?

3 What are monoclonal antibodies used for?

Common misconception

Students may think it doesn't take very long to develop a drug.

Answers: 1 Antibodies made by a single clone of cells *2* A cell made by combining a lymphocyte which makes the required antibody with a tumour cell/ a cell which both divides and makes a required antibody *3* Pregnancy tests; diagnostic tests for diseases; measuring and monitoring hormones; research to locate or identify specific molecules; treating diseases

Trigger interest

Monoclonal antibodies (5 mins)

Ask students to write down bullet points showing what they remember about monoclonal antibodies from the last lesson. Suggest they group them under the headings: What, Why, and How?

Exposition

Treating cancer (15 mins)

Explain that cancer cells and non-cancerous cells have different surface antigens.

Use a diagram or image to reinforce your explanation. Suggest that students carry out an Internet search for images, if required.

Ask students to explain why monoclonal antibodies are used to target cancer cells, and how they would be made.

Tell students that when monoclonal antibodies were first developed, they were described as a 'magic bullet' for disease. Ask groups of students to discuss what they think was meant by this.

Main activity

Treating cancer (30 mins)

Divide the class into groups and assign each group one of the three methods by which monoclonal antibodies are used to treat cancer, as outlined in the Student Book.

Ask each group to conduct online research into this information to complete the Activity: *Treating cancer*.

This can be a very sensitive subject for students who may have relatives or friends who have/had cancer, so this should be considered when delivering this sensitive topic.

Review and practice

Uses of monoclonal antibodies (5 mins)

Students use this interactive to answer a multiple-choice question and then select true or false in a series of statements about monoclonal antibodies.

B6 Preventing and treating disease

Introduction to checkpoint intervention

The checkpoint intervention provides suggestions for a lesson to follow up the *B6 Preventing and treating disease* checkpoint assessment. Depending on the outcomes of the assessment, these suggestions can be used to move a class towards secure, to target specific learning outcomes, or to consolidate knowledge for students achieving secure outcomes. Use students' outcomes from the checkpoint assessment to address general misconceptions from the content covered. Be prepared to re-visit content as required; three key concepts from the chapter are covered in more detail below.

Checkpoint secure learning outcomes

- Describe, with reference to the human immune system, how vaccination works.
- Describe what is meant by antibiotic-resistant bacteria.
- Explain why a placebo is used during drug trialling.
- Describe the sequence of events in the discovery of penicillin from discovery to industrial production and explain why the latter step was so important.
- Explain why each procedure in drugs testing and trialling is used.
- Describe some ways in which monoclonal antibodies are used.

Student reflection

Encourage students to reflect on whether there were any checkpoint questions they found difficult or straightforward as well as their preparation for the checkpoint assessment, including how they revised, what they revised, the time spent on revision, and what areas they could improve upon. For more strategies, see the Metacognition spread.

Intervention activity: Describe, with reference to the human immune system, how vaccination works.

Ask students to create a public information poster to explain how a vaccine works. They should include information on how it would both protect them as an individual, and the wider population through herd immunity. Remind students to adapt their language style for a lay audience.

Intervention activity: Describe what is meant by antibiotic-resistant bacteria.

Go through and recall the following terms with students: resistance, mutation, overuse, antibiotics, competition. Students should then use these words to create a storyboard to show how a bacterium can become resistant to multiple antibiotics. Some students may need help with this so ask them to read the Student Book first and then create the storyboard with the book closed.

Intervention activity: Describe some ways in which monoclonal antibodies are used.

Remind students how monoclonal antibodies are used in pregnancy tests. This could be through a video or reading through previous slides. Tell students to use this knowledge and apply it to explain how the COVID-19 lateral flow tests work. Remind students that these also use monoclonal antibodies.

B7 Non-communicable diseases

B7.1 Non-communicable diseases

AQA GCSE Biology specification points

4.2.2.6 The effect of lifestyle on some non-communicable diseases

Smart Curriculum Narrative

Health and disease

Theme/subtheme

Health and disease/Non-communicable diseases

Disciplinary knowledge

- WS 1.4, 1.5
- MS 2c, 2d, 2g, 4a

Activity

How big is the risk?

	Learning outcomes		
Learning objectives	**Developing** **Aiming for 3**	**Secure** **Aiming for 5**	**Extending** **Aiming for 8**
Classify a range of diseases as communicable or non-communicable and describe some impacts of non-communicable diseases on individuals	Define non-communicable diseases and name some examples	**Classify a range of diseases as communicable or non-communicable and describe some impacts of non-communicable diseases on individuals**	Discuss the human and financial cost of non-communicable diseases to a local community, a nation, and globally
Define sampling and explain why it is necessary in the scientific investigation of risk factors for non-communicable diseases	List some risk factors linked to an increased rate of a non-communicable disease	**Define sampling and explain why it is necessary in the scientific investigation of risk factors for non-communicable diseases**	Describe examples of good and bad sampling techniques
Describe the difference between a correlation and a causal mechanism	Identify correlations in data	**Describe the difference between a correlation and a causal mechanism**	Explain the role of sampling in determining a causal mechanism for a non-communicable disease

Tier 2 vocabulary	Tier 3 vocabulary
causal mechanism, correlation, diet, gene, ionising radiation, obesity, UV light	cancer, carcinogen, lifestyle factor, non-communicable disease, risk factor, smoking

Lesson plan

Reactivate knowledge

1 What is a non-communicable disease?
2 Give two diseases linked to smoking.
3 What is a recreational drug?

Common misconception

Students may think that a correlation between a risk factor and incidence of disease means that risk factor alone caused the disease.

Answers: 1 A disease which cannot be transmitted from one person to another *2* Lung cancer/heart disease *3* A substance taken for reasons other than treatment of disease

Trigger interest

Communicable or non-communicable? (5 mins)

Display a list of diseases and ask the class to state whether they are communicable or non-communicable. Suitable examples of diseases include: HIV, asthma, diabetes, heart disease, and lung infection. Students may want to recap definitions of communicable and non-communicable diseases first.

Exposition

Correlation doesn't mean causation (15 mins)

Draw, and describe if necessary, a scatter graph with number of universities in an area on the x-axis and number of religious places of worship in an area on the y-axis. Add scatter points to form a positive correlation. Show students this graph and ask them to describe the data. Students should point out a positive correlation between the number of universities and the number of places of worship in an area. Ask students whether this means that having more universities causes more places of worship to be built, or vice versa. Discuss their ideas, and then emphasise that correlation doesn't prove causation and there could be other factors that cause or impact the correlation seen. In this example, it could be that the area has a larger population, and that is why there are more places of worship and also more universities, or it has better transport links, etc.

Emphasise that this is the same in science, and when looking at data students must be aware that a correlation with a factor doesn't necessarily mean the factor caused the pattern they see. Address the misconception here.

Main activity

How big is the risk? (30 mins)

Students complete Working Scientifically: *How big is the risk?* They place a range of risk factors on a continuum line showing effect on health, and explain their choices. Students then complete the questions on the sheet and discuss answers to self-mark.

Provide groups of students with a range of data showing links between risk factors and different diseases. Ask students to interpret the data and look for correlations and causal links.

To support, go through how to interpret data on risk factors by analysing one piece of data as a class.

As an extension, students discuss how reducing or limiting risk factors that result in higher incidences of non-communicable diseases could have a benefit to the local and national community. For example, this could cut NHS costs, waiting times, and benefit the economy.

Review and practice

Tips for healthy living (5 mins)

Ask students to write down their top five tips to minimise the risk of developing a non-communicable disease and explain the science behind their tips.

B7.2 Cancer

AQA GCSE Biology specification points

4.2.2.7 Cancer

Theme/subtheme

Health and disease/Non-communicable diseases

Smart Curriculum Narrative

Health and disease

Activity

Risk of cancer

	Learning outcomes		
Learning objectives	**Developing Aiming for 3**	**Secure Aiming for 5**	**Extending Aiming for 8**
Describe the difference between benign and malignant tumours	Describe cancer as the result of changes in cells that lead to uncontrolled growth and division and define a tumour as a mass of abnormally growing cells	**Describe the difference between benign and malignant tumours**	Explain how both benign and malignant tumours can be life-threatening
Describe why chemical carcinogens and ionising radiation increase the risk of tumours forming	State some causes of cancer	**Describe why chemical carcinogens and ionising radiation increase the risk of tumours forming**	Link a lack of control in the cell cycle to tumour formation and explain that this may be the result of many factors interacting
Describe how radiotherapy and chemotherapy act to treat cancers	Describe current methods of treating cancer as surgery, radiotherapy, and chemotherapy	**Describe how radiotherapy and chemotherapy act to treat cancers**	Discuss the risks and benefits of radiotherapy for both the individual and for society

Tier 2 vocabulary	Tier 3 vocabulary
cell cycle, cell division, ionising radiation, mitosis, UV light	benign, cancer, carcinogen, chemotherapy, genetic risk factor, human papillomavirus (HPV), lifestyle risk factor, malignant, melanoma, mutation, radiotherapy, tumour

Lesson plan

Reactivate knowledge

1 Which risk factors for disease can't you change?
2 List two lifestyle risk factors for non-communicable diseases.
3 What is difference between correlation and causation?

Common misconception

Students may think that benign tumours, being non-cancerous, do not cause any harm.

Answers: 1 Inherited genes; age *2* Any two from: smoking; drinking alcohol; lack of exercise; poor nutrition/obesity *3* Correlation is an apparent link between two factors; causation is when there is a biological mechanism which explains how one factor affects another

Trigger interest

Cell cycle order (5 mins)

Provide students with the stages of the cell cycle from B2.1. Students put the stages in the correct order.

Discuss how cells may sometimes stop responding to the normal mechanisms that control the cell cycle and start to divide rapidly.

Exposition

Cancer (20 mins)

Introduce the term tumour as a mass of abnormally dividing cells. This is a sensitive topic and could be upsetting for students who have experienced cancer. This should be considered when delivering this content.

Link this to the cell cycle order activity and point out that the cells divide uncontrollably to form tumours when the cell cycle isn't controlled.

On their mini-whiteboards, ask students to list any examples they might know of carcinogens (chemicals that increase the risk of cancer) or types of radiation that increase the risk of cancer. Talk through how carcinogens and ionising radiation increase the risk of the cell cycle not being controlled, and therefore increase the likelihood of tumours forming.

Explain the difference between benign and malignant tumours, and suggest how they cause health problems. Address the misconception here, and give the example of benign brain tumours still being very harmful.

Main activity

Risk of cancer (25 mins)

Students create a leaflet or a poster to educate the public about the causes of cancer (e.g., carcinogens, ionising radiation, some viruses) and some of the risks and benefits of cancer treatments (e.g., radiotherapy, chemotherapy).

They can use information from the Student Book and the Internet if resources are available.

Review and practice

Cancer risk factors (5 mins)

Read out some lifestyle factors, for example, smoking, a stressful job, a diet consisting of saturated fats, exercising regularly, using sunbeds, etc.

Ask students to write on their mini-whiteboards whether they think each lifestyle factor greatly increases the risk of cancer, if they think there is a link, or if they think there is no link. Ask students to write a justification for their choice.

To finish, discuss the progress being made in cancer research, especially in extending the lives of those living with cancer so it becomes a manageable disease. Mention some cancer research institutions and what they do.

B7.3 Smoking and the risk of disease

AQA GCSE Biology specification points

4.2.2.6 The effect of lifestyle on some non-communicable diseases

Smart Curriculum Narrative

Health and disease

Theme/subtheme

Health and disease/Health and diet

Disciplinary knowledge

- WS 1.5
- MS 2c, 2d, 2g, 4a

Practical

The effects of smoking

	Learning outcomes		
Learning objectives	**Developing** **Aiming for 3**	**Secure** **Aiming for 5**	**Extending** **Aiming for 8**
Describe the effects of the harmful substances found in tobacco smoke on the cardiovascular system, the lungs, and on unborn babies	State that tobacco smoke is a risk factor for cardiovascular disease and lung diseases including lung cancer and is harmful to unborn babies, and name three harmful substances found in tobacco smoke	**Describe the effects of the harmful substances found in tobacco smoke on the cardiovascular system, the lungs, and on unborn babies**	Explain in detail the effects of the harmful substances found in tobacco smoke on the tissues and organs of the body
Analyse data to describe evidence for the link between smoking and lung disease, smoking and low birth weight babies, and smoking and cardiovascular disease	Plot a bar chart of data showing the link between cigarette smoking and different diseases	**Analyse data to describe evidence for the link between smoking and lung disease, smoking and low birth weight babies, and smoking and cardiovascular disease**	Suggest possible causal mechanisms to explain smoking-related health trends such as lung cancer, low birth weight babies, stillbirths, and cardiovascular disease, and explain how the causal link between smoking and lung cancer was identified using data and data analysis

Tier 2 vocabulary	Tier 3 vocabulary
alveoli, bronchus, carbon monoxide, cilia, foetus, oxygen, placenta, trachea	addictive, bronchitis, carcinogen, cardiovascular disease, chronic obstructive pulmonary disease (COPD), coronary heart disease, low birth weight, lung cancer, lung disease, nicotine, premature birth, smoking, stillbirth, tar, tobacco

Lesson plan

Reactivate knowledge

1 What is a benign tumour?
2 What is a malignant tumour?
3 What are the three most common ways of treating cancer?

Common misconception

Students may think that a correlation between a risk factor and incidence of disease means the risk factor alone caused the disease.

Answers: 1 Growth of abnormal cells contained in one place – does not spread to other parts of the body *2* Growth of abnormal cells when some cells spread around the body in the bloodstream invading other tissues *3* Surgery, radiotherapy, chemotherapy

Trigger interest

Smoking links (10 mins)

Show students a graph that indicates the link between smoking and lung cancer for them to interpret verbally.

Support students by providing them with a list of possible conclusions to choose from.

Students with higher prior attainment can be extended here and use their knowledge about the causes of cancer to suggest a causal mechanism.

Exposition

Demonstration: The effects of smoking (20 mins)

Carry out the demonstration to show some of the products of smoking cigarettes. Ask students to observe and write down what the experiment showed.

Students then create a table to show how nicotine and carbon monoxide in cigarette smoke affect health. Students can use the Student Book to support them.

Main activity

The effects of smoking (20 mins)

Students complete the questions on the practical worksheet to analyse data on smoking and risk factors. They describe how the data provide evidence for causal links between smoking and various non-communicable diseases.

As an extension, students should explain how smoking during pregnancy can lead to unborn babies obtaining less oxygen and having a low birth weight.

Students can self-mark their answers.

Review and practice

Trends in smoking (5 mins)

Students complete the Interactive: *Trends in smoking* to analyse a graph and answer questions about how the number of people smoking in the UK has changed over the last 50 years.

B7.4 Diet, exercise, and disease

AQA GCSE Biology specification points

4.2.2.6 The effect of lifestyle on some non-communicable diseases

Smart Curriculum Narrative

Health and disease

Theme/subtheme

Health and disease/Health and diet

Disciplinary knowledge

- WS 1.4
- MS 2c, 2d, 2g, 4a

Activity

Preventing the diabetes epidemic

	Learning outcomes		
Learning objectives	**Developing** **Aiming for 3**	**Secure** **Aiming for 5**	**Extending** **Aiming for 8**
Describe causal mechanisms for the link between exercise and health	Describe some health problems caused by a poor diet and lack of exercise	**Describe causal mechanisms for the link between exercise and health**	Suggest reasons for the correlation between exercise and health, and decide which are causal
Suggest measures to prevent a further rise in the number of people with Type 2 diabetes	List ways in which people can avoid becoming overweight	**Suggest measures to prevent a further rise in the number of people with Type 2 diabetes**	Explain in detail why eating a poor diet can lead to health problems
Translate information about the impacts of diet and exercise on health between graphical and numerical forms	Describe the evidence presented in data on the links between diet or exercise and health	**Translate information about the impacts of diet and exercise on health between graphical and numerical forms**	Extract and interpret information about risk factors relating to diet, exercise, and health from charts, graphs, and tables

Tier 2 vocabulary	Tier 3 vocabulary
balanced diet, causal mechanism, cholesterol, diet, exercise, insulin, metabolic rate, muscle	body mass index (BMI), cardiovascular disease, heart disease, obesity, overweight, Type 2 diabetes

Lesson plan

Reactivate knowledge

1 What is the effect of carbon monoxide in tobacco smoke?

2 What are the three main risks linked to smoking during pregnancy?

3 Which substance in tobacco smoke mainly increases the risks of developing COPD, lung, and throat cancer?

Answers: *1* Binds to red blood cells, stops them carrying oxygen *2* Higher rate of premature births, low birth weight babies, stillbirths *3* Tar

Common misconception

Students may think that a correlation between a risk factor and incidence of disease means the risk factor alone caused the disease.

Trigger interest

Exercise and health (10 mins)

Tell the class the statement that people who exercise are less likely to develop non-communicable diseases. Ask students to list possible reasons why this is true, and list reasons when this isn't true.

Exposition

Food and exercise (15 mins)

Provide groups of students with some sticky notes. Ask them to write down information from the Student Book about the effects on the body of eating too much unhealthy food and not taking enough exercise. Students should stick the notes down on the desk to create flow charts. These should contain several paths describing how an unhealthy diet and lack of exercise lead to problems such as Type 2 diabetes and heart disease.

Main activity

Preventing the diabetes epidemic (25 mins)

Tell the class that the number of people being diagnosed with Type 2 diabetes is rising. Students should explain why, using what they have learnt, the Student Book, or research on the Internet. Ask students to work as a government advisory group and come up with suggestions to stop this epidemic. Students should write these suggestions in their books. Students can discuss and debate their ideas.

Review and practice

Obesity data (5 mins)

Show students a piece of data that shows a correlation between a factor and obesity, for example, country of residence, age, sex, ethnicity, or income. They then describe and explain the pattern.

B7.5 Alcohol and other carcinogens

AQA GCSE Biology specification points

4.2.2.6 The effect of lifestyle on some non-communicable diseases

Smart Curriculum Narrative

Health and disease

Theme/subtheme

Health and disease/Health and diet

Disciplinary knowledge

- WS 1.4
- MS 2c, 2d, 2g, 4a

Activity

The effects of alcohol

	Learning outcomes		
Learning objectives	**Developing** **Aiming for 3**	**Secure** **Aiming for 5**	**Extending** **Aiming for 8**
Describe the short-term and long-term effects of drinking alcohol on the liver and the brain	State that drinking too much alcohol can affect liver and brain function	**Describe the short-term and long-term effects of drinking alcohol on the liver and the brain**	Explain in detail why the liver and the nervous system are particularly vulnerable to the short-term and longer-term impacts of drinking too much alcohol
Describe the effects of alcohol on unborn babies	State that alcohol can affect unborn babies	**Describe the effects of alcohol on unborn babies**	Evaluate evidence on the effects of alcohol on a developing baby
Describe the link between ionising radiation and cancer	Define the term carcinogen	**Describe the link between ionising radiation and cancer**	Explain the link between ionising radiation and cancer and evaluate evidence for the impact of different levels of exposure to ionising radiation on cancer risks

Tier 2 vocabulary	Tier 3 vocabulary
alcohol, brain, ethanol, ionising radiation, liver, nervous system, placenta, radioactive, UV light, X-ray	addictive, alcoholic, cancer, carcinogen, chromosome, cirrhosis, DNA, drug, foetal alcohol syndrome (FAS), liver cancer, low birth weight, melanoma, miscarriage, mutation, premature birth, stillbirth

Lesson plan

Reactivate knowledge

1 What percentage of a healthy food intake is used for the basic reactions of life?

2 What health benefits have been observed in studies from regular exercise?

3 Give three factors that increase the risk of developing Type 2 diabetes.

Common misconception

Students may think that Class A drugs, such as heroin, are the most harmful drugs to society, rather than the correct answer, alcohol.

Answers: 1 60–75% *2* Reduced risk of heart disease, osteoarthritis, Type 2 diabetes, high blood pressure *3* Any three from: obesity; genetics; age; lack of exercise

Trigger interest

Which drug is the most harmful? (10 mins)

Ask students to write on their mini-whiteboards which recreational drug they think is the most harmful to society and reveal their answers.

Show students data that claims that alcohol is the most harmful recreational drug. Discuss why this is by considering the effects of alcohol on the individual and society and the fact that it is legal for people over the age of 18.

Exposition

Ionising radiation (10 mins)

Explain that ionising radiation is a carcinogen. Ask students for suggestions of sources of ionising radiation that they may come into contact with. After hearing their ideas, allow students to use the Student Book to help them list the sources.

Discuss the fact that coming into contact with ionising radiation increases the risk of developing cancer. Ask students to list lifestyle factors that will increase this risk, for example, using tanning beds or not wearing a high SPF suncream.

Main activity

The effects of alcohol (30 mins)

Students look at the effects of alcohol on the body and on baby development using the activity Working Scientifically: *The effects of alcohol*. They analyse the given evidence and use it to come to their own conclusion on the risks of consuming alcohol during pregnancy.

Review and practice

Alcohol statements (5 mins)

Use the interactive, which asks students to type in answers to complete sentences about the effects of alcohol on the body.

B7 Non-communicable diseases

Introduction to checkpoint intervention

The checkpoint intervention provides suggestions for a lesson to follow up the *B7 Non-communicable diseases* assessment. Depending on the outcomes of the assessment, these suggestions can be used to move a class towards secure, to target specific learning outcomes, or to consolidate knowledge for students achieving secure outcomes. Use students' outcomes from the checkpoint assessment to address general misconceptions from the content covered. Be prepared to re-cover content as required; three key concepts from the chapter are covered in more detail below.

Checkpoint secure learning outcomes

- Classify diseases as communicable or non-communicable.
- Analyse data on risk factors for non-communicable diseases and draw conclusions from the data.
- Describe the difference between benign and malignant tumours.
- Suggest measures to prevent a further rise in the number of people with Type 2 diabetes.
- Analyse data to describe evidence for the link between smoking and lung disease, smoking and low birthweight babies and stillbirths, and smoking and cardiovascular disease.
- Describe the effects of alcohol on unborn babies.

Student reflection

Encourage students to reflect on whether there were any checkpoint questions they found difficult or straightforward as well as their preparation for the checkpoint assessment, including how they revised, what they revised, the time spent on revision, and what areas they could improve upon. For more strategies, see the Metacognition spread.

Intervention activity: Classify diseases as communicable or non-communicable.

Ask students to write definitions for communicable and non-communicable diseases.

Students then list as many communicable and non-communicable diseases as they can think of. They can look through their notes to try and find examples in the Student Book if they struggle to think of any.

To support students, this activity could be modified so that the teacher provides a list of diseases and students sort them into communicable and non-communicable.

As an extension, for each communicable disease, students could state how it is transmitted, and for the non-communicable diseases they could state associated risk factors.

Intervention activity: Describe the difference between benign and malignant tumours.

Students complete a table to summarise the differences between benign and malignant tumours.

They should consider whether the tumour:

- can spread to other tissues in the body (metastasis)
- has its own blood supply
- is cancer
- can cause harm.

Intervention activity: Suggest measures to prevent a further rise in the number of people with Type 2 diabetes.

Provide students with the following three strategies for how to reduce the risk of developing Type 2 diabetes:

- Controlled carbohydrate diet
- Losing weight
- Exercise

Ask students to rank these in order of what they believe to be the most important down to the least important strategy. Ask them to justify their choice.

There is no correct ranking, but the activity should get students debating and therefore considering the impact of each strategy in more detail.

B8 Photosynthesis

B8.1 Photosynthesis

AQA GCSE Biology specification points

4.4.1.1 Photosynthetic reaction

Smart Curriculum Narrative

How organisms work

Theme/subtheme

Organisation/Plant organ systems

Disciplinary knowledge

- WS 1.4

Activity

The process of photosynthesis

	Learning outcomes		
Learning objectives	**Developing Aiming for 3**	**Secure Aiming for 5**	**Extending Aiming for 8**
Describe how the leaf is adapted for photosynthesis	Describe how plants get the materials they need for growth	**Describe how the leaf is adapted for photosynthesis**	Explain how adaptations of the leaf make photosynthesis efficient
Describe photosynthesis as an endothermic reaction in which energy is transferred from the environment to the chloroplasts by light and recognise the chemical symbols for the molecules involved in photosynthesis	State the word equation for photosynthesis	**Describe photosynthesis as an endothermic reaction in which energy is transferred from the environment to the chloroplasts by light and recognise the chemical symbols for the molecules involved in photosynthesis**	Write a balanced chemical equation for photosynthesis and define an endothermic reaction
Describe a method to demonstrate that plants need light to carry out photosynthesis	Describe why plants need light to carry out photosynthesis	**Describe a method to demonstrate that plants need light to carry out photosynthesis**	Explain why chlorophyll is needed for photosynthesis

Tier 2 vocabulary	Tier 3 vocabulary
endothermic, energy, food, glucose	chlorophyll, chloroplasts, diffusion, *Elodea*, xylem

Lesson plan

Reactivate knowledge

Common misconception

Students can sometimes confuse the equation for photosynthesis with respiration.

1 What is photosynthesis?
2 Which is the main organ of photosynthesis in most plants?
3 Where in the plant cell does photosynthesis take place?

Answers: 1 The process by which plants make food using light *2* Leaf *3* Chloroplast

Trigger interest

Leaf structure (5 mins)

Ask students to think about the plants in their house or around the school and what the plants need to grow and survive. Show students an image of a leaf and ask them to describe the adaptations it has to maximise the rate of photosynthesis. Discuss these ideas as a class.

To support, remind students that carbon dioxide, water, and light are required for photosynthesis (which they will have covered at KS3).

Extend students by asking how they think each feature they described can alter the rate of photosynthesis (i.e., increase or decrease it).

Exposition

Photosynthesis equations (15 mins)

Discuss photosynthesis as a reaction that plants and algae use to produce food. Students should write out the word equation in their notes. Discuss that carbon dioxide and water are reactants, and glucose and oxygen are products. Provide students with the chemical formula for glucose ($C_6H_{12}O_6$) and ask them to complete the equation by writing out the chemical formula for water, carbon dioxide, and oxygen.

Ask students if they can balance the equation. Discuss that photosynthesis is an example of an endothermic reaction as it takes in energy. Then ask which part of the word equation they think shows that the reaction is endothermic.

Main activity

The process of photosynthesis (25 mins)

Students will need access to the internet and the Student Book to work through the worksheet. Task 1 is to produce an information poster about the process of photosynthesis. Make students aware of using appropriate terminology and imagery for a lay audience. Then work through the questions in Task 2.

Support students by reminding them that glucose is a product of photosynthesis, and is required by the plant to grow (glucose is used in respiration to provide the energy for growth).

Review and practice

The photosynthesis equations (10 mins)

Students can complete the *Bump up your Grade* worksheet where they consolidate the word equation and balanced symbol equation for photosynthesis.

B8.2 The rate of photosynthesis

AQA GCSE Biology specification points

4.4.1.2 Factors affecting the rate of photosynthesis in plants

Smart Curriculum Narrative

How organisms work

Theme/subtheme

Bioenergetics/Photosynthesis

Disciplinary knowledge

- MS 1a, 1c, 2c, 3a, 3d, 4a, 4c
- WS 2.1, 2.2, 2.5, 2.6, 3.1, 3.2

Required Practical 6

Photosynthesis

	Learning outcomes		
Learning objectives	**Developing** **Aiming for 3**	**Secure** **Aiming for 5**	**Extending** **Aiming for 8**
Describe the effect of changes in temperature, carbon dioxide concentration, light intensity, and shortage of chlorophyll on the rate of photosynthesis	List the factors that affect the rate of photosynthesis (temperature, carbon dioxide concentration, light intensity, amount of chlorophyll)	**Describe the effect of changes in temperature, carbon dioxide concentration, light intensity, and shortage of chlorophyll on the rate of photosynthesis**	Apply scientific understanding to explain the effect of changes in temperature, carbon dioxide concentration, light intensity, and amount of chlorophyll on the rate of photosynthesis
Describe how a particular factor limits the rate of photosynthesis in a given situation	Give simple statements of the relationship between each of these factors and the rate of photosynthesis	**Describe how a particular factor limits the rate of photosynthesis in a given situation**	Predict how the rate of photosynthesis will be affected when there is more than one limiting factor
Plan and carry out an investigation into the effect of light intensity on photosynthesis	Follow instructions to carry out an investigation into the effect of light intensity on the rate of photosynthesis	**Plan and carry out an investigation into the effect of light intensity on photosynthesis**	Explain the use of a heat shield when investigating the effect of light intensity of the rate of photosynthesis and explain why the rate of photosynthesis eventually becomes constant even when light intensity continues to increase
Interpret and explain graphs of the rate of photosynthesis involving one limiting factor	Plot a line graph and write a simple conclusion	**Interpret and explain graphs of the rate of photosynthesis involving one limiting factor**	Understand and use the inverse square law and light intensity in the context of photosynthesis

Tier 2 vocabulary	Tier 3 vocabulary
CO_2, enzymes, light, temperature	inverse square law, light intensity, limiting factors

Lesson plan

Reactivate knowledge

Common misconception

Students can sometimes confuse the equation for photosynthesis with respiration.

1 What is the word equation for photosynthesis?
2 Name two adaptations of a leaf for photosynthesis.
3 Photosynthesis is an endothermic reaction. What is an endothermic reaction?

Answers: 1 Carbon dioxide + water → glucose + oxygen (using sunlight) *2* Thin, lots of chlorophyll *3* More energy is taken in than given out

Trigger interest

Factors that affect the rate of photosynthesis (5 mins)

Tell students they are now growing plants and need to sell them much faster. What factors would speed up the rate of photosynthesis and what could they do as the farmers? Have the word equation for photosynthesis written on the board and ask students to list the factors that would speed up the rate of this reaction. To extend, ask students why this is the case.

Exposition

Limiting factors graphs (10 mins)

On the board, draw line graphs showing how carbon dioxide, light intensity, concentration, and temperature affect the rate of photosynthesis.

Ask pairs to describe the shape of, and explain, each graph.

You could prompt them to consider enzymes, successful collisions, and endothermic reactions.

Main activity

Photosynthesis (35 mins)

Use Required Practical: *Photosynthesis* worksheet.

The practical element can be time-consuming so two lessons may be needed to enable time for analysis.

Students use *Elodea* in water with a lamp shining on it to investigate the effect of light intensity on the rate of photosynthesis. Measure and record the different volumes of oxygen produced when the lamp is placed at varying distances away from the plant. As a class, discuss how and why it is difficult to control temperature.

After the practical, instruct students to draw line graphs of their results and to write an analysis of their results.

Higher attaining students should plan a suitable method and choose their own ranges and intervals. Encourage students to use the inverse square to plot points on the graph during analysis.

Review and practice

Data handling skills (5 mins)

Interactive where students answer multiple choice questions about supplied graphs showing the rate of photosynthesis.

B8.3 How plants use glucose

AQA GCSE Biology specification points

4.4.1.3 Use of glucose from photosynthesis

Smart Curriculum Narrative

How organisms work

Theme/subtheme

Bioenergetics/Photosynthesis

Disciplinary knowledge

- WS 2.4, 2.6

Practical

Testing for starch

	Learning outcomes		
Learning objectives	**Developing Aiming for 3**	**Secure Aiming for 5**	**Extending Aiming for 3**
Describe five ways in which plants use glucose, including how they make proteins	List some ways in which plants use glucose	**Describe five ways in which plants use glucose, including how they make proteins**	Explain how carnivorous plants are adapted to their environment
Give reasons for the different steps when testing a leaf for starch, including the safety precautions	Describe how to test a leaf for starch and state some safety rules	**Give reasons for the different steps when testing a leaf for starch, including the safety precautions**	Explain how and why plants convert glucose to starch for storage

Tier 2 vocabulary	Tier 3 vocabulary
glucose, proteins, starch, sugars	cellulose, iodine solution, nitrate ions

Lesson plan

Reactivate knowledge

1 What four factors affect the rate of photosynthesis in a leaf?

2 What effect do you expect an increase in light intensity to have on the rate of photosynthesis?

3 Why does photosynthesis stop permanently if the temperature of a plant goes up too high?

Common misconceptions

Students may think that both photosynthesis and respiration stop in the dark. Clarify it is *only* photosynthesis that doesn't happen in the dark because plants require light to photosynthesise.

Students may also think plants don't respire, and that they only photosynthesise.

Answers: 1 Light; carbon dioxide concentration; temperature; chlorophyll *2* Increases *3* Enzymes controlling photosynthesis are denatured

Trigger interest

Food test reminder (10 mins)

Remind students about the food tests carried out in B3.3. With help from the class, demonstrate the tests for glucose, starch, and protein on a range of plant substances (e.g., potato, peas, eggs, bananas). Check for any allergies students may have before selecting safe and appropriate foods to test.

Exposition

Use of glucose (15 mins)

Ask students to list what plants use glucose for. Allow them to use information from the Student Book to complete their lists, *if required*. It is now a good time to clarify the misconception and outline that plants do respire.

Describe starch as a long molecule made up of many repeating units of glucose.

Demonstrate that glucose is soluble in water but starch isn't, by adding both to water and stirring. Ask students to write down why plants store glucose in their cells as starch.

To deepen their understanding, ask students to use this concept and explain how plants make proteins.

Main activity

Testing for starch (25 mins)

Students can follow the testing for starch worksheet and carry out the test on leaves that have been kept in the light and leaves that have been kept in the dark.

Before they start, ask students to read through the method, write down any potential hazards, and state how they will minimise any risks.

They could identify the following hazards:

- Ethanol – it is flammable and therefore should be kept away from naked flames.
- Glass test tubes – any breakages could break and cut the skin. Test tubes should be placed in a test-tube rack to reduce the chance of this happening.
- Hot water – could cause burns to the skin/eyes. Students to stand during any experiments with hot water, ready to quickly step back in the event of any spillages.

Some students may need support working independently on the practical so it may be a good idea to go through each stage of the experiment with them first.

Review and practice

Match the uses (5 mins)

Students complete the interactive and match the name of a product made from glucose with its use in the plant.

Ⓗ B8.4 Making the most of photosynthesis

AQA GCSE Biology specification points

4.4.1.2 Limiting factors and their effect on photosynthesis

Smart Curriculum Narrative

How organisms work

Theme/subtheme

Bioenergetics/Photosynthesis

Disciplinary knowledge

- WS 1.4

Activity

Hydroponics

	Learning outcomes		
Learning objectives	**Developing Aiming for 3**	**Secure Aiming for 5**	**Extending Aiming for 8**
Describe how two limiting factors may interact to affect the rate of photosynthesis	State that temperature, carbon dioxide concentration, light intensity, and the amount of chlorophyll all affect the rate of photosynthesis	**Describe how two limiting factors may interact to affect the rate of photosynthesis**	Explain how interactions between several different factors affect, and may limit, the rate of photosynthesis
Interpret and explain graphs of photosynthesis rate involving two limiting factors	Describe a graph of the rate of photosynthesis involving one limiting factor	**Interpret and explain graphs of photosynthesis rate involving two limiting factors**	Interpret and explain graphs of photosynthesis rate involving three limiting factors
Explain how changing factors such as carbon dioxide concentration and temperature in greenhouses increases the rate of photosynthesis	State that plants grow faster in greenhouses	**Explain how changing factors such as carbon dioxide concentration and temperature in greenhouses increases the rate of photosynthesis**	Explain, in detail, how controlling limiting factors in greenhouses increases the rate of growth of the plants
Comment on the cost-effectiveness of adding heat, light, or carbon dioxide to greenhouses	State that there is an economic cost to changing conditions in a greenhouse	**Comment on the cost-effectiveness of adding heat, light, or carbon dioxide to greenhouses**	Use data to comment on the economics of enhancing conditions in greenhouses to gain the maximum rate of photosynthesis whilst still maintaining profit

Tier 2 vocabulary	Tier 3 vocabulary
carbon dioxide, *control*, greenhouse, light, temperature	biomass, hydroponic, insulation, limiting factors, profit, rate of photosynthesis

Lesson plan

Reactivate knowledge

1 What compound built from glucose is the main storage compound in plant cells?

2 What yellow-brown solution is used to test from the presence of starch?

3 Which important compounds are made in plants using glucose from photosynthesis and nitrates from the soil?

Answers: 1 Starch *2* Iodine *3* Proteins

Common misconceptions

Students may think that both photosynthesis and respiration stop in the dark. Clarify it is *only* photosynthesis that doesn't happen in the dark because plants require light to photosynthesise.

Students may also think plants don't respire, and that they only photosynthesise.

Trigger interest

Multiple factors (10 mins)

Students select the correct limiting factors for graphs which show the rate of photosynthesis when more than one limiting factor is acting.

Exposition

Greenhouse economics (20 mins)

Use the worksheet Calculations: *Cost effectiveness* to introduce students to the idea of using greenhouses to increase plant growth and therefore increase profits for growers. Link each feature of a greenhouse to the limiting factors of photosynthesis.

Students are then presented with data looking at the costs of running a greenhouse and potential profits. Students use this data to decide on the cost-effectiveness of adding extra carbon dioxide, light, and heat to the greenhouse.

Review the answers with students self-marking their work.

Main activity

Hydroponics (20 mins)

Introduce students to the use of hydroponics to increase the growth of plants and therefore increase profits for growers. Many students will not have heard of hydroponics and therefore will need thorough explanation before completing the worksheet. Students then complete the literacy worksheet to analyse given text on hydroponics.

Students should then evaluate the economics of hydroponics (i.e., potential profit versus costs).

Alternatively, or time permitting, students can work through Activity: *Using greenhouses* as a follow-up activity from the exposition.

Review and practice

Plants on Mars (5 mins)

Introduce the idea of people potentially living on Mars – they will need to grow plants for food!

Supply groups of students with a list of problems that may affect any plant growth on Mars (e.g., low temperature, lack of soil minerals, no liquid water) and ask them to discuss possible ways to overcome these issues.

Highlight the importance of biology in space exploration and how important it is if humans are to one day become interplanetary.

B8 Photosynthesis

Introduction to checkpoint intervention

The checkpoint intervention provides suggestions for a lesson to follow up the *B8 Photosynthesis* assessment. Depending on the outcomes of the assessment, these suggestions can be used to move a class towards secure, to target specific learning outcomes, or to consolidate knowledge for students achieving secure outcomes. Use students' outcomes from the checkpoint assessment to address general misconceptions from the content covered. Be prepared to re-cover content as required; three key concepts from the chapter are covered in more detail below.

Checkpoint secure learning outcomes

- Describe how the leaf is adapted for photosynthesis.
- Describe photosynthesis as an endothermic reaction in which energy is transferred from the environment to the chloroplasts by light and recognise the chemical symbols for the molecules involved in photosynthesis.
- Describe the effect of changes in temperature, carbon dioxide concentration, light intensity, and shortage of chlorophyll on the rate of photosynthesis.
- (H) Suggest which factor limits the rate of photosynthesis in a given situation.
- Comment on the cost-effectiveness of adding heat, light, or carbon dioxide to greenhouses.
- Describe a method to demonstrate that plants need light to carry out photosynthesis.

Student reflection

Encourage students to reflect on whether there were any checkpoint questions they found difficult or straightforward as well as their preparation for the checkpoint assessment, including how they revised, what they revised, the time spent on revision, and what areas they could improve upon. For more strategies, see the Metacognition spread.

Intervention activity: Describe how the leaf is adapted for photosynthesis.

Instruct students to draw a leaf. Students should annotate their leaf to identify and describe the structural features that increase the rate of photosynthesis.

Students should then recall and explain how each leaf feature is an adaptation for photosynthesis.

Intervention activity: Describe photosynthesis as an endothermic reaction in which energy is transferred from the environment to the chloroplasts by light and recognise the chemical symbols for the molecules involved in photosynthesis.

Ask students to write down everything they can remember about photosynthesis based on the following four prompts. Students must do this from memory and by themselves in five minutes.

i. What is the word equation for photosynthesis?

ii. What is the balanced symbol equation for photosynthesis?

iii. What is the definition of an endothermic reaction?

iv. Explain why photosynthesis is an endothermic reaction.

Then, in pairs, give students five minutes to compare what they wrote, and in a different coloured pen or in brackets, add in anything they forgot that their partner remembered.

Finally, show students the answers to each question, and in a bright coloured pen, students correct their answers or add in any missing information.

H Intervention activity: Suggest which factor limits the rate of photosynthesis in a given situation.

Show students graph from B8.4 which shows the effect of light intensity, carbon dioxide concentration, and temperature.

Ask students to copy this graph into their notes and to label the point on the graph where:

- temperature is the limiting factor
- carbon dioxide concentration is the limiting factor
- light intensity is the limiting factor.

B9 Respiration

B9.1 Aerobic respiration

AQA GCSE Biology specification points

4.4.2.1 Aerobic and anaerobic respiration

***Smart Curriculum* Narrative**

How organisms work

Theme/subtheme

Bioenergetics/Respiration

Practical

Investigating respiration

	Learning outcomes		
Learning objectives	**Developing Aiming for 3**	**Secure Aiming for 5**	**Extending Aiming for 8**
Describe cellular respiration as an exothermic reaction occurring continuously in the mitochondria of cells, transferring the energy needed for living processes. Recognise the chemical symbols for the molecules involved in respiration	State the word equation for aerobic cellular respiration	**Describe cellular respiration as an exothermic reaction occurring continuously in the mitochondria of cells, transferring the energy needed for living processes. Recognise the chemical symbols for the molecules involved in respiration**	Write the balanced symbol equation for respiration and define an exothermic reaction
Describe four ways the energy transferred during aerobic respiration is used in organisms	State three reasons why organisms need energy	**Describe four ways the energy transferred during aerobic respiration is used in organisms**	Apply understanding of aerobic respiration to explain the distribution of mitochondria in different types of cells
Describe two ways of indicating that aerobic respiration is taking place	State the colour change in limewater that identifies the presence of carbon dioxide	**Describe two ways of indicating that aerobic respiration is taking place**	Explain the scientific basis of using limewater and temperature changes to indicate aerobic respiration is taking place in an organism
Describe an investigation into aerobic respiration in a named organism, including a control	Identify a control in a given investigation into aerobic respiration in an organism	**Describe an investigation into aerobic respiration in a named organism, including a control**	Explain why a control is included in many scientific investigations

Tier 2 vocabulary	Tier 3 vocabulary
carbon dioxide, *concentration*, energy, glucose, limewater, organism, oxygen, products, temperature, transfer, *waste*, water	aerobic, *cellular*, exothermic reaction, mitochondria, organelle, respiration

Lesson plan

Reactivate knowledge

Common misconception

Students may think that only animals respire.

1 Which are the organs of gas exchange in the human body?
2 Which gases are exchanged?
3 In which cell organelle do the reactions of respiration take place?

Answers: *1* Lungs *2* Oxygen in, carbon dioxide out *3* Mitochondria

Trigger interest

Energy gels (5 mins)

Show the class some picture examples of energy gels. Ask them to discuss in pairs why cyclists in events such as the Tour de France use them. Draw out the idea that they are high in glucose, which is the source of energy for the body.

To extend, ask students to explain why the cyclists will have a breakfast high in starch, such as porridge.

Exposition

Respiration paired blurt (15 mins)

Give the students the following questions about respiration and ask them to use their KS3 and chemistry knowledge to write out the answers on mini-whiteboards or in their books.

Questions can include:

- What are the reactants/products in aerobic respiration?
- Where does respiration take place in the cell?
- Respiration is exothermic – what does this mean?
- Why is respiration an exothermic reaction?
- Why do living things need to carry out respiration?

Review answers and address the misconception to highlight that all organisms respire.

If students are struggling to recall respiration from KS3, they could use the Student Book to find the answers.

Once students have completed this, challenge them to write the balanced symbol equation for respiration. Tell them that glucose is $C_6H_{12}O_6$ to enable them to do this.

Main activity

Investigating respiration (30 mins)

Remind students of the word equation for aerobic respiration.

Students then use the Practical: *Investigating respiration* worksheet to plan an investigation to prove that living organisms carry out respiration. They will need to use controls in their plan.

If you have time, you can allow students to carry out the investigation in the next lesson.

To support, students should focus on measuring thermal energy.

As an extension, students should plan two investigations, one to measure thermal energy and one to measure production of carbon dioxide, including an explanation of setting up the controls.

Review and practice

Sperm cells (5 mins)

Show students a diagram of a sperm cell to show the large number of mitochondria in the middle section. Ask students to write down an explanation for why sperm cells contain so many mitochondria.

Students may need a reminder of specialised cells so discuss the role of the sperm cell before students write their explanation.

B9.2 The response to exercise

AQA GCSE Biology specification points

4.4.2.2 Response to exercise

Theme/subtheme

Bioenergetics/Respiration

***Smart Curriculum* Narrative**

How organisms work

Activity

Measuring the effects of exercise

Learning objectives	Learning outcomes: Developing – Aiming for 3	Learning outcomes: Secure – Aiming for 5	Learning outcomes: Extending – Aiming for 8
Explain why heart rate, breathing rate, and breath volume change with exercise	Describe how heart rate, breathing rate, and breath volume change with exercise	**Explain why heart rate, breathing rate, and breath volume change with exercise**	Explain why stores of glycogen change with exercise
Choose the best way to display given data and calculate percentage changes	Draw a suitable chart/graph to display data with some support	**Choose the best way to display given data and calculate percentage changes**	Justify the choice of chart/graph used to display data

Tier 2 vocabulary	Tier 3 vocabulary
arteries, breathe, contraction, depth, energy, exercise, *fibre*, movement, muscle, *rate*, *volume*	aerobic, glycogen, mitochondria, respiration, oxygenated

Lesson plan

Reactivate knowledge

Common misconception

Students may think that only animals respire.

1. What is the word equation for aerobic respiration?
2. Why is respiration needed?
3. Aerobic respiration is an exothermic reaction. What is an exothermic reaction?

Answers: 1 Glucose + oxygen → carbon dioxide + water *2* Carry out basic functions of life; in animals muscle contraction; birds and mammals maintain body temperature; active transport in plants and animals *3* A reaction where more energy is transferred to the surroundings when new bonds are formed than is taken in when initial bonds are broken

Trigger interest

Changes during exercise (5 mins)

Show students an online map showing where exercise might happen in the local area. Students could suggest their local gym, leisure centre, parks, tennis court, etc. Ask the class what changes they think were happening to the participants' bodies during the exercise.

Exposition

The effects of exercise (10 mins)

Discuss with students how they think the following change with exercise:

- Heart rate
- Breathing rate
- Breath volume

For each change, explain why the change occurs and link it back to the word equation for respiration.

Then ask students why stores of glycogen change with exercise. Inform students that glycogen is a store of glucose in the muscles.

Main activity

Measuring the effects of exercise (35 mins)

Demonstrate how to measure the pulse at the wrist or neck.

Students carry out an investigation where they measure their own pulse rate before, during, and after exercise. Consider the needs of the students in your class when suggesting which vigorous activity to do, to ensure it is inclusive for all.

Students record results in a table then analyse the data by:

- calculating the percentage change in resting heart rate after exercise. Inform students that to calculate percentage change they must use the formula:

$$\frac{\text{(final heart rate – initial heart rate)}}{\text{initial heart rate}} \times 100$$

- drawing suitable charts to display their data (higher tier students should be able to justify their choice)
- ask students to use the information in the Student Book to write a scientific explanation for the change in heart rate.

Demonstrate how to calculate percentage changes if students are struggling.

Review and practice

Exercise effects (5 mins)

Students complete the interactive of questions and activities on how exercise affects the body.

B9.3 Anaerobic respiration

AQA GCSE Biology specification points

4.2.1 Aerobic and anaerobic respiration

4.2.2 Response to exercise

Smart Curriculum Narrative

How organisms work

Theme/subtheme

Bioenergetics/Respiration

Activity

Anaerobic respiration

	Learning outcomes		
Learning objectives	**Developing Aiming for 3**	**Secure Aiming for 5**	**Extending Aiming for 8**
State the word equation for anaerobic respiration in animals and give examples of when and where anaerobic respiration occurs	Define anaerobic respiration	**State the word equation for anaerobic respiration in animals and give examples of when and where anaerobic respiration occurs**	Explain why anaerobic respiration transfers less energy than aerobic respiration
State the word equation for anaerobic respiration in plant and yeast cells and describe the economic and social importance of fermentation by yeast	State that anaerobic respiration in yeast cells is known as fermentation and give two ways people use fermentation	**State the word equation for anaerobic respiration in plant and yeast cells and describe the economic and social importance of fermentation by yeast**	Compare and contrast aerobic and anaerobic respiration in animals, plants, and yeast cells
Describe what happens in the muscles during prolonged exercise	State that muscle fibres become fatigued and stop working efficiently after a long period of exercise	**Describe what happens in the muscles during prolonged exercise**	(H) Define the term oxygen debt and explain in detail why heart and breathing rate continue to be high for a period of time after exercise

Tier 2 vocabulary	Tier 3 vocabulary
carbon dioxide, energy, ethanol, fatigue, incomplete, muscle, oxygen, transfer, waste	anaerobic, fermentation, lactic acid, respiration, oxygen debt

Lesson plan

Reactivate knowledge

Common misconception

Students may think that only animals respire.

1 Name a type of cell where you would expect to find many mitochondria.

2 What happens to the heart during exercise?

3 What is glycogen?

Answers: 1 Muscle cell/sperm cell, etc. *2* Pumps faster; pumps more blood at each beat *3* A carbohydrate energy store in the muscles

Trigger interest

Muscle contraction (10 mins)

Ask the class to stand up on tiptoes or to clench and relax their hands repeatedly (consider the needs of the class when the suggestion is made to ensure it is an inclusive activity for all).

Whilst they are doing so, talk about their muscles having to contract to keep them in this position, or contract and relax for the movement of their hands.

Question them on how their muscles feel and why they could not do this for very long.

Exposition

Making lactic acid (15 mins)

Discuss that muscles start to feel tired and burn after a period of intense exercise. Explain why by showing the formula for respiration in animals and linking the burning pain to the lactic acid. Inform students that glucose isn't fully broken down and this is why lactic acid forms, and it is also why less energy is released from anaerobic respiration compared to aerobic respiration.

Students then repeatedly clench and relax their fist. They compare how long they can do this for with their arm at different positions, for example, down by their side, held horizontally in front of them, and held up in the air. They should find that they tire most quickly with their arm in the air, because blood cannot travel so quickly to the arm muscles so less oxygen is being delivered to respiring cells.

Define the term oxygen debt and then ask students to explain in detail why the heart rate and breathing rate continues to be high for a period of time after exercise.

Main activity

Anaerobic respiration (20 mins)

Show students images of useful products of fermentation (bread, wine, soy sauce, etc.) and ask them what they have in common.

Introduce the word equation for anaerobic respiration in plants and microorganisms, and the balanced symbol equation for higher tier students. Address the misconception here.

Ask students to create a comparison table for aerobic and anaerobic respiration.

They should compare which releases more energy, which releases energy more rapidly, whether each occurs in the presence of oxygen, and the products.

Students with higher prior attainment may not need to be told the comparative points, other students can use the Student Book if they are struggling.

Review and practice

Testing fitness (10 mins)

Ask for two volunteers to measure their resting heart rate and breathing rate. Explain that the fitter a person is, the lower the resting heart and breathing rate will be.

Then ask students to exercise for one minute, (e.g., star jumps on the spot) measure the increase in heart and breathing rate, and measure how long it takes for them to return to normal. This can be done by measuring the pulse on the student's wrist and counting how many times the student breathes in. Explain how all three of these measurements can be used to gauge fitness levels.

B9.4 Metabolism and the liver

AQA GCSE Biology specification points

4.4.2.3 Metabolism

***Smart Curriculum* Narrative**

How organisms work

Theme/subtheme

Bioenergetics/Metabolism

Activity

The liver

	Learning outcomes		
Learning objectives	**Developing Aiming for 3**	**Secure Aiming for 5**	**Extending Aiming for 8**
Describe the role of enzymes in synthesis and breakdown reactions in metabolism	Define metabolism as the sum of all reactions in a cell or the body	**Describe the role of enzymes in synthesis and breakdown reactions in metabolism**	Describe the importance of the energy transferred by respiration in metabolism
Describe a range of metabolic reactions including both breakdown and synthesis reactions	List some metabolic reactions	**Describe a range of metabolic reactions including both breakdown and synthesis reactions**	(H) Describe the metabolic role of the liver in repaying the oxygen debt

Tier 2 vocabulary	Tier 3 vocabulary
detoxify, energy, glucose, liver, photosynthesis, protein, reaction	cellulose, excretion, glycerol, glycogen, lactic acid, lipid, metabolism, oxygen debt, respiration, urea

Lesson plan

Reactivate knowledge

Common misconception

Students may think enzymes are only involved in digestion as this was where the topic was previously covered.

1. What is anaerobic respiration?
2. Why do people breathe harder and faster after they finish exercising?
3. Give the word summary for anaerobic respiration in yeast.

Answers: 1 Breakdown of glucose without oxygen present *2* Paying off their oxygen debt/converting the lactic acid that builds up during anaerobic respiration back to glucose in the liver and using it for aerobic respiration *3* Glucose → ethanol + carbon dioxide

Trigger interest

Enzymes recap (10 mins)

Ask students to write down everything they can remember about enzymes on their mini-whiteboards. Support students with the hints below, then collate key ideas and address the misconception.

- What molecule are they made of?
- What do they do to a reaction?
- What is an enzyme-substrate complex?
- Which reactions do they catalyse?
- What is the part of an enzyme that the substrate binds to?
- Why do enzymes only catalyse one reaction?

Exposition

Metabolism (15 mins)

Ask students to work in pairs to write down a list of reactions that occur inside living organisms. This can be on the mini-whiteboards. Collate these on the board.

Introduce the term metabolism as a sum of all the reactions that happen in a cell or the body. Let students know that enzymes are involved in the synthesis and breakdown of substances in metabolism. Ask students if they have heard the term before, for example, someone saying that they have a fast metabolism. Discuss what metabolism means when used in this sense.

Extend students by asking them to consider what the energy released from respiration could be used for. Students should be able to think of active transport, for example, in plants to absorb mineral ions into the root hair cell.

Main activity

The liver (20 mins)

Remind students that drinking too much alcohol can lead to liver disease. Ask them to discuss in pairs why they think this is. Discuss the fact that one role of the liver is to detoxify poisonous substances such as ethanol.

Students complete Literacy worksheet: *The liver* to understand the role of the liver in metabolism. Provide students with the Student Book to support them.

Review and practice

Liver adaptations (10 mins)

Provide students with a list of adaptations of the liver, for example, cells have a lot of mitochondria and cells grow and regenerate rapidly. Ask students to suggest why the liver has these adaptations.

B9 Respiration

Introduction to checkpoint intervention

The checkpoint intervention provides suggestions for a lesson to follow up the *B9 Respiration* checkpoint assessment. Depending on the outcomes of the assessment, these suggestions can be used to move a class towards secure, to target specific learning outcomes, or to consolidate knowledge for students achieving secure outcomes. Use students' outcomes from the checkpoint assessment to address general misconceptions from the content covered. Be prepared to re-cover content as required; three key concepts from the chapter are covered in more detail below.

Checkpoint secure learning outcomes

- Describe cellular respiration as an exothermic reaction occurring continuously in the mitochondria of cells, transferring the energy needed for living processes. Recognise the chemical symbols for the molecules involved in respiration.
- State the word equation for anaerobic respiration in animals and give examples of when and where anaerobic respiration occurs.
- State the word equation for anaerobic respiration in plant and yeast cells and describe the economic and social importance of fermentation by yeast.
- Describe a range of metabolic reactions including both breakdown and synthesis reactions.
- Describe the role of enzymes in synthesis and breakdown reactions in metabolism.
- Describe what happens in muscles during prolonged exercise.

Student reflection

Encourage students to reflect on whether there were any checkpoint questions they found difficult or straightforward as well as their preparation for the checkpoint assessment, including how they revised, what they revised, the time spent on revision, and what areas they could improve upon. For more strategies, see the Metacognition spread.

Intervention activity: Describe cellular respiration as an exothermic reaction occurring continuously in the mitochondria of cells, transferring the energy needed for living processes. Recognise the chemical symbols for the molecules involved in respiration.

Students should write out the word or symbol equation for aerobic respiration in the middle of a page.

Students then annotate around the equation to describe where the oxygen and glucose originate from and how they reach the respiring cells. They should also annotate how the carbon dioxide is removed.

They should show that energy is released and use this to demonstrate it is an exothermic reaction.

Intervention activity: Describe a range of metabolic reactions including both breakdown and synthesis reactions.

Ask students to list a range of metabolic reactions. They can use their previous notes or the Student Book to support with this.

Students should then split the reactions into two columns to indicate whether the reaction is an example of breakdown or synthesis.

Intervention activity: Describe the role of enzymes in synthesis and breakdown reactions in metabolism.

Students create a storyboard showing how an enzyme catalyses a reaction. They apply this to respiration, showing the glucose as the substrate and the products as water and carbon dioxide. Students label the diagram in their storyboard with the following key terms:

active site, enzyme, enzyme-substrate complex, product, substrate.

3 Biological responses

Introduction to unit

Students will already be familiar with multicellular organisms and the organisation of cells. This unit focuses on endocrine and nervous mechanisms which allow organisms to control their internal environment within narrow limits. Students will also study the eye, control of body temperature, control of water and nitrogen, and how plants have mechanisms for sensing and responding to stimuli.

Assessment

Suggested minimum guided learning hours: 21 hours

Exam: Paper 2

Prerequisite knowledge from KS3 (chapter references from *Oxford Smart* Activate)

B1.2.6 Muscles

- Muscles contract.

B1.3.2 Reproductive systems

- Structure of the female reproductive system.
- Human development is affected by hormones.
- Changes to the female body during puberty

B3.3.3 Apply: Preventing pregnancy

- Contraception prevents pregnancy.

B3.2 Sexual reproduction

- Main structures found in plants (roots, stem, leaves, flower).

AQA GCSE Biology specification links

4.5 Homeostasis and response

- 4.5.1 Homeostasis
- 4.5.2 The human nervous system
- 4.5.3 Hormonal coordination in humans
- 4.5.4 Plant hormones (biology only)

Identity, awe, and wonder

With most of the topic being based around human physiology, students are likely to show a keen interest in this unit. Students will be affected by their hormones throughout life and will be very conscious of the changes brought about during puberty. This unit will allow students to become more familiar with the workings of their own body and understand the functions of those changes. Infertility is thought to affect over 3.5 million people in the UK and students may have experienced this within their family. Advances in fertility treatments such as IVF can be discussed in this unit. The speed of nervous transmission is something that students should marvel at. Every day activities involve thousands of decisions and actions, and they feel instantaneous to us. Sports scientists will have an investment in maximising the output of human physiology. There is constant research into neuroscience from the medical world, looking for new drugs or drug targets to help people suffering from neurological issues.

A class is very likely to have students with a range of eyesight issues and most will have experienced an eye test, which can be discussed. There are now stem cell treatments for age-related macular degeneration, laser surgeries for correcting the lens rather than adding a corrective lens in spectacles, corneal transplants, as well as lens replacement cataract surgery.

Required practicals

AQA required practicals	Practical skills	Topic
Required practical 7: Reaction time Plan and carry out an investigation into the effect of a factor on human reaction time.	AT 1 – use appropriate apparatus to record time. AT 3 – selecting appropriate apparatus and techniques to measure the process of reaction time. AT 4 – safe and ethical use of humans to measure physiological function of reaction time and responses to a chosen factor.	B10.2
Required practical 8 (biology only): Plant responses Record results both as length measurements and as accurate, labelled biological drawings to show the effects.	AT 1 – use appropriate apparatus to record length and time. AT 3 – selecting appropriate apparatus and techniques to measure the growth of shoots or roots. AT 4 – safe and ethical use of plants to measure physiological function of growth in response to light or gravity. AT 7 – observations of biological specimens to produce labelled scientific drawings.	B11.9

Working scientifically

Building on what they already know of reproduction and fertilisation, students will begin to critically assess information about methods of contraception and IVF treatments. Through this, they'll be able to discuss some of the social and ethical issues associated with IVF. There are good opportunities to discuss experimental design and the effects of different factors on plant growth in B11.9 and B11.10, as well as problem solving with commercial applications of plant hormones. Finally, B12 is a good opportunity to discuss scientific methods, associated health risks, and making predictions based on scientific data related to removing waste products.

Maths skills

Students will handle and process data from experiments, using an appropriate number of significant figures, calculate means and percentage change, interpret data in a variety of forms (including graphs), and draw tangents to calculate rates of reaction.

Literacy links

Any of the practical investigations could be used to develop scientific/instructional writing skills such as writing a step-by-step method. B11.1 includes a comprehension exercise for students to test their knowledge of terminology used in this chapter. B11.3 includes writing a 'webpage' where students can discuss the differences between type 1 and type 2 diabetes, evaluate different treatments, and discuss the work being done to discover a cure.

B10 The human nervous system

B10.1 Principles of homeostasis

AQA GCSE Biology specification points

4.5.1 Homeostasis

***Smart Curriculum* Narrative**

How organisms work

Theme/subtheme

Hormone functions/Homeostasis

Disciplinary knowledge

- MS 2c, 4a

Activity

Controlling internal conditions

	Learning outcomes		
Learning objectives	**Developing Aiming for 3**	**Secure Aiming for 5**	**Extending Aiming for 8**
Describe the internal body conditions that need to be controlled for cells to work effectively	State that the cells of the body only function within narrow limits	**Describe the internal body conditions that need to be controlled for cells to work effectively**	Apply knowledge of enzymes and osmosis to explain in detail why internal conditions need to be maintained
Describe the basic principles of homeostasis	Define homeostasis	**Describe the basic principles of homeostasis**	Suggest external and internal factors that may change the internal environment
Identify some of the key conditions that must be controlled for effective homeostasis and identify some of the receptors, coordination centres, and effectors involved in the process	Describe the pathway of a control system as receptor → coordination centre → effector	**Identify some of the key conditions that must be controlled for effective homeostasis and identify some of the receptors, coordination centres, and effectors involved in the process**	Describe autonomic control and explain why it is important in homeostasis

Tier 2 vocabulary	Tier 3 vocabulary
coordination centre, enzyme, nervous system, receptor	autonomic control, effector, homeostasis

Lesson plan

Reactivate knowledge

1 What is a specialised nerve cell called?

2 How does pH affect enzyme action?

3 What is a large multicellular organism made up of?

Common misconception

Students may think homeostasis *only* occurs when there is a big change in their internal or external environment – rather than homeostasis constantly occurring to maintain a constant internal environment.

Answers: 1 A neurone *2* Each enzyme has an optimum pH where it works best. Changes in pH may increase or lower the rate of an enzyme-controlled reaction *3* A number of organ systems working together in a coordinated way

Trigger interest

Lost in the desert (10 mins)

Tell students to imagine that they are lost in the middle of a desert. In pairs, ask them to discuss what problems they are likely to face. Discuss their ideas, which might include lack of water, high temperature in the day, low temperature at night, and lack of food.

To extend students, ask them to think about mechanisms in the body which combat these problems.

Support students by asking them to think about changes they would expect to happen in the body when they are too hot.

Exposition

Homeostasis (15 mins)

Ask students to list examples of conditions inside the body that are kept constant (they may suggest temperature, blood glucose, or water levels).

Ask students to suggest any factors – internally or externally – that may change the levels of these conditions in the body (they may consider the outside temperature, exercise, consumption of food and drink, chemical reactions happening in the body, e.g., respiration).

Introduce the term homeostasis and explain what it means, including examples of conditions that need to be controlled, such as temperature and blood glucose levels.

Describe the pathway of a control system as receptor → coordination centre → effector. Indicate that the effector is a muscle or a gland and this brings about a response, such as those seen in homeostasis (e.g., when cold, muscles contract and relax to cause shivering, which increases internal body temperature). Inform students that autonomic control means 'without conscious thought'. Ask students to suggest why this is important in homeostasis.

Main activity

Controlling internal conditions (25 mins)

Students complete this activity using their notes and the Student Book to explain why these conditions need to be kept constant. Use one example of an internal condition that needs to be kept constant to guide students through the pathway of a control system (receptor → coordination centre → effector). Students repeat this for different examples.

Review and instruct students to mark their own answers within the lesson.

Review and practice

Key points (5 mins)

Ask students to write five bullet points to indicate the five things they have learnt from the lesson. Ask volunteers to read their sentences to the class.

Students should include the key terms:

Homeostasis, enzymes, chemical response, nervous response, thermoregulation, blood glucose control, receptors, effectors, coordination centre.

B10.2 The structure and function of the human nervous system

AQA GCSE Biology specification points

4.5.1 Homeostasis

Smart Curriculum Narrative

How organisms work

Theme/subtheme

Organisation/Nervous system

Disciplinary knowledge

- MS 2c, 4a

Required Practical 7

Reaction time

	Learning outcomes		
Learning objectives	**Developing** **Aiming for 3**	**Secure** **Aiming for 5**	**Extending** **Aiming for 8**
Identify and describe the main features of the nervous system, defining the terms neurone and nerve, and give a word summary of the main sequence of events in a nervous response	State that the nervous system enables humans to react to their surroundings and coordinate their behaviour quickly	**Identify and describe the main features of the nervous system, defining the terms neurone and nerve, and give a word summary of the main sequence of events in a nervous response**	Explain in detail how the nervous system coordinates a response
Describe the main sense organs and link each to the stimulus/stimuli it responds to	Name the main sense organs in a mammal	**Describe the main sense organs and link each to the stimulus/stimuli it responds to**	Explain clearly the difference between sensory receptors and the sense organs where they are found
Display data on the impact of different factors on reaction times in an appropriate form	Translate measurements on reaction times into a simple graph, with support	**Display data on the impact of different factors on reaction times in an appropriate form**	Evaluate results in detail in order to discuss precision and accuracy

Tier 2 vocabulary	Tier 3 vocabulary
nervous system, receptor	nerve, neurone, reaction time

Lesson plan

Reactivate knowledge

1 What is homeostasis?

2 List three internal conditions which must be controlled.

3 What are the key features of a control system?

Common misconception

Students may confuse receptors and sense organs and use these terms interchangeably.

Answers: 1 The regulation of the internal environment of the body to maintain the optimum conditions for functioning *2* Body temperature, water content, blood glucose concentration *3* Receptors; a coordination centre; effectors

Trigger interest

Sense organs (5 mins)

Students use the interactive to identify the external stimulus that a range of sense organs detect.

Some students may need to be supported here and so provide the class with some useful key words, such as light, eyes, receptors, neurones, impulses, brain, and muscles.

Exposition

The nervous system (10 mins)

Show students a labelled diagram indicating the different parts of the nervous system. Refer students back to the previous lesson where they were introduced to responses and how they are triggered.

stimulus → receptor → coordinator → effector → response

Address the misconception that receptors are cells that detect a stimulus, whereas sense organs contain these receptors. Give the example of light receptors (rods and cones in the retina) found in the sense organ, the eye.

Main activity

Reaction time (35 mins)

In this required practical, students should work in pairs to choose one factor to investigate. Some examples include the consumption of cola (which contains caffeine), listening to music, or talking to someone. If using cola, check if any students are sensitive to cola/caffeine. If so, choose a different variable.

Students can also choose a suitable method of measuring reaction times. This can be simply dropping a metre rule or using an online reaction tester. Discuss with students the importance of repetition and remind them that their plan should include a results table.

Tell students to include independent, dependent, control variables, prediction, and method in their plan. Check methods and, in the next lesson, allow students to carry out their investigation. They should analyse their results and write a conclusion and an evaluation.

To support students, provide them with a suitable method for measuring reaction times.

Challenge students by asking them to evaluate their results in greater detail and discuss how precise their measurements were (showing the range of repeat sets of readings on their graph). They should also include ways they could possibly change their method to make their results more accurate.

Review and practice

Nervous system ordering (5 mins)

Write the different parts of the nervous system (effector, stimulus, receptor, sensory neurone, coordinator, motor neurone) on the board and ask students to put them into the correct order.

To extend, students should use this to write an example of an event that uses this pathway, for example, picking up a book.

B10.3 Reflex actions

AQA GCSE Biology specification points

4.5.2.1 Structure and function

Smart Curriculum Narrative

How organisms work

Theme/subtheme

Organisation/Nervous system

Disciplinary knowledge

- MS 2c, 4a

Activity

Reflexes

	Learning outcomes		
Learning objectives	Developing Aiming for 3	Secure Aiming for 5	Extending Aiming for 8
Describe the importance of reflex actions	Define reflex reactions and give examples	**Describe the importance of reflex actions**	Compare reflex responses with conscious responses
Describe the events of a reflex arc	Order the events involved in a reflex action	**Describe the events of a reflex arc**	Explain in detail the route of an impulse in a reflex arc, highlighting the differences with a conscious reaction
Describe how an impulse crosses from one neurone to another across a synapse	Define a synapse as a gap between two neurones	**Describe how an impulse crosses from one neurone to another across a synapse**	Apply knowledge of synapses to explain the effects of drugs on the speed of reaction times or reflex responses

Tier 2 vocabulary	Tier 3 vocabulary
impulse, neurone, synapse	conscious reaction, reflex arc

Lesson plan

Reactivate knowledge

1 What do the initials CNS stand for?

2 Name two different types of neurones and what they do.

3 Identify two different types of effectors.

Answers: 1 Central nervous system *2* Sensory neurones – carry impulses from the receptors to the CNS/motor neurones – carry impulses from the CNS to the effectors *3* Muscles, glands

Common misconception

Students often get confused about where in the body the parts of a reflex are.

Trigger interest

Voluntary or involuntary? (10 mins)

Explain to students what is meant by voluntary and involuntary (reflex) actions. Then, read out a list of examples that is a mixture of each (e.g., picking up a ringing phone, moving hand away from a hot pan, talking, sneezing, blinking, shivering, swallowing). Ask them to indicate on the mini-whiteboards if the action is involuntary or voluntary.

After using a few examples, invite students to think of their own examples to share with the class.

Exposition

The reflex arc (15 mins)

Use a diagram of the reflex arc to go through the pathway from stimulus to response. Students often get confused about the location of these so it may be helpful to use a model skeleton (if you have one available) to show the pathway using a piece of coloured string.

Ask students to write down what happens during a reflex action and to describe why it is fast and automatic. Introduce what a synapse is, a chemical transmission instead of an electrical transmission, that occurs in the neurones and explain the importance of it in a reflex action.

Some students may need support writing the events in order, so provide them with the events in the *incorrect* order and ask them to sequence it correctly.

Students should explain in detail what happens at a synapse and why synapses are needed. They can then apply their knowledge by explaining how drugs can affect the synapses.

Main activity

Reflexes (25 mins)

Provide students with instructions on how to carry out a range of reflex tests. Some very simple yet effective tests include: the knee jerk, blink, pupil, and ankle reflex.

Using the Reflexes activity, ask for volunteers to have their reflexes tested. Students can work in pairs or groups, and write down a stimulus and response. They should then explain why they are carrying out these tests, and why each reflex might be useful.

To extend, encourage students to carry out their own research to find some more examples of reflexes and why they are useful. You could steer them in the direction of babies and the unique responses that they possess.

Review and practice

Reflex keywords (5 mins)

Students use the interactive to match each part of the reflex arc with its definition. Students then arrange the parts of the reflex arc in the correct order.

B10.4 The brain

AQA GCSE Biology specification points

4.5.2.2 The brain (Biology only)

Smart Curriculum Narrative

How organisms work

Theme/subtheme

Organisation/Nervous system

Disciplinary knowledge

- WS 1.5

Practical

The structure and function of the brain

	Learning outcomes		
Learning objectives	**Developing Aiming for 3**	**Secure Aiming for 5**	**Extending Aiming for 8**
Describe the brain as made up of billions of interconnected neurones, with different regions that carry out different functions	State that the brain controls complex behaviour	**Describe the brain as made up of billions of interconnected neurones, with different regions that carry out different functions**	H Describe some of the difficulties of investigating brain functions
Describe the functions of the cerebral cortex, cerebellum, and medulla	Identify the cerebral cortex, cerebellum and medulla on a diagram of the brain	**Describe the functions of the cerebral cortex, cerebellum, and medulla**	H Explain how neuroscientists have mapped regions of the brain to particular functions, both historically and today
H Describe some of the difficulties of investigating and treating brain disorders	H State that the complexity and delicacy of the brain makes investigating and treating brain disorders very difficult	**H Describe some of the difficulties of investigating and treating brain disorders**	H Consider ethical dilemmas surrounding brain research, and consider the benefits and risks of investigating and treating brain disorders

Tier 2 vocabulary	Tier 3 vocabulary
brain, disorder, research	cerebellum, cerebral cortex, medulla

Lesson plan

Reactivate knowledge

1 What are reflexes?

2 What is a synapse?

3 Why are reflexes so important?

Common misconception

Students often get confused about where in the body the parts of a reflex are, specifically visualising the placement of the sensory, relay, and motor neurone, with the relay neurone running through the spine or brain.

Answers: 1 Rapid, automatic responses which do not involve the conscious brain *2* Junctions/physical gaps between two neurones *3* Help avoid danger or harm; control basic body functions so don't have to think consciously about them

Trigger interest

Brain function (5 mins)

In pairs, ask students to discuss the function of the brain, write down their answers, and feed back to the class.

Students may be able to use their Student Book to help with ideas.

Exposition

The structure and function of the brain (15 mins)

Provide students with an unlabelled diagram of the brain. Students use the Student Book or any explanations you can give to colour in the different regions of the diagram. Students then create a key to go with the diagram that includes the name of the brain regions and their corresponding functions.

(H) Explain how neuroscientists have mapped regions of the brain to particular functions, both historically and today, giving the historical example of Phineas Gage. Encourage students to carry out some research about Phineas Gage. If time is an issue, encourage them to conduct this research for homework.

Inform students about the complexity of the brain, with its billions of interconnected neurones. Ask students why investigating and treating brain disorders is so difficult, or why we even need to carry out such investigations.

To support students with such questions, ask them to consider what might happen if a neurone were accidently cut. Time permitting, this is a good opportunity to encourage some further research.

Main activity

The structure and function of the brain (25 mins)

Students complete the practical where they carry out a co-ordination task several times. Students then analyse their results by drawing a graph or chart. Ask them if repeating the co-ordination task has any benefits and if so, what they were.

To support students, provide them with a results table and discuss which type of graph is the most suitable to record their results.

To extend, students should plan another method to test a hypothesis, for example, they change the independent variable (reaction with the left hand versus right hand).

Review and practice

Areas of the brain (10 mins)

Students complete the interactive to match an area of the brain to its function. Higher tier students can then do some online research about some cutting-edge breakthroughs in neuroscience. For example, using AI to treat brain cancer or brain-computer interfaces. Students can then present their findings to the class.

B10.5 The eye

AQA GCSE Biology specification points

4.5.2.3 The eye (Biology only)

Smart Curriculum Narrative

How organisms work

Theme/subtheme

Organisation/Nervous system

Disciplinary knowledge

- WS 1.2

Activity

Finding the blind spot

	Learning outcomes		
Learning objectives	**Developing Aiming for 3**	**Secure Aiming for 5**	**Extending Aiming for 8**
Describe the main structures of the eye and their functions	State that the eye is a sense organ containing receptors sensitive to light intensity and colour	**Describe the main structures of the eye and their functions**	Relate the structures of the eye to their functions
Describe how the eye focuses light	Describe how the amount of light entering the eye is controlled by the size of the pupil	**Describe how the eye focuses light**	Draw an accurate ray diagram to show how the structures of the eye focus light on the retina

Tier 2 vocabulary	Tier 3 vocabulary
eye, light, sense organ	pupil, ray diagram, receptor

Lesson plan

Reactivate knowledge

1 What is the role of the cerebral cortex of the brain?

2 Which part of the brain controls unconscious activities such as heartbeat, gut movements and breathing?

3 Give three ways in which scientists find out about the way the brain works.

Answers: 1 Consciousness/intelligence/memory/language *2* Medulla *3* For example, studying people whose brains are damaged/ electrical stimulation of brain areas/MRI scans

Common misconception

Students may sometimes confuse receptors and sense organs and use these terms interchangeably.

Trigger interest

Can you believe your eyes? (10 mins)

Show the class a series of optical illusions (e.g., the blue/black or gold/white dress that sparked global discussion in 2015). Ask students what they can see, being mindful of students with visual impairments. Nominate one student to record the findings on the board and to total them for each optical illusion – this way everybody has the same information. Each student should then individually calculate the percentage of students who saw each optical illusion.

Use this as an opportunity to link what was learnt in the last lesson about the brain to explain vision.

Exposition

Structure of the eye demonstration (15 mins)

Carry out an eye dissection as a demonstration, or show a video of one. Alternatively, if you have a 3D model of the eye, you can use this.

Provide students with a diagram of the eye (Figure 1 in the Student Book) to label the different parts and their functions.

This is a good opportunity to address the misconception that the eye is a sense organ and that the light receptors are located in the retina.

Main activity

Finding the blind spot (20 mins)

Discuss how the light that enters the eye is focused on the retina, and how this initiates an impulse. Use a ray box, or a diagram of this demonstration, to show how light is focused onto the retina.

Students can also try and find their blind spot (using the Student Book) and explain why you have one.

To support students, use cloze sentences and provide them with a diagram of the eye for them to add rays of light to.

Students then build a 2D model of the eye following the instructions from the activity.

Review and practice

Explaining the pupil reflex (10 mins)

Students may have carried out the pupil reflex activity in B10.3. Ask them to use what they learnt in this lesson to explain why we have the pupil reflex. Ask students to provide some examples.

B10.6 Common problems of the eye

AQA GCSE Biology specification points

4.5.2.3 The eye (Biology only)

Smart Curriculum Narrative

How organisms work

Theme/subtheme

Organisation/Nervous system

Disciplinary knowledge

- WS 1.2, 1.4

Activity

Measuring range of accommodation

Learning objectives	Learning outcomes: Developing Aiming for 3	Secure Aiming for 5	Extending Aiming for 8
Describe how the lens changes shape to focus on near or distant objects	State that the lens changes shape to focus on near or distant objects	**Describe how the lens changes shape to focus on near or distant objects**	Draw accurate ray diagrams to explain what happens during accommodation
Describe several common causes for myopia (short-sightedness) and hyperopia (long-sightedness) and interpret ray diagrams of these conditions	Define short-sightedness and long-sightedness where rays of light focus either in front or behind the retina	**Describe several common causes for myopia (short-sightedness) and hyperopia (long-sightedness) and interpret ray diagrams of these conditions**	Draw accurate ray diagrams to explain what causes long and short-sightedness
Describe the use of spectacle and contact lenses, laser surgery and lens replacement surgery to correct long and short-sightedness, and interpret ray diagrams to show how spectacle lenses work	Describe the use of spectacle lenses to correct short and long sight	**Describe the use of spectacle and contact lenses, laser surgery, and lens replacement surgery to correct long and short-sightedness, and interpret ray diagrams to show how spectacle lenses work**	Draw accurate ray diagrams to show how concave and convex lenses are used to correct short- and long-sightedness, and evaluate the risks and benefits of surgery to treat these conditions

Tier 2 vocabulary	Tier 3 vocabulary
long-sightedness, short-sightedness	concave, convex, ray diagram

Lesson plan

Reactivate knowledge

1 Which part of the eye controls the amount of light that gets in through the pupil?

2 What is the blind spot?

3 How are the ciliary muscles and suspensory ligaments involved in focusing light on the retina?

Answers: 1 Muscles of the iris *2* Where the optic nerve leaves the eye – no retina *3* Change the shape of the lens to bend light more or less

Common misconception

Students sometimes confuse receptors and sense organs and use these terms interchangeably.

Trigger interest

Near and far (10 mins)

Ask students to hold out their forefinger in front of them, at arm's length and focus on it. Then, focus on an object a long way away (out of a window if possible) and then back to their finger. Ask them to talk about what happened – why their finger went out of focus when they looked at the distant object, and vice versa. This can be done in groups, time permitting.

To extend students, encourage them to apply what they know about the eye and how it focuses light.

Exposition

Accommodation and eye defects (15 mins)

Firstly, hand out diagrams of the eye focusing on near and distant objects. Simultaneously, introduce students to the term 'accommodation' – the ability of the eye to see both at a distance and close-up.

Ask students to annotate the diagram to show the changes that occur in the eye, and why these changes are needed in order to form an image on the retina. Encourage students to refer to the Student Book, if necessary.

To support students, provide diagrams with rays of light drawn on them.

To challenge students, provide diagrams without rays of light drawn on them and ask students to add them.

Main activity

Measuring range of accommodation (20 mins)

Students should now measure their range of accommodation. Ask students to focus on the furthest distance they can see, this would ideally be through an open window into the distance outside. If students are able to see objects clearly in focus at very far distances, we state that they can see as far as 'infinity'. If students are not able to see objects clearly in focus in the distance, then they should look at various points in the distance until an object is clearly in focus. Students who wear glasses can remove them for this activity if they are comfortable to.

Students should then hold out their arms in front of them and look at their hands. Students should slowly bring their hands closer to their faces until they can no longer see the creases and details on their hands clearly and measure this distance, another student should use a rule to measure the distance between their hands to their eyes, taking care not to touch their eyes.

The range is in dioptres (1/distance in m) and is calculated as:

$$\frac{1}{\text{near distance}} - \frac{1}{\text{far point distance}}$$

The far point distance will be infinity, so their range will simply be:

$$\frac{1}{\text{near distance}}$$

Show students a graph that plots how the range of accommodation changes with age. Students can analyse the information in pairs.

Introduce myopia and hyperopia and relate these conditions to the process of accommodation.

As an extension, draw accurate ray diagrams to show how concave and convex lenses are used to correct short- and long-sightedness. Evaluate the risks and benefits of surgery to treat these conditions.

Review and practice

Eye health (10 mins)

Show students a picture of an optician and discuss the types of things that opticians check for to measure good eye health. For example, checking the structure of the eye and detecting early signs of disease. This could lead onto a discussion about the importance of optometry and careers in the field such as optical engineering or laser-eye correction.

Expand by introducing the class to Dr Patricia Bath, a scientist who transformed the nature of laser eye surgery and was the first Black woman physician to receive a patent for a device she invented to remove cataracts in the eye.

B10 The human nervous system

Introduction to checkpoint intervention

The checkpoint intervention provides suggestions for a lesson to follow up the *B10 The human nervous system* assessment. Depending on the outcomes of the assessment, these suggestions can be used to move a class towards secure, to target specific learning outcomes, or to consolidate knowledge for students achieving secure outcomes. Use students' outcomes from the checkpoint assessment to address general misconceptions from the content covered. Be prepared to re-cover content as required; three key concepts from the chapter are covered in more detail below.

Checkpoint secure learning outcomes

- Describe the basic principles of homeostasis.
- Describe how an impulse crosses from one neurone to another across a synapse.
- Describe the importance of reflex actions.
- Describe the brain as made up of billions of interconnected neurones, with different regions that carry out different functions.
- Describe how the lens changes shape to focus on near or distant objects.
- Describe the use of spectacle and contact lenses, laser surgery, and lens replacement surgery to correct long and short sightedness, and interpret ray diagrams to show how spectacle lenses work.

Student reflection

Encourage students to reflect on whether there were any checkpoint questions they found difficult or straightforward as well as their preparation for the checkpoint assessment, including how they revised, what they revised, the time spent on revision, and what areas they could improve upon. For more strategies, see the Metacognition spread.

Intervention activity: Describe the importance of reflex actions.

Students use Figure 3 in B10.3 of the Student Book to explain why a reflex action is fast and automatic.

To support students in understanding why reflexes are fast:

- remind students that impulses travel faster as electricity through the neurones than as chemicals at a synapse
- ask students to look at how many synapses are in a reflex. Compare this to non-reflex responses, which could contain far more than two synapses
- ask students to use this information, to consider the speed.

To support students in understanding why reflexes are automatic:

- ask students to name the location of the relay neurone
- inform students that reflexes travel through the spine or unconscious part of the brain, so thought is not involved.

Intervention activity: Describe how the lens changes shape to focus on near or distant objects.

A student is looking at their book, which is near to them on a table. Describe what happens to the ciliary muscles, suspensory ligaments, and the lens to focus on this near object.

If students require support, provide them with the following hints:

- The ciliary muscles and suspensory ligaments are antagonistic (as one contracts the other relaxes).
- To focus on a near object, the light needs to be bent more to focus on the retina, so the lens must be fatter and rounded.

A student then looks out the window at a tree (or any object) in the distance. Describe what happens to the ciliary muscles, suspensory ligaments, and the lens when focussing on this distant object.

If students require support, provide them with the following hints:

- The ciliary muscles and suspensory ligaments are antagonistic (as one contracts the other relaxes).
- To focus on a distant object, the light needs to be bent less to focus on the retina, so the lens must be longer and flatter.

Intervention activity: Describe the use of spectacle and contact lenses, laser surgery and lens replacement surgery to correct long and short sightedness, and interpret ray diagrams to show how spectacle lenses work.

Provide students with a ray image (e.g., Figures 3–4 from B10.6 of the Student Book) and ask them to write an explanation of how a concave lens corrects myopia and how a convex lens corrects hyperopia.

If students require support, provide them with the following hints:

- Myopia is when the light focuses in front the retina. To correct this, the light needs to be refracted (bent) less.
- Hyperopia is when the light focuses behind the retina. To correct this, the light needs to be refracted (bent) more.

B11 Hormonal coordination

B11.1 Principles of hormonal control

AQA GCSE Biology specification points
4.5.3.1 Human endocrine system

Theme/subtheme
Organisation/Endocrine system

***Smart Curriculum* Narrative**
How organisms work

Activity
Endocrine glands

	Learning outcomes		
Learning objectives	**Developing Aiming for 3**	**Secure Aiming for 5**	**Extending Aiming for 8**
Describe the principles of hormonal coordination and control, and compare coordination and control by the endocrine system and the nervous system	State that the endocrine system is composed of glands which secrete hormones directly into the bloodstream, and the blood carries the hormone to its target organ where it has an effect	**Describe the principles of hormonal coordination and control, and compare coordination and control by the endocrine system and the nervous system**	Explain how hormones have their effects on the body
Describe the main roles of the hormones released by named endocrine glands	Label the pituitary gland, pancreas, thyroid, adrenal gland, ovary, and testes on a diagram of the human body	**Describe the main roles of the hormones released by named endocrine glands**	Explain why the pituitary gland is known as a 'master gland'

Tier 2 vocabulary	Tier 3 vocabulary
hormone, ovary, pancreas	ADH, adrenal gland, adrenaline, endocrine gland, endocrine system, FSH, growth hormone, insulin, oestrogen, pituitary gland, receptor, testosterone, thyroid gland, TSH

Lesson plan

Reactivate knowledge

1 Maintaining the internal conditions of the body in a stable state is known as....?

2 Which two systems are involved in maintaining the internal environment of the body?

3 Where is the pituitary gland found?

Answers: 1 Homeostasis *2* Nervous system and hormone system *3* In the brain

Common misconception

Students often think that hormones are only involved in the changes to the body during puberty, and are not aware of the wide range of hormones produced and their varied roles.

Trigger interest

Hormones: *think, pair, share* (10 mins)

Inform students that hormones are chemical messengers that are released by glands and transported in the blood to the target organ. Ask students to work alone and write down any hormones they have heard of. Then ask them to share their lists in pairs and discuss whether they know any extra information about these hormones; for example, where in the body they are produced, or what they do. Ask pairs to share what they know with the rest of the class to build up a class list on the board.

Exposition

Endocrine glands diagram (10 mins)

Supply students with an outline of the human body and ask them to draw on the endocrine glands, which they can label with their names and functions. Students can use information from the Student Book to help them do this.

This is a good opportunity to address the misconception and talk through the role of each hormone to students.

After labelling the diagram, ask students to suggest why the pituitary gland is named the 'master gland'.

Main activity

Endocrine glands (25 mins)

Students answer a series of questions from the activity, which looks at the hormones that each gland produces, how hormones travel to their target organs, and the differences between hormonal and nervous control.

Students can use the Student Book to help with this, or this could be a computer room research lesson.

Review answers with the class so students can self-mark their work.

As an extension, students should suggest the effects on the body if glands make too much or too little of the named hormone.

Students can then complete the worksheet Go further: *Hormones and cell signalling*.

Review and practice

Nerves vs hormones (10 mins)

Create a comparison table to compare coordination and control by the endocrine system and the nervous system.

Model answer:

Feature	Nervous system	Endocrine system
Nature of signals	Electrical impulses (nerve impulses)	Chemical messengers (hormones)
Speed of response	Rapid (milliseconds)	Slower (seconds to minutes)
Duration of response	Short-lived (milliseconds to seconds)	Longer-lasting (minutes to days)
Mode of transmission	Neurones and synapses	Bloodstream
Specificity of response	Precise and specific	General and widespread
Target cells/organs	Muscle cells, glands, other neurones	Cells or organs with appropriate receptors
Major control centres	Brain and spinal cord	Endocrine glands (e.g., pituitary, thyroid)

B11.2 The control of blood glucose levels

AQA GCSE Biology specification points

4.5.3.2 Control of blood glucose concentration

Smart Curriculum Narrative

How organisms work

Theme/subtheme

Hormone functions/Hormonal control in human systems

Disciplinary knowledge

- WS 1.3
- MS 2c

Activity

Control of blood glucose levels

	Learning outcomes		
Learning objectives	**Developing Aiming for 3**	**Secure Aiming for 5**	**Extending Aiming for 8**
Describe how insulin produced by the pancreas controls blood glucose levels in the body	State that blood glucose (sugar) concentration is monitored and controlled by the pancreas using the hormone insulin	**Describe how insulin produced by the pancreas controls blood glucose levels in the body**	(H) Explain how glucagon interacts with insulin to control blood glucose levels
Describe the difference in the causes of Type 1 and Type 2 diabetes	State that there are two types of diabetes	**Describe the difference in the causes of Type 1 and Type 2 diabetes**	Explain why it is important to control the level of glucose in the blood

Tier 2 vocabulary	Tier 3 vocabulary
glucose, lipid, liver, muscle, obese, pancreas, respiration	blood glucose concentration, diabetes, (H) glucagon, glycogen, homeostasis, insulin, (H) negative feedback control, Type 1 diabetes, Type 2 diabetes

Lesson plan

Reactivate knowledge

1 What is the scientific term for the hormone system?

2 Name three human endocrine glands.

3 Give one example of a hormone and state its function in the body.

Common misconception

Students might think that blood sugar levels can only be affected by the amount of sugar you eat, rather than by eating or drinking anything containing carbohydrates.

Answers: 1 Endocrine system *2* Any from: pituitary gland; thyroid gland; adrenal gland; pancreas; ovary (female); testis (male) *3* For example, insulin – controls blood sugar levels; adrenaline – prepares body for fight or flight; thyroxine – controls rate of metabolism; growth hormone – controls rate of growth in children etc

Trigger interest

Digesting carbohydrates (10 mins)

Ask students to use what they have learnt about digestion to write down what happens in their body after eating a slice of sugary cake.

Prompt students to think about the enzyme produced in the salivary glands and pancreas.

Students with prior higher attainment could also discuss why blood glucose levels will increase after eating the cake.

Exposition

Control of blood glucose levels (15 mins)

Ask students to write on their mini-whiteboards what might cause the amount of sugar in their blood to increase and decrease. Discuss the answers and explain that food and drink containing carbohydrates are digested and the sugar is absorbed from the small intestines into the blood, increasing blood glucose levels. Glucose is a respiratory substrate, therefore exercising will cause a decrease in blood glucose levels as it requires more respiration.

Ask students to suggest on their mini-whiteboards why it may be harmful for blood sugar levels to become higher or lower than the normal range. Discuss their responses and link the dangers of low blood sugar to a lack of respiratory substrate and high levels of water from surrounding cells moving into the blood by osmosis, as the blood is more concentrated due to the sugar, resulting in shrivelled cells and a higher blood pressure.

Explain to students how the body controls blood glucose levels. Use a circular flow diagram, and Figure 1 from B11.2 of the Student Book, to help explain a homeostasis example that involves negative feedback. Students only need to know what happens when blood glucose levels get too high.

Challenge students by asking them to explain why it is important to control the level of glucose in the blood.

Ask students to consider how a blood glucose level that is too high or too low would affect respiration and osmosis.

Main activity

Control of blood glucose levels (25 mins)

Students complete the activity, including using the data to plot a graph and analyse what it shows about diabetes. Students then use information in the Student Book to describe what causes Type 1 and Type 2 diabetes.

Review and practice

Glucose, glycogen, and glucagon (5 mins)

Ask students to write down definitions for glucose and glycogen to make sure they understand the difference.

Higher tier students should also write a definition for glucagon.

B11.3 Treating diabetes

AQA GCSE Biology specification points

4.5.3.2 Control of blood glucose concentration

Smart Curriculum Narrative

How organisms work

Theme/subtheme

Hormone functions/Hormonal control in human systems

Disciplinary knowledge

- MS 2c

Activity

Diabetes webpage

Interpreting graphs on blood glucose levels

	Learning outcomes		
Learning objectives	**Developing** **Aiming for 3**	**Secure** **Aiming for 5**	**Extending** **Aiming for 8**
Explain why Type 1 diabetes is treated with insulin injections	State that Type 1 diabetes is usually treated with insulin injections	**Explain why Type 1 diabetes is treated with insulin injections**	Evaluate different treatments and potential cures for Type 1 diabetes
Explain how a carbohydrate-controlled diet and an exercise regime are often effective treatments for Type 2 diabetes. State that different drugs and even insulin may be needed to control Type 2 diabetes in some people	State that Type 2 diabetes is commonly treated by changes to diet and exercise	**Explain how a carbohydrate-controlled diet and an exercise regime are often effective treatments for Type 2 diabetes. State that different drugs and even insulin may be needed to control Type 2 diabetes in some people**	Explain in detail how lifestyle factors and genetics affect the risk of developing Type 2 diabetes
Extract information and interpret data from graphs showing the blood sugar levels in both people with diabetes and people without diabetes	Describe graphs showing the effect of insulin on blood sugar levels in both people with diabetes and people without diabetes	**Extract information and interpret data from graphs showing the blood sugar levels in both people with diabetes and people without diabetes**	Extract information and interpret data from graphs showing blood sugar levels and the effect of different treatments on people with Type 1 and Type 2 diabetes

Tier 2 vocabulary	Tier 3 vocabulary
carbohydrate, genetic engineering, glucose, liver, obesity, pancreas, stem cell	blood glucose concentration, carbohydrate-controlled diet, diabetes, glycogen, immunosuppressant, insulin, insulin injection, insulin pump, transplant, Type 1 diabetes, Type 2 diabetes

Lesson plan

Reactivate knowledge

1 Which endocrine organ makes the hormone insulin?
2 What is the name of the carbohydrate stored in the human liver and muscles?
3 What is Type 1 diabetes?

Common misconception

Students might think that blood sugar levels can only be affected by the amount of sugar you eat, rather than by eating or drinking anything containing carbohydrates.

Answers: 1 Pancreas 2 Glycogen 3 When the pancreas does not make enough or any insulin and control of the blood sugar levels is lost

Trigger interest

Foods containing sugar (5 mins)

Show students an image of a cake, a sugary drink, potato, and pasta. Ask the students to identify which of those would result in an increase in blood sugar. Use this activity to address the misconception.

Exposition

Treatments for diabetes (10 mins)

Explain that diabetes is a condition where the body cannot control blood glucose levels. Measuring blood glucose levels after a sugary meal is a way of diagnosing diabetes. Summarise the differences between the causes of Type 1 and Type 2 diabetes.

Explain that Type 1 diabetes is due to the pancreas producing little to no insulin, and therefore the treatment is injections of insulin.

Explain that people with Type 2 diabetes do produce insulin, but the body no longer responds to it. Therefore, a carbohydrate-controlled diet and an exercise regime are often effective treatments for Type 2 diabetes. State that different drugs or insulin may be needed to control Type 2 diabetes in some people.

Main activity

Diabetes webpage (15 mins)

Students will research online, using, for example, the NHS website, to discover how both types of diabetes can be treated and how treatments have changed over time.

Students should use their research to complete the summary comparison table to compare Type 1 and Type 2 diabetes and their causes.

Then, ask students to make informed opinions of their findings. For example, considering why they think diabetes in adults and children is rising.

Support students by showing them an example of a real page from NHS Choices that explains the treatment of an illness.

As an extension, students should include information on recent attempts to find a cure for diabetes and evaluate different treatments for Type 1 diabetes.

Interpreting graphs on blood glucose levels (20 mins)

Students then complete this activity by extracting information and interpreting data from graphs on diabetes.

Review and practice

Lifestyle factors (5 mins)

Students use the interactive to assess a series of lifestyle factors (eating lots of sugary food, taking plenty of exercise, smoking, etc.) to decide if they will increase the risk of developing Type 2 diabetes. Students can also explain why certain lifestyle factors increase the risk.

Ⓗ B11.4 The role of negative feedback

AQA GCSE Biology specification points

Ⓗ 4.5.3.7 Negative feedback

Smart Curriculum Narrative

How organisms work

Theme/subtheme

Hormone functions/Hormonal control in human systems

Disciplinary knowledge

- WS 1.2
- MS 2c

Activity

Adrenaline

	Learning outcomes		
Learning objectives	**Developing** Aiming for 3	**Secure** Aiming for 5	**Extending** Aiming for 8
Describe the roles of adrenaline and thyroxine in the body	Recall that adrenaline is produced by the adrenal glands and thyroxine is produced by the thyroid gland	**Describe the roles of adrenaline and thyroxine in the body**	Explain in detail how adrenaline prepares the body for 'fight or flight'
Describe the negative feedback loop that controls the levels of thyroxine in the body	State that many hormones are controlled as part of a negative feedback system	**Describe the negative feedback loop that controls the levels of thyroxine in the body**	Explain the principles of negative feedback control in hormonal systems in response to changes in the internal environment

Tier 2 vocabulary	Tier 3 vocabulary
glucose, hormone	adrenal gland, adrenaline, basal metabolic rate, 'flight or fight', negative feedback, pituitary gland, thyroid gland, thyroid stimulating hormone (TSH), thyroxine

Lesson plan

Reactivate knowledge

1 Name the hormone that controls the release of thyroxine from the thyroid gland.

2 Where is this hormone produced and how does it reach the thyroid gland?

3 Where are the adrenal glands situated in the body?

Common misconception

Students may think that negative feedback is a bad thing for humans because of the term 'negative'.

Answers: 1 TSH/thyroid stimulating hormone *2* Produced in the pituitary gland, carried in the blood *3* On top of the kidneys

Trigger interest

Start the lesson with a bang! (5 mins)

During the usual start to the lesson, suddenly make a loud noise, for example, by slamming a book on a desk or turning on loud music. Invite students to comment on how they felt when they heard the noise and write a list of the effects of a shock like this has on the body. (Consider whether any students in your class have hypersensitivity to sound before doing this activity.)

Students should think about what hormone was released to bring about these changes, and why the changes are useful.

Exposition

Thermostat interactive (15 mins)

Explain to students how a thermostat works in a similar way to a negative feedback system. Students then use the interactive to complete a paragraph explaining this. They then complete sentences to describe what a negative feedback system is.

Provide students with the flow diagrams from the worksheet Literacy: *Adrenaline and thyroxine*, and use these to describe the roles of adrenaline and thyroxine and how thyroxine is an example of negative feedback.

Main activity

Adrenaline (25 mins)

Students complete the activity by using the Student Book to help them complete the table. Review the answers with students so they can self-mark in the lesson. The follow-up activity could be set as homework.

Ask students to discuss in pairs what the possible effects of an underactive or overactive thyroid gland might be. Show them the answers by searching on the NHS Choices website for the conditions.

Review and practice

Negative feedback summary (10 mins)

Students consolidate their understanding from the last two lessons to explain how the control of blood glucose levels and thyroxine are all examples of negative feedback.

Students should then draw a diagram to explain how blood glucose control is an example of negative feedback.

B11.5 Human reproduction

AQA GCSE Biology specification points

4.5.3.4 Hormones in human reproduction

Smart Curriculum Narrative

Growth and development

Theme/subtheme

Reproduction/Menstrual cycle and hormonal control of reproduction

Activity

The menstrual cycle

	Learning outcomes		
Learning objectives	**Developing** **Aiming for 3**	**Secure** **Aiming for 5**	**Extending** **Aiming for 8**
Describe the development of female and male secondary sex characteristics during puberty, including the hormones involved and physical changes	Identify oestrogen and testosterone as the main female and male reproductive hormones	**Describe the development of female and male secondary sex characteristics during puberty, including the hormones involved and physical changes**	Compare and contrast the process of puberty in females and males including the hormones involved and physical changes
Describe the changes that take place in the ovaries and the uterus during the menstrual cycle and name the hormones involved at each stage in the cycle	Describe the main events of the menstrual cycle	**Describe the changes that take place in the ovaries and the uterus during the menstrual cycle and name the hormones involved at each stage in the cycle**	Explain the main role of each hormone in the menstrual cycle

Tier 2 vocabulary	Tier 3 vocabulary
egg, gamete, hormone, ovary, reproduction, reproductive organ, sperm, testis	follicle stimulating hormone (FSH), luteinising hormone (LH), menstrual cycle, menstruation, oestrogen, ovulation, *period*, primary sexual characteristic, progesterone, puberty, reproductive hormone, secondary sexual characteristic, testosterone, uterus

Lesson plan

Reactivate knowledge

1 Which endocrine gland produces the female hormone oestrogen?

2 Which hormone is produced by the male testes?

3 Oestrogen has two main functions in the female body. What are they?

Common misconception

Students may think that biological females only release hormones during their period, rather than different hormones at different levels continually throughout the menstrual cycle.

Answers: *1* Ovary/ovaries *2* Testosterone *3* Development of the female secondary sexual characteristics; control of the menstrual cycle

Trigger interest

Scrambled hormones (5 mins)

Write anagrams of oestrogen and testosterone on the board. Tell the class that they are examples of the hormones they will be learning about today, and ask them to unscramble them. Throughout this lesson, use the terms biological female/male.

Support students by writing the first letter of each hormone as a clue.

Challenge students to suggest one action of each hormone on the body.

Exposition

Secondary sexual characteristics (15 mins)

Describe primary sexual characteristics as the ones you are born with, and secondary sexual characteristics as ones that develop during puberty.

Ask students to write a list of changes that happen to biological females and biological males during puberty. Discuss students' answers, and students can add any missing changes to their lists. Prompt further discussion to highlight secondary sexual characteristics that both biological males and females share.

Main activity

The menstrual cycle (25 mins)

Students should complete the first task of the activity where they label a diagram of a biological female human reproductive system. Show students the correct answers.

Go through the events of the menstrual cycle with the class using Figure 3 and information from the Student Book. The students label key events during the menstrual cycle on the diagram in Task 1 of the activity.

Students then complete the rest of the menstrual cycle activity. Review the answers with the class so students can self-mark.

Review and practice

Menstrual cycle hormones (10 mins)

Students use the interactive to match each key term from the lesson to a question that could have the key term as an answer.

Students complete the final part of the interactive activity in their books, summarising FSH, LH, oestrogen, and progesterone.

Ⓗ B11.6 Hormones and the menstrual cycle

AQA GCSE Biology specification points

Ⓗ 4.5.3.4 Hormones in human reproduction

Smart Curriculum Narrative

Growth and development

Theme/subtheme

Reproduction/Menstrual cycle and hormonal control of reproduction

Disciplinary knowledge

- MS 2c

Activity

Hormones and the menstrual cycle

Learning objectives	Learning outcomes: Developing – Aiming for 3	Learning outcomes: Secure – Aiming for 5	Learning outcomes: Extending – Aiming for 8
Name the glands that produce the hormones oestrogen, progesterone, LH, and FSH	Recall the main female reproductive hormones	**Name the glands that produce the hormones oestrogen, progesterone, LH, and FSH**	Describe the effect of the reproductive hormones produced by the pituitary gland on the ovary during the menstrual cycle
Describe the function of the hormones that control the menstrual cycle	Recall the main events of the menstrual cycle	**Describe the function of the hormones that control the menstrual cycle**	Explain the interactions of the pituitary and ovarian hormones in the control of the menstrual cycle
Extract data from graphs showing hormone levels during the menstrual cycle	Describe a graph showing hormone levels during the menstrual cycle	**Extract data from graphs showing hormone levels during the menstrual cycle**	Extract and interpret data from graphs showing hormone levels during the menstrual cycle

Tier 2 vocabulary	Tier 3 vocabulary
egg, embryo, fertilise, hormone, ovary, pregnancy, sperm	follicle stimulating hormone (FSH), luteinising hormone (LH), menstrual cycle, menopause, menstruation, oestrogen, ovulation, *period*, pituitary gland, progesterone, uterus, uterus lining

Lesson plan

Reactivate knowledge

1 Give four secondary sexual characteristics which develop during puberty in both females and males.

2 How often is a mature egg released from the ovary during the female menstrual cycle?

3 What is the monthly period?

Common misconception

Students may think that biological females only release hormones during their period, rather than different hormones at different levels continually throughout the menstrual cycle.

Answers: 1 Growth spurt/growth of underarm hair and pubic hair/growth of external genitalia/brain structure changes and matures *2* Once a month/once a cycle/every 24–38 days *3* When the lining of the uterus is shed if the released ovum is not fertilised

Trigger interest

Menstrual cycle review (5 mins)

Display Figure 3 from B11.5, the diagram of the menstrual cycle. Students use the Interactive: *Menstrual cycle review* to answer a series of questions to check their understanding.

Exposition

Hormone levels in the menstrual cycle (15 mins)

Provide students with copies of Figure 3 from B11.5 of the Student Book, which shows the changing levels of the female sex hormones throughout the menstrual cycle.

Ask students to write a description of what the graph shows. For example, on which day does ovulation occur in this cycle? What happens to the uterus lining as levels of oestrogen increase?

Main activity

Hormones and the menstrual cycle (25 mins)

Students explain the interaction of the hormones in controlling the menstrual cycle using this activity.

Discuss the roles of each hormone, as shown on the sheet, and students then use this information to complete the flow diagram showing how the hormones interact.

Review and practice

Which hormone? (10 mins)

Provide students with cards showing the names of hormones, or have students write the name of the hormone each time on a mini-whiteboard. Read a series of statements and ask students to hold up the hormone (or hormones) you are describing.

Example statements:

- It causes ovulation.
- It causes an egg to mature.
- It stimulates the release of oestrogen.
- It is secreted by the pituitary gland. (this has more than one answer)

B11.7 The artificial control of fertility

AQA GCSE Biology specification points

4.5.3.5 Contraception

Smart Curriculum Narrative

Growth and development

Theme/subtheme

Reproduction/Menstrual cycle and hormonal control of reproduction

Disciplinary knowledge

- WS 1.3, 1.4

Activity

Contraceptive advice

Evaluating contraceptive methods

	Learning outcomes		
Learning objectives	**Developing Aiming for 3**	**Secure Aiming for 5**	**Extending Aiming for 8**
Describe the main categories of contraception and explain simply how the different types of non-hormonal contraception work	Define contraception and list examples	**Describe the main categories of contraception and explain simply how the different types of non-hormonal contraception work**	Discuss the advantages and disadvantages of the main non-hormonal methods of contraception
Describe the main types of hormonal contraception and explain simply how they work	Categorise contraceptives as hormonal and non-hormonal	**Describe the main types of hormonal contraception and explain simply how they work**	Apply knowledge of hormones in the natural menstrual cycle to explain how hormonal contraceptives work
Use data to help discuss the advantages and disadvantages of different methods of contraception and explain why issues around contraception cannot be answered by science alone	List the main advantages and disadvantages of the main methods of contraception	**Use data to help discuss the advantages and disadvantages of different methods of contraception and explain why issues around contraception cannot be answered by science alone**	Evaluate personal, social, economic, and environmental implications of the different methods of contraception based on evidence and discussion

Tier 2 vocabulary	Tier 3 vocabulary
egg, hormone, ovary, pregnancy, sperm	barrier method, condom, contraception, contraceptive implant, contraceptive patch, contraceptive pill, *diaphragm*, hormonal contraception, intrauterine device, mixed pill, non-hormonal contraception, oral contraceptive, oviduct, sexually transmitted disease, spermicide, surgical sterilisation, vasectomy

Lesson plan

Reactivate knowledge

1 What must happen to the egg released at ovulation for pregnancy to occur?

2 Which hormone stimulates the eggs in the ovary to mature?

3 Which hormone triggers ovulation?

Common misconception

Students may think that condoms and 'the pill' are the only forms of contraception.

Answers: 1 Must be fertilised by a sperm *2* FSH (follicle stimulating hormone) *3* LH (luteinising hormone)

Trigger interest

What is contraception? (5 mins)

Ask students to write down their answer to this question. Use images of examples of contraceptives to initiate ideas. Discuss potential definitions with the class and ask students to write down the following in their books: Contraception methods are interventions that can stop the spread of sexually transmitted infections and prevent pregnancy.

Exposition

The contraceptive pill (10 mins)

Introduce the contraceptive pill as a contraceptive. Tell the class that one type of the contraceptive pill contains oestrogen. Ask students to use what they have learnt about the function of oestrogen to suggest why.

Main activity

Contraceptive advice (10 mins)

Discuss the fact that contraception is a way of preventing pregnancy. Make a class list of all the contraceptive methods that the students are aware of. Allow students to use the Student Book to add any extras that were not mentioned. Students then categorise the contraceptives as hormonal or non-hormonal, and explain what the difference is.

Evaluating contraceptive methods (25 mins)

Discuss the fact that there are many different forms of contraception and there are advantages and disadvantages for each. Students complete the evaluation worksheet and consider data to help them evaluate each type of contraceptive method.

Review the answers with the class so they can self-mark.

Review and practice

Advantages and disadvantages (5 mins)

Students use the interactive to match up the advantages and disadvantages of different forms of contraception. The interactive can be done individually or presented on the board to do as a class.

As an extension, discuss the idea of the latest developments from scientists for a male contraceptive pill which will prevent the sperm's ability to swim.

Ⓗ B11.8 Hormones and infertility treatments

AQA GCSE Biology specification points

Ⓗ 4.5.3.6 The use of hormones to treat infertility

Disciplinary knowledge

- WS 1.1, 1.3, 1.4

Smart Curriculum Narrative

Growth and development

Activity

Solving fertility problems

Theme/subtheme

Reproduction/Human reproductive systems

	Learning outcomes		
Learning objectives	**Developing** Aiming for 3	**Secure** Aiming for 5	**Extending** Aiming for 8
Describe some of the common reasons for female and male infertility and define what is meant by a fertility drug	Define infertility	**Describe some of the common reasons for female and male infertility and define what is meant by a fertility drug**	Explain some of the known causes of infertility and explain how LH and FSH may be used to overcome some forms of female infertility
Describe the main steps in the IVF process	Define *in vitro* fertilisation	**Describe the main steps in the IVF process**	Evaluate the impact of age on the success of IVF treatments
Outline the main biological, social, and ethical issues surrounding IVF	List some of the advantages and disadvantages of using IVF to overcome infertility	**Outline the main biological, social, and ethical issues surrounding IVF**	Evaluate the advantages and disadvantages of IVF treatment and discuss some of the social and ethical issues associated with the technology

Tier 2 vocabulary	Tier 3 vocabulary
egg, embryo, fertilise, hormone, ovary, pregnancy, sperm	ethical issue, fertility drug, fertility treatment, FSH, infertility, *in vitro* fertilisation (IVF), LH, oestrogen, oviduct, ovulation, premature birth, stillbirth, uterus

Lesson plan

Reactivate knowledge

1 List the main different approaches to contraception.

2 Which two hormones are in the mixed contraceptive pill?

3 How does the mixed contraceptive pill work?

Answers: 1 Hormone-based; chemical; barrier; intrauterine devices; abstinence; surgical *2* Oestrogen and progesterone *3* Stops ovulation, stops buildup of uterus lining (endometrium), thickens mucus – all preventing pregnancy

Common misconception

Students may think that only biological females can be infertile as the treatments discussed focus on females.

Trigger interest

What is infertility? (5 mins)

Ask pairs to discuss this question and come up with a definition of infertility that they can share with the class.

If students seem to be unsure, provide the definition as well as two incorrect definitions and ask pairs to discuss which they think is correct.

As an extension, students could suggest causes of infertility linked to the reproductive system.

Exposition

Causes of infertility (5 mins)

Provide pairs of students with diagrams of the male and female reproductive systems, using Figure 1 and Figure 2 from B11.5 of the Student Book. Ask students to discuss possible reasons why someone may be infertile. Students share their ideas on mini-whiteboards.

Fertility Challenges (10 mins)

Introduce the ways in which males and females can have fertility difficulties. Ask groups to continue the discussion to come up with ways of helping people with each of the fertility problems identified. For example, one suggestion might be that an egg is not regularly released from the female's ovaries. One solution would be to give FSH injections to stimulate the eggs in the ovaries to mature.

Listen to the ideas of the class on solving fertility problems and introduce *in vitro* fertilisation (IVF) as one method. Describe the process of IVF.

Main activity

Solving fertility problems (25 mins)

Students complete the activity and review the answers as a class for self-marking.

Students should then evaluate the impact of age and consider the ethics when completing the student follow-up in this activity.

Review and practice

IVF: Advantages and disadvantages (10 mins)

Ask students to create a table to summarise the advantages and disadvantages of IVF.

If students are struggling, provide statements and ask students to work in groups to categorise them as advantages or disadvantages.

B11.9 Plant hormones and responses

AQA GCSE Biology specification points

4.5.4.1 Control and coordination

Smart Curriculum Narrative

How organisms work

Theme/subtheme

Hormone functions/Plant hormones

Disciplinary knowledge

- WS 2.2, 2.3, 2.6, 2.7, 3.1

Required Practical 8

Plant responses

Learning objectives	Learning outcomes: Developing – Aiming for 3	Learning outcomes: Secure – Aiming for 5	Learning outcomes: Extending – Aiming for 8
Use diagrams and descriptions to explain how plant shoots and roots respond to light and gravity (phototropisms and geotropisms)	Define plant hormones and describe the growth responses of plant shoots and roots in response to light and gravity	**Use diagrams and descriptions to explain how plant shoots and roots respond to light and gravity (phototropisms and geotropisms)**	Explain in detail how the production and diffusion of auxin affects the growth responses of shoots and roots in phototropisms and geotropisms
Explain the importance of tropisms in the life of plants	Identify plant responses as phototropisms or gravitropisms	**Explain the importance of tropisms in the life of plants**	Explain the terms positive and negative phototropism or geotropism and identify examples in roots and shoots
Independently plan an investigation into the effect of light or gravity on plant growth	Plan an investigation into the effect of light on plant growth, given support	**Independently plan an investigation into the effect of light or gravity on plant growth**	Predict the results of an investigation into plant tropisms, with detailed scientific reasons

Tier 2 vocabulary	Tier 3 vocabulary
gravity, hormone, root, *shoot*	auxin, gravitropism, phototropism, plant hormone, tropism

Lesson plan

Reactivate knowledge

1 What is a hormone?

2 What do plants need from the environment to grow?

3 Why are the leaves and stems of most plants green?

Common misconception

Students often think that plants just respond to light. Even in a given scenario where no light is present, they may discuss the plant responding to light, rather than gravity.

Answers: 1 A substance made in one part of an organism, carried around the body and has an effect in a different, specific part of the organism *2* Light, water, carbon dioxide, mineral ions *3* Contain chloroplasts/chlorophyll to capture light for photosynthesis

Trigger interest

What do plants need? (10 mins)

Ask students to write a list of everything plants need in order to grow healthily. As a class, discuss how plants get each of these things. Students should be able to link this to photosynthesis.

Exposition

Phototropism (15 mins)

Show the class an image of plant stems bending towards the light (or set up a plant in advance so it displays this growth). Ask students why the plant has grown like this.

Introduce students to the concepts of phototropism and gravitropism. Build on the earlier activity to discuss why it is important for plant growth that plant shoots grow upwards and towards light, and that roots grow downwards. Provide students with unlabelled diagrams of seedlings and ask them to annotate the diagrams and write captions in their own words to explain what is happening. You could use Figure 1 from B11.9 in the Student Book for this.

To support students, provide them with a range of key words that they must include (phototropism, light, gravitropism, gravity).

Main activity

Plant responses (25 mins)

In this required practical, students work in small groups to write a hypothesis on the effect of light on the growth of seedlings, and then plan a method to allow them to investigate it. They may need to be reminded to set one or more controls so they can accurately describe the effect of the independent variable. After checking their method, allow students to set up their investigation, and leave it for at least five days for the seedlings to grow. Some time should be allocated in another lesson to review the results of the practical.

If students are struggling, they can simply set up an experiment using germinated cress seeds in a box with a hole cut in the side to investigate the effect of light from one direction.

Review and practice

Which way? (5 mins)

Students use the interactive to look at a diagram of an investigation into the effect of light on newly germinated shoots and complete sentences about what is happening.

Ⓗ B11.10 Using plant hormones

AQA GCSE Biology specification points

Ⓗ 4.5.4.1 Control and coordination

Ⓗ 4.5.4.2 Use of plant hormones

Smart Curriculum Narrative

How organisms work

Theme/subtheme

Hormone functions/Plant hormones

Disciplinary knowledge

- WS 1.3, 1.4

Practical

Plant growth hormones

Learning objectives	Learning outcomes: Developing Aiming for 3	Learning outcomes: Secure Aiming for 5	Learning outcomes: Extending Aiming for 8
Describe the effect of gibberellins and ethene in plants	List auxin, gibberellins, and ethene as major plant hormones	**Describe the effect of gibberellins and ethene in plants**	Compare and contrast auxins, gibberellins, and ethene
Describe some uses of plant hormones (gibberellins, ethene, and auxins) in agriculture, horticulture, and the food industry	State that plant hormones are used in agriculture and horticulture	**Describe some uses of plant hormones (gibberellins, ethene, and auxins) in agriculture, horticulture, and the food industry**	Explain in detail how plant hormones, such as auxins, act as weedkillers and discuss how the use of hormones as weedkillers affects biodiversity

Tier 2 vocabulary	Tier 3 vocabulary
agriculture, biodiversity, cell division, cloning, germination, hormone, horticulture, tissue culture, weedkiller	auxin, *cutting*, dicotyledon, ethene, gibberellin, monocotyledon, plant hormone, ripening, rooting powder, selective herbicide

Lesson plan

Reactivate knowledge

1 How do plant roots respond to gravity?
2 Which plant hormone controls the response of plants to light and gravity?
3 Which is a phototropism?

Common misconception

Students often think that plants just respond to light. Even in a given scenario where no light is present, they may discuss the plant responding to light, rather than gravity.

Answers: 1 Grow down in the direction of gravity *2* Auxin *3* The response of a plant to light

Trigger interest

What's the problem? (10 mins)

Ask the class a series of questions. For example, what's the problem with:

- picking ripe fruit to transport to the supermarket?
- weeds growing in crop fields?
- trees that produce small fruit?

Gather responses and introduce synthetic plant hormones as a solution to these problems.

Exposition

Plant hormones (15 mins)

Describe the effects of gibberellins and ethene in plants. Students could create a summary table for the hormones auxins, gibberellins, and ethene, describing the effect each hormone has.

From this, describe the uses of the plant hormones gibberellins, ethene, and auxins in agriculture, horticulture, and the food industry. Ask students to use their knowledge of each hormone to explain why each hormone can be used in this way.

Main activity

Plant growth hormones (20 mins)

Ask students to work in pairs. They carry out one (or both depending on timing) experiment where they observe and evaluate the effects of plant hormones. In the first experiment, they take two sets of cuttings and use hormone rooting powder on one set. They can then observe the differences in the roots after two weeks.

In the second experiment they use a weedkiller on a selected area of the school playing field to assess its effectiveness at removing the weeds. They need to repeat the observations once a week for a fortnight. Students can then compare the plant roots in the treated area to the control area. Additional time will need to be given in a different lesson to accommodate this and assess the results.

Ask students to consider the drawbacks of herbicides for biodiversity.

Review and practice

Writing predictions (10 mins)

Students write predictions for the two experiments they set up. They should then use their scientific knowledge to explain their predictions.

B11 Hormonal coordination

Introduction to checkpoint intervention

The checkpoint intervention provides suggestions for a lesson to follow up the *B11 Hormonal coordination* assessment. Depending on the outcomes of the assessment, these suggestions can be used to move a class towards secure, to target specific learning outcomes, or to consolidate knowledge for students achieving secure outcomes. Use students' outcomes from the checkpoint assessment to address general misconceptions from the content covered. Be prepared to re-cover content as required; three key concepts from the chapter are covered in more detail below.

Checkpoint secure learning outcomes

- Describe the role of hormones released by endocrine glands.
- Explain why Type 1 diabetes is treated with insulin injections.
- Describe the negative feedback loop that controls the levels of thyroxine in the body.
- Use data to help discuss the advantages and disadvantages of different methods of contraception and explain why issues around contraception cannot be answered by science alone.
- Outline the main biological, social and ethical issues surrounding IVF.
- Independently plan and carry out an investigation into the effect of light or gravity on plant growth.

Student reflection

Encourage students to reflect on whether there were any checkpoint questions they found difficult or straightforward as well as their preparation for the checkpoint assessment, including how they revised, what they revised, the time spent on revision, and what areas they could improve upon. For more strategies, see the Metacognition spread.

Intervention activity: Describe the role of hormones released by endocrine glands.

Ask students to complete the following summary table, outlining all the hormones they have learnt about in this chapter. If students are struggling, you could provide a table with some of the columns filled in as a hint.

Hormone	Where is the hormone released from?	Role of the hormone

Intervention activity: Use data to help discuss the advantages and disadvantages of different methods of contraception and explain why issues around contraception cannot be answered by science alone.

Ask students to complete the following summary table, outlining all the different types of contraceptions they have learnt about in this chapter.

If students, are struggling, you could provide a table with some of the columns filled in as a hint.

Contraceptive	Advantages	Disadvantages

Intervention activity: Independently plan and carry out an investigation into the effect of light or gravity on plant growth.

Ask students to create a story board with images and descriptions outlining how unidirectional light sources result in positive phototropism in plants. Their storyboards should include how to carry out this investigation. To support students, play a short video or animation showing this and ask students to describe and explain what is happening.

B12 Maintaining balance in the body

B12.1 Controlling body temperature

AQA GCSE Biology specification points
4.5.2.4 Control of body temperature

Theme/subtheme
Hormone functions/Homeostasis

***Smart Curriculum* Narrative**
How organisms work

Activity
Controlling body temperature

	Learning outcomes		
Learning objectives	**Developing Aiming for 3**	**Secure Aiming for 5**	**Extending Aiming for 8**
Describe how body temperature is monitored and controlled by the thermoregulatory centre in the brain	State that it is important to maintain the human body temperature at around 37 °C	**Describe how body temperature is monitored and controlled by the thermoregulatory centre in the brain**	Explain why it is dangerous if the core body temperature is too high or too low
Describe the mechanisms that cool the body down if the core temperature is getting too hot, or warm it up if the core temperature is too low	Predict whether a given activity or situation will tend to raise or lower core body temperature	**Describe the mechanisms that cool the body down if the core temperature is getting too hot, or warm it up if the core temperature is too low**	(H) Explain in detail how the mechanisms to lower or raise body temperature have their effect, in terms of energy transfers

Tier 2 vocabulary	Tier 3 vocabulary
capillary, *constrict*, dilate, enzyme, evaporate, exothermic, nerve impulse, radiation, receptor, respiration, skeletal muscle, skin, temperature	core body temperature, hypothalamus, shivering, sweat gland, temperature sensor, thermoregulatory centre, vasoconstriction, vasodilation

Lesson plan

Reactivate knowledge

1 What are enzymes?

2 What happens to enzymes if the temperature goes very high?

3 Which hormone is involved in controlling blood glucose levels in the body?

Common misconception

Students often think that some people appear to go red in warm environments because their blood vessels move toward the surface of the skin, to allow heat to escape. In fact, this actually occurs because the blood vessels have dilated, so there is increased blood flow to the surface of the skin.

Answers: 1 Biological catalysts/proteins that speed up reactions in cells but are not changed/can be reused *2* They denature/the protein structure unravels/they do not work *3* Insulin

Trigger interest

Taking your temperature (10 mins)

Provide students with forehead thermometers and ask them to take their temperature. Collect results and calculate the mean of the class.

As an extension, discuss how much variation there was, and possible reasons why. Do people have different temperatures, or is it due to the resolution of the thermometers? Discuss how you could decide.

Exposition

Maintaining a constant body temperature (15 mins)

Discuss the fact that our bodies maintain a body temperature of 37 °C.

Either as a discussion, or a mini-whiteboard quiz, ask students the following questions:

- Why would it be dangerous if our body temperature went above 37 °C?
- Why would it be dangerous if our body temperature went below 37 °C?

(For these first two questions, you could support students by prompting them to consider chemical reactions and enzymes.)

- What mechanisms cool you down when you are too hot?
- What mechanisms warm you up if you are too cold?

(For these final two questions, you could prompt students to consider what happens to the amount of sweating, muscle contractions, colour of your skin, etc.)

After the discussion, describe and explain the mechanisms that cool the body down if the core temperature is getting too hot, or warm it up if the core temperature is too low.

Ask students to explain how the mechanisms to lower or raise body temperature have their effect, in terms of energy transfers. For example, they could explain how increased blood to the surface of the skin, due to dilated blood vessels, results in heat loss.

Main activity

Controlling body temperature (20 mins)

Students complete the activity using the information from the class discussion and the Student Book to support them, if needed.

Discuss the answers and students self-mark their work.

Support students by providing the stages in temperature regulation and ask students to arrange them in the correct order.

If students complete this task quickly, they can complete the 'Student follow-up' section of the worksheet to explain how these mechanisms work.

Review and practice

Misconceptions (10 mins)

Present the class with a series of misconceptions about controlling body temperature:

- Blood capillaries dilate or constrict to control body temperature. (Explain that vasodilation and vasoconstriction increase and decrease the volume and speed of blood flowing to the skin's surface. This changes the amount of heat energy that can radiate from the blood and skin.)
- In order to cool the body down, blood vessels move towards the surface of the skin to allow more heat to escape. (Explain that blood vessels do not move closer or further away. Blood vessels dilate so that more blood can flow closer to the surface of the skin, or constrict to reduce the blood flow to the skin.)
- Sweating cools the body down because sweat is cold. (Explain that sweat is not cold. The water in sweat evaporates from the skin's surface, and heat energy from the skin is used to evaporate this water, therefore removing heat energy from the skin as the sweat evaporates.)

Students should say why the statements are incorrect, and write out corrected versions.

B12.2 Removing waste products

AQA GCSE Biology specification points

4.5.3.3 Maintaining water and nitrogen balance in the body

Smart Curriculum Narrative

How organisms work

Theme/subtheme

Organisation/Urinary and respiratory system

Activity

Removing waste

Learning objectives	Learning outcomes: Developing – Aiming for 3	Secure – Aiming for 5	Extending – Aiming for 8
Explain why the body needs to get rid of carbon dioxide, urea, excess mineral ions, and water	State that the internal conditions of the body must be controlled to remain as stable as possible for the cells to function efficiently	**Explain why the body needs to get rid of carbon dioxide, urea, excess mineral ions, and water**	Explain the importance of maintaining the water balance of the body in terms of the impact of osmotic changes in the body fluids on the cells
Describe how carbon dioxide is removed from the body in the lungs	State that carbon dioxide is a waste product of aerobic respiration and is removed from the body by the lungs. Water is lost at the same time	**Describe how carbon dioxide is removed from the body in the lungs**	Explain how removing carbon dioxide also results in loss of water from the body
Clarify the difference between urea and urine, and describe how urea is lost from the body through the skin and by the kidneys	State that excess water, mineral ions, and urea are removed from the body in the sweat and by the kidneys	**Clarify the difference between urea and urine, and describe how urea is lost from the body through the skin and by the kidneys**	(H) Explain the link between high levels of protein in the diet and an increase in urea excreted by the kidneys

Tier 2 vocabulary	Tier 3 vocabulary
alveolus, amino acid, carbon dioxide, concentration gradient, diffusion, exhalation, kidney, liver, lung, mineral ion, osmosis, protein, respiration, waste product, water	amino group, ammonia, deamination, homeostasis, nitrogenous waste, sweat, urea, urine

Lesson plan

Reactivate knowledge

1 What is the typical human core body temperature?

2 How does sweating cool you down?

3 What is vasoconstriction?

Common misconceptions

Students may think that only carbon dioxide is lost when you exhale. Students may also think that the kidneys are the only organs involved in removing urea from the body.

Answers: *1* 37 °C *2* When the water in the sweat evaporates it transfers energy to the environment and cools you down *3* When the blood vessels that supply your surface skin capillaries dilate (open wider)

Trigger interest

What do they have in common? (5 mins)

Show the class images of the kidneys, skin, and lungs. Ask students to discuss in groups what they think the organs have in common.

Discuss as a class the fact that they all have a role in removing waste products from the body and give examples such as: kidneys remove urea, skin removes water, and lungs remove carbon dioxide.

Then ask students if they can recall how each waste product is made.

Exposition

Removing waste from the body (15 mins)

Introduce students to how excess water and other waste products are removed from the body. Explain that carbon dioxide produced in respiration is a waste product. Carbon dioxide dissolves in the blood and becomes acidic. This is harmful as it could denature enzymes and damage proteins. The carbon dioxide is removed when the blood travels to the alveoli in the lungs and the carbon dioxide diffuses from the blood to the alveoli and is exhaled. Students then answer questions one and two from the Student Book.

Main activity

Removing waste (25 mins)

Students should use the information discussed in the first part of the lesson to label which waste products are removed from the lungs, kidney, skin, and liver on the diagram on the worksheet.

Students then complete the series of exam-style questions using their notes, and the Student Book if needed.

Review the answers as a class as students self-mark.

Review and practice

My role (10 mins)

Ask students to work in groups of three. Designate each student as the skin, lungs, or kidneys. Ask them to present an argument to each other that their role in removing waste products is the most vital in the body. If there is a group of four, the liver can be included. Facilitate discussions around the class and ask groups to present their arguments to the rest of the class.

B12.3 The human kidney

AQA GCSE Biology specification points

4.5.3.3 Maintaining water and nitrogen balance in the body

Smart Curriculum Narrative

How organisms work

Theme/subtheme

Organisation/Urinary system

Disciplinary knowledge

- MS 4a

Practical

Urine testing

Learning objectives	Learning outcomes: Developing Aiming for 3	Secure Aiming for 5	Extending Aiming for 8
Describe the processes of filtering and selective reabsorption in the kidneys	State the main functions of the kidneys	**Describe the processes of filtering and selective reabsorption in the kidneys**	Describe how the kidneys interact with other systems to play a homeostatic role in maintaining the water and mineral ion balance in the body
Suggest how the composition of the urine will change in given situations	List the substances found in urine	**Suggest how the composition of the urine will change in given situations**	Explain why the volume and colour of urine are useful indicators of hydration levels in the body
State that the kidneys are made up of tubules and describe the effect of ADH on the permeability of the kidney tubules	Describe factors that affect the amount of urine produced by the kidneys	**State that the kidneys are made up of tubules and describe the effect of ADH on the permeability of the kidney tubules**	Explain how the production of ADH will change in given situations and the effect this has on the water balance of the body as an example of negative feedback control
Translate tables and bar charts of glucose, mineral ion, and urea concentration of the blood before and after filtration	Produce bar charts of glucose, mineral ions, and urea concentrations in the blood from given tables with support	**Translate tables and bar charts of glucose, mineral ion, and urea concentration of the blood before and after filtration**	Analyse data from tables and bar charts on blood and urine concentrations before and after filtration

Tier 2 vocabulary	Tier 3 vocabulary
active transport, blood, *concentrated*, diffusion, dilute, excretion, *filter*, glucose, homeostasis, hormone, kidney, negative feedback, protein, water	(H) ADH, bladder, pituitary gland, selective reabsorption, (H) tubule

Lesson plan

Reactivate knowledge

1 What are the main waste products of the body?
2 Which organ produces urea?
3 How do you lose water from your body?

Common misconceptions

Students may think that only carbon dioxide is lost when you exhale. Students may also think that the kidneys are the only organs involved in removing urea from the body.

Answers: 1 Carbon dioxide, urea *2* Liver *3* Breathing out, sweating, urine

Trigger interest

Urea and urine (5 mins)

Ask students to write down the difference between urea and urine. They then peer assess each other's answers.

Exposition

The kidneys and urine (15 mins)

The kidneys

Show students a diagram of the kidney (showing internal structures) and the nephron (students do not need to know details of the structure of the nephron). Describe the processes of filtering and selective reabsorption in the kidneys.

Ask students to consider other systems in the body that would have an impact on the amount of sugar and water in the blood (e.g., digestive system and respiratory system).

Urine

Show students four beakers of 'urine' (which will later be used in the practical):

Urine sample A – pale colour

Urine sample B – dark colour

Urine sample C – pale colour

Urine sample D – dark colour

Ask students to suggest why urine B and urine D are darker in colour. Prompt students to consider what this indicates about how much water is in the urine, and why there may be less water in the urine.

For higher tier students, use this to lead into a discussion of how the composition of the urine will change in given situations, and how ADH affects the permeability of tubules in the kidney to cause these changes.

Main activity

Urine testing (25 mins)

Students complete the practical and discuss the results at the end to conclude which urine was from a person with diabetes, a person with kidney failure, a dehydrated person, and a healthy person.

Review and practice

Waste analysis (10 mins)

Students complete the activity where they analyse the data provided or use it to draw a graph, depending on the level of support needed.

B12.4 Dialysis – an artificial kidney

AQA GCSE Biology specification points

4.5.3.3 Maintaining water and nitrogen balance in the body

Smart Curriculum Narrative

How organisms work

Theme/subtheme

Organisation/Urinary system

Disciplinary knowledge

- WS 1.4, 1.5

Practical

Modelling kidney dialysis

	Learning outcomes		
Learning objectives	Developing Aiming for 3	Secure Aiming for 5	Extending Aiming for 8
Describe some reasons for kidney failure and why it is a threat to life	State what kidney failure means	**Describe some reasons for kidney failure and why it is a threat to life**	Describe the expected impact of kidney failure on the concentrations of urea, mineral ions, and water in the blood
Describe the process of kidney dialysis, using a diagram	State the two main ways of treating kidney failure	**Describe the process of kidney dialysis, using a diagram**	Using knowledge on factors which affect diffusion rates, suggest and explain suitable concentrations of substances in dialysis fluid compared to the blood
Describe some of the advantages and disadvantages of kidney dialysis	State that a kidney dialysis machine is a model of a kidney	**Describe some of the advantages and disadvantages of kidney dialysis**	Compare and contrast a kidney dialysis machine with a working kidney

Tier 2 vocabulary	Tier 3 vocabulary
concentration gradient, diffusion, kidney, mineral ion, osmosis, partially permeable membrane	dialysis, dialysis fluid, dialysis machine, kidney failure, kidney transplant, urea, water balance

Lesson plan

Reactivate knowledge

1 What are the functions of the kidneys?

2 What type of urine is produced if you are short of water?

3 How much urine is typically produced in 24 hours?

Common misconception

Students may not realise that the kidneys are constantly filtering the blood, and therefore filtering the blood using dialysis only provides short-term effects.

Answers: 1 Removal of urea, which is a toxin, balancing water and mineral ion concentration in the blood *2* Small amount, concentrated *3* 1500 cm^3

Trigger interest

Kidney questions (5 mins)

Ask the class a series of multiple choice questions covering the function of the kidney to recap learning from B12.3 (using the Interactive: *Kidney questions*).

Discuss what would happen to the body if the kidneys stopped working.

Exposition

Kidney dialysis (10 mins)

Introduce how a kidney can fail.

Ask students to suggest ways that a failing kidney can be treated.

Introduce students to the idea of dialysis, and ask students to suggest how they think a kidney dialysis machine works. Build on their suggestions to explain how a kidney dialysis machine works, linking this back to the function of the kidney.

Main activity

Modelling kidney dialysis (35 mins)

Students set up the equipment for a practical where they model how kidney dialysis works. This is used to show how a partially permeable membrane can allow the passage of some molecules but not others.

Provide students with a diagram of a kidney dialysis machine and ask them to use information from the Student Book to label it and explain how it works. This can be done whilst the kidney dialysis model is in the water.

Students then test the water surrounding the Visking tubing for the presence of starch or glucose, and record their results. They then write a conclusion and explain how the model is similar to what happens in kidney dialysis.

If required, support students by setting up the Visking tubing prior to the lesson.

As a challenge, ask students to write a prediction based on their scientific knowledge.

Review and practice

Quiz (5 mins)

Ask students to answer the following questions in a mini-whiteboard quiz:

- Give one similarity and one difference between how the kidney filters the blood and how a dialysis machine filters the blood.
- What benefits does a dialysis machine provide?
- What are some disadvantages of using a dialysis machine?

Prompt students to consider how diet might affect the concentration of blood and how often a person would need dialysis?

B12.5 Kidney transplants

AQA GCSE Biology specification points

4.5.3.3 Maintaining water and nitrogen balance in the body

Smart Curriculum Narrative

How organisms work

Theme/subtheme

Organisation/Urinary system

Disciplinary knowledge

- WS 1.4, 1.5

Activity

Transplant versus dialysis

	Learning outcomes		
Learning objectives	**Developing** **Aiming for 3**	**Secure** **Aiming for 5**	**Extending** **Aiming for 8**
Describe a kidney transplant and list the two currently available sources of potential donors	Define an organ transplant	**Describe a kidney transplant and list the two currently available sources of potential donors**	Explain why family members are often but not always a good choice for a living organ donation
Compare the advantages and disadvantages of treating kidney failure using dialysis or kidney transplant	List some advantages and disadvantages of kidney transplants	**Compare the advantages and disadvantages of treating kidney failure using dialysis or kidney transplant**	Use economic, social, and ethical arguments to evaluate treating organ failure by mechanical device or by transplant

Tier 2 vocabulary	Tier 3 vocabulary
antibody, antigen, genetic engineering, immune system, kidney, stem cell	dialysis, dialysis machine, *donor*, immunosuppressant drug, kidney transplant, live donor transplant, recipient, rejection, tissue type, urea, xenotransplantation

Lesson plan

Reactivate knowledge

1 What are the main problems for the body if the kidneys fail?

2 What process is used in a dialysis machine to remove urea from the blood?

3 What is the alternative to dialysis to overcome kidney failure.

Common misconception

Students may not realise that organ transplants typically last for ten years, and not a lifetime.

Answers: 1 Build up of urea, loss of water and mineral ion balance *2* Diffusion down a concentration gradient *3* A kidney transplant

Trigger interest

Organ transplants (10 mins)

Ask students to write down as much as they can in five minutes about organ transplants, considering the following:

- Where do the organs come from for transplants?
- What are the risks of having a transplant?
- Are organs readily available for transplants?
- Can you accept a transplanted organ from anyone?

Gather key ideas from the class and list them on the board.

Exposition

Kidney transplants (15 mins)

Leading on from the starter questions, talk students through what a kidney transplant is. Then talk through the following considerations and risks:

- The kidney is replaced through surgery, which poses risks of infection.
- There must be a very close match of antigens between the donor and the recipient to prevent the organ being rejected.
- The availability of kidneys does not meet the demand.
- A transplanted kidney doesn't work forever.
- There is the potential to use stem cells to grow organs for transplants in the future.

Main activity

Transplant versus dialysis (20 mins)

Students complete the activity to help them consider the advantages and disadvantages of kidney transplants and dialysis.

The activity provides students with information on the costs of kidney dialysis and kidney transplants in the UK. Students then analyse this data and comment on which treatment is the best economically.

Students could also use information from the Student Book to look at social and ethical considerations and write down advantages and disadvantages of each method. Discuss the answers to this activity to conclude which they think is the best choice overall.

To support students, read through the information as a class and list the advantages and disadvantages together.

Students should then write a letter of recommendation to the government, commenting on the financial benefits of transplants versus dialysis. They should use calculations to support their argument.

Review and practice

The future of transplants (10 mins)

Present two possibilities for kidney transplants in the future – the production of new kidneys using a patient's own cells, and the use of animals that have been genetically modified to grow human kidneys. Ask students to discuss the benefits and drawbacks of each method.

B12 Maintaining balance in the body

Introduction to checkpoint intervention

The checkpoint intervention provides suggestions for a lesson to follow up the *B12 Maintaining balance in the body* assessment. Depending on the outcomes of the assessment, these suggestions can be used to move a class towards secure, to target specific learning outcomes, or to consolidate knowledge for students achieving secure outcomes. Use students' outcomes from the checkpoint assessment to address general misconceptions from the content covered. Be prepared to re-cover content as required; three key concepts from the chapter are covered in more detail below.

Checkpoint secure learning outcomes

- Describe the mechanisms that cool the body down if the core temperature is getting too hot, or warm it up if it the core temperature is too low.
- Explain why the body needs to get rid of carbon dioxide, urea, excess ions, and water.
- Clarify the difference between urea and urine, and describe how urea is lost from the body through the skin and by the kidneys.
- State that the kidneys are made up of tubules and describe the effect of ADH on the permeability of the kidney tubules.
- Compare the advantages and disadvantages of treating kidney failure using dialysis or kidney transplant.
- Use a diagram to show how kidney dialysis works.

Student reflection

Encourage students to reflect on whether there were any checkpoint questions they found difficult or straightforward as well as their preparation for the checkpoint assessment, including how they revised, what they revised, the time spent on revision, and what areas they could improve upon. For more strategies, see the Metacognition spread.

Intervention activity: Explain why the body needs to get rid of carbon dioxide, urea, excess ions, and water.

Students create a table showing the substances and the impact each has on the body if not removed:

Substance	Why the body must remove this substance
Carbon dioxide	
Urea	
Excess ions	
Water	

If students need additional support, provide them with a series of statements to add into the table:

- If the cells of your body lose too much water by osmosis to the fluids surrounding them, they do not function efficiently. (water)
- If too much water enters a cell, it could cause the cell to lyse. (water)
- This dissolves to produce an acidic solution. This change in pH would affect the working of all the enzymes in the cells. (carbon dioxide)
- This is the nitrogenous waste produced by the breakdown of excess amino acids in your liver. (urea)
- This is poisonous and if levels build up in your body, it will cause extensive damage to cells. (urea)
- Excess salt can affect the concentration of the blood and will impact osmosis. This could cause more water to leave cells by osmosis and prevent the cells functioning efficiently. (excess ions)

Intervention activity: Clarify the difference between urea and urine, and describe how urea is lost from the body through the skin and by the kidneys.

Students sort the statements below into two columns. One column is for statements that are about urea, and the other is for statements that are about urine:

- This is the nitrogenous waste produced by the breakdown of excess amino acids in your liver.
- This is poisonous and if levels build up in your body, it will cause extensive damage to the cells.
- It is produced by the kidneys.
- It contains excess water and salt.
- It is passed out of the body via the skin and kidneys.
- Ammonia is immediately converted into this.

Intervention activity: Compare the advantages and disadvantages of treating kidney failure using dialysis or kidney transplant.

Students should complete the fishbone evaluation diagram, as shown below. The top two boxes should contain two advantages of dialysis, and the bottom two boxes should be filled with two disadvantages. The final box is for students to add their conclusion.

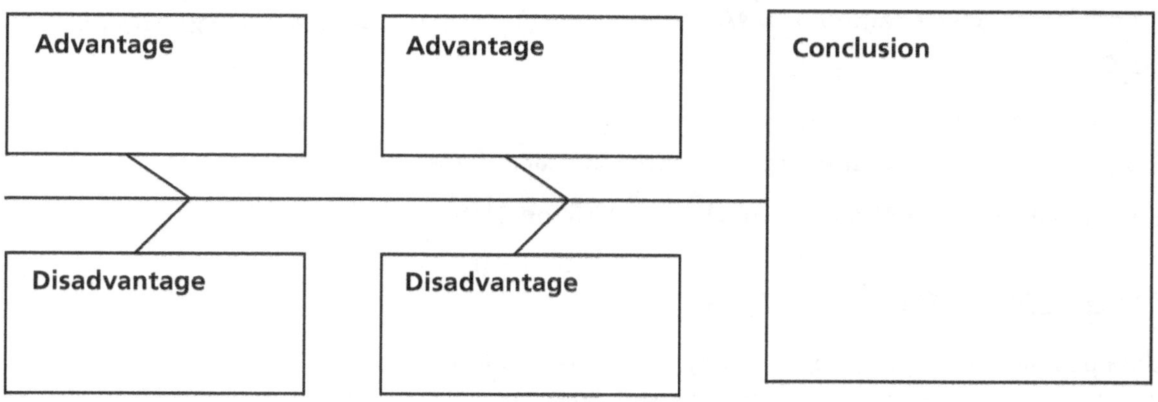

To support students, you could provide them with the following statements to select from:

- The person must follow a very carefully controlled diet.
- The person must spend regular, long sessions connected to the dialysis machine.
- There is only a relatively short time when the balance of chemicals in the blood is at the right level, and so the person may feel tired and unwell as the next dialysis session approaches.
- With long-term use of dialysis, the balance of substances in the blood can become more difficult to control.
- Dialysis acts as an artificial kidney and keeps the person alive.
- Dialysis removes toxins and restores the concentrations of dissolved substances to healthy levels.
- Dialysis is readily available.

4 Genetics and evolution

Introduction to unit

This unit will give students a deeper understanding of the principles of genetics, and how variation, adaptation, and selection can lead to the process of evolution. It will bring students up to date with the latest advances in biology, such as cloning, in-vitro fertilisation, and the use of genetic engineering as well as research into gene therapy to treat inherited disorders.

Assessment

Suggested minimum guided learning hours: 27 hours

Exam: Paper 2

Prerequisite knowledge from KS3 (chapter references from *Oxford Smart* Activate)

B1.3.2 Reproductive systems

- Sperm and egg are sex cells.

B3.3.6 Reach: Inheriting characteristics

- DNA is found in the nucleus of cells.
- DNA gives organisms inherited characteristics.
- Differences in DNA give differences in inherited characteristics.
- Not all characteristics are caused by DNA.

B3.4.2 Apply: Variation

- There is competition between organisms for resources.

B3.4.5 Apply: Natural selection and antibiotic-resistant bacteria

- Extinction means there are no more of a species alive on Earth.

→

AQA GCSE Biology specification links

4.6 Inheritance, variation, and evolution

- 4.6.1 Reproduction
- 4.6.2 Variation and evolution
- 4.6.3 The development of understanding of genetics and evolution
- 4.6.4 Classification of living organisms

Identity, awe, and wonder

The world is becoming increasingly affected by genetic technologies as our understanding of molecular biology processes grows.

In an age where genetic testing and personalised medicine are becoming more prevalent, understanding genetics will empower students to make informed decisions about their health as they will be better equipped to understand risk factors, interpret test results, and engage in discussions about ethical considerations in genetic research and technology. Will they be comfortable eating genetically modified foods? Will they want laws to govern research on genetic modification?

Genetics is a rapidly growing field with diverse career opportunities for students to explore.

Required practicals

There are no required practicals in this unit.

Working scientifically

In this unit, students will learn the different theories of evolution and inheritance and how they were developed by scientists through observation and data handling. Students are usually keen to discuss why theories were accepted or rejected due to social, economic, or environmental limitations. These can easily lead to written evaluation tasks or debate-style activities. Genetic engineering or gene editing will also be of interest to students where they will be able to use descriptive and representational models to explain scientific developments.

Maths skills

Students will be given the opportunity to work with a variety of graphs and frequency tables in the main lesson activities within this unit. The maths skills worksheets will cover a range of questions including working with ratios, fractions, percentages, and probabilities. Students should also be able to understand and use different mathematical symbols.

Literacy links

Written skills can be developed through answering questions on tasks in the main activities. The GM crops literacy worksheet in B14.4 will allow students to read and analyse debate-style arguments for and against GM crops. Further literacy skills can be developed in the main activity in B15.2 which provides students with the opportunity to write a narration for a storyboard on the work of Charles Darwin.

B13 Reproduction and genetic inheritance

B13.1 Types of reproduction

AQA GCSE Biology specification points

4.6.1.1 Sexual and asexual reproduction

4.6.1.3 Advantages and disadvantages of sexual and asexual reproduction

***Smart Curriculum* Narrative**

Growth and development

Theme/subtheme

Reproduction/Sexual and asexual reproduction

Practical

Modelling reproduction

Learning objectives	Learning outcomes: Developing Aiming for 3	Learning outcomes: Secure Aiming for 5	Learning outcomes: Extending Aiming for 8
Describe the key elements in sexual and asexual reproduction, including the role of meiosis and mitosis	Define sexual and asexual reproduction	**Describe the key elements in sexual and asexual reproduction, including the role of meiosis and mitosis**	Compare and contrast the processes of asexual and sexual reproduction
Describe the effect of meiosis on the chromosome number of gametes and name the gametes found in plants and animals	State that sexual reproduction involves special sex cells (gametes) and asexual reproduction does not	**Describe the effect of meiosis on the chromosome number of gametes and name the gametes found in plants and animals**	Explain in detail why meiosis is important for sexual reproduction
Describe why variation is produced in offspring from sexual reproduction but not from asexual reproduction	State why variation is produced in offspring from sexual reproduction but not from asexual reproduction	**Describe why variation is produced in offspring from sexual reproduction but not from asexual reproduction**	Explain the importance of the variation produced in offspring from sexual reproduction but not from asexual reproduction

Tier 2 vocabulary	Tier 3 vocabulary
bacterium, cell division, chromosome, clone, DNA, egg cell, gamete, mitosis, offspring, reproduction, sperm, variation	asexual reproduction, meiosis, sexual reproduction, zygote

Lesson plan

Reactivate knowledge

Common misconception

Students may think that bacteria reproduce asexually by mitosis, rather than binary fission.

1 What is a gamete?
2 How long is a typical human pregnancy?
3 What are the two main ways in which plants are pollinated?

Answers: 1 A specialised sex cell *2* Nine months *3* Wind pollination, insect pollination

Trigger interest

Bacterial reproduction (10 mins)

Ask students to write a paragraph that describes how bacteria reproduce. As a class, discuss how this compares with how animals reproduce. Use this activity to address the misconception.

If students find this tricky, ask them to refer back to their notes from *B2 Cell division* or show them a video of bacteria dividing (there are many available online).

Exposition

Sexual and asexual reproduction (15 mins)

Recap what students can remember from KS2 and KS3 about sexual and asexual reproduction, asking them to write down any facts they remember. Prompt them to consider:

- the number of parents involved
- whether gametes fuse
- whether identical offspring are made
- whether the cells divide by mitosis or meiosis.

Discuss students' responses. They should summarise the key comparisons in a table in their notes.

Meiosis is covered in the next lesson but, at this point, identify that there are some key differences between mitosis and meiosis. This includes that meiosis involves two rounds of division, resulting in only one copy of each chromosome being present in the gametes, and also introducing genetic variation.

Let students know that in animals, the gametes are sperm and egg cells, whereas in plants, the gametes are pollen and egg cells.

Main activity

Modelling reproduction (20 mins)

Ask students to read the instructions in the practical worksheet and then to carry out the practical in pairs to model why sexual reproduction produces variation in offspring, but asexual reproduction does not.

Students complete the questions and discuss the answers as a class.

Review and practice

Definitions (10 mins)

Ask pairs of students to write definitions for each type of reproduction and then review their answers with another pair.

B13.2 Cell division in sexual reproduction

AQA GCSE Biology specification points

4.6.1.2 Meiosis

Smart Curriculum Narrative

Growth and development

Theme/subtheme

Reproduction/Meiosis and mitosis

Disciplinary knowledge

- WS 1.2

Activity

Which one?

	Learning outcomes		
Learning objectives	**Developing** **Aiming for 3**	**Secure** **Aiming for 5**	**Extending** **Aiming for 8**
Describe the processes of meiosis and mitosis	State that gametes (sex cells) are formed by meiosis	**Describe the processes of meiosis and mitosis**	Compare and contrast mitosis and meiosis
Explain how meiosis halves the number of chromosomes in gametes and fertilisation restores the full number	State that meiosis halves the number of chromosomes in gametes	**Explain how meiosis halves the number of chromosomes in gametes and fertilisation restores the full number**	Explain in detail why gametes are all genetically different to each other
Describe what happens after the gametes fuse at fertilisation in terms of cell division	State that when gametes fuse at fertilisation, the normal chromosome number is restored	**Describe what happens after the gametes fuse at fertilisation in terms of cell division**	Discuss the roles of meiosis and mitosis in sexual reproduction

Tier 2 vocabulary	Tier 3 vocabulary
cell division, chromosome, *differentiate*, egg cell, fertilisation, gamete, gene, mitosis, ovary, ovum, sperm, testis, variation	asexual reproduction, chromatid, meiosis, sexual reproduction

Lesson plan

Reactivate knowledge

Common misconception

Students may think meiosis happens only in animals to make sperm and eggs, and not realise it also occurs in plants to make pollen and eggs.

1 What are the two main types of reproduction?

2 How many parents are involved in asexual reproduction?

3 What is the main difference between the offspring produced by asexual reproduction and the offspring produced by sexual reproduction?

Answers: *1* Sexual and asexual *2* One *3* Asexual: offspring identical to parent/no variation. Sexual: offspring contain information from both parents so show variation

Trigger interest

Passing on instructions (5 mins)

Ask students to write a short answer to the question: *Why do children look like their biological parents?* Go through their answers and discuss the fact that during fertilisation, chromosomes from both parents pair up. We inherit genes from them and, as a result, express those genes.

If necessary, prompt students with the following key words: egg, sperm, genes, chromosomes, fertilisation.

Exposition

Meiosis (15 mins)

Ask students how many chromosomes there are in a human body cell and a human gamete (sex cell), and why the two numbers are different.

Show students a simple animation that shows the process of meiosis, or go through Figure 1 in the Student Book with them, describing what happens at each stage.

Explain that each gamete will contain a different combination of maternal and paternal chromosomes, and that it is random how these are split between the gametes. This is why siblings are not identical.

Challenge students to use their maths skills to calculate how many different combinations of unique gametes are possible in humans (2^{23}).

An easy way students can remember the difference between mitosis and meiosis is:

MITosis – **M**aking **I**dentical **T**wo

MEiosis – **M**aking **E**ggs (and sperm)

Main activity

Which one? (25 mins)

Explain that it is important that gametes contain half the usual number of chromosomes, so that the full number is restored when gametes fuse during fertilisation. After fertilisation, a zygote is made. The zygote then divides by mitosis to make identical copies, forming an embryo. The cells of the embryo continue to divide and start to differentiate into specialised cells to form a foetus.

Students complete the activity using their notes from the lesson and the Student Book if needed. Review the answers at the end for students to self-mark.

Review and practice

Inheritance (10 mins)

In this interactive, students check their understanding by answering questions on mitosis, meiosis, and the differences between sexual and asexual reproduction.

B13.3 The best of both worlds

AQA GCSE Biology specification points

4.6.1.3 Advantages and disadvantages of sexual and asexual reproduction

Smart Curriculum Narrative

Growth and development

Theme/subtheme

Reproduction/Sexual and asexual reproduction

Disciplinary knowledge

- WS 1.1, 2.4

Practical

Exploring plants and fungi

	Learning outcomes		
Learning objectives	**Developing** **Aiming for 3**	**Secure** **Aiming for 5**	**Extending** **Aiming for 8**
Describe the advantages and disadvantages of sexual and asexual reproduction	State that some organisms reproduce asexually, some reproduce sexually, and some use both methods	**Describe the advantages and disadvantages of sexual and asexual reproduction**	Explain in detail the advantages and disadvantages of asexual and sexual reproduction in a stable environment and in a changing environment
Describe how malarial parasites and fungi reproduce both asexually and sexually	Name three types of organism that reproduce both sexually or asexually	**Describe how malarial parasites and fungi reproduce both asexually and sexually**	Compare the variation in the offspring produced by asexual and sexual reproduction in fungi and malarial parasites and discuss the implications of these differences
Describe some of the ways in which flowering plants reproduce asexually in addition to sexual reproduction	State that flowers contain the reproductive organs of flowering plants	**Describe some of the ways in which flowering plants reproduce asexually in addition to sexual reproduction**	Explain the advantages and disadvantages to an organism of the different types of reproduction

Tier 2 vocabulary	Tier 3 vocabulary
chromosome, clone, egg cell, fungus, gamete, life cycle, malaria, mitosis, natural selection, offspring, pollen, red blood cell, selective breeding, variation	asexual reproduction, *bulb*, fruiting body, hypha, meiosis, mould, nectar, parasite, petal, pollination, pollinator, *runner*, sexual reproduction, spore, zygote

Lesson plan

Reactivate knowledge

1 Which type of cell division is involved when body cells divide?
2 Which type of cell division is involved in the formation of gametes?
3 What happens to the chromosome number when gametes are formed?

Answers: 1 Mitosis *2* Meiosis *3* It is halved

Common misconception

Students may think meiosis happens only in animals to make sperm and eggs, and not realise it also occurs in plants to make pollen and eggs.

Trigger interest

Plant reproduction (5 mins)

Ask students to consider whether plants reproduce sexually or asexually. Students can share their ideas with the class.

Then ask students if they can give examples of plants that can reproduce sexually or asexually. Tell students that flowering plants reproduce sexually, but others, such as strawberries and daffodils can use both methods.

Exposition

The advantages (15 mins)

Ask students to list the advantages of reproducing asexually and the advantages of reproducing sexually. They could do this on mini-whiteboards, or write their ideas directly into their books. Discuss as a class, asking students for contributions. Write these as a table on the white-board and ask students to write down what they missed for their own lists.

Reproduction in malaria parasites

Discuss that some organisms reproduce using both sexual and asexual reproduction, and give malaria parasites as an example. Provide students with the worksheet which has empty boxes for students to label the diagram. Ask students to discuss in pairs where the parasite is using asexual and sexual reproduction, and why it uses both methods. Then ask students to check their answers and add further information about the advantages and disadvantages of each method by referring to information in the Student Book.

Main activity

Exploring plants and fungi (25 mins)

Students carry out a range of practical tasks looking at reproduction in plants and fungi. They carry out a flower dissection to identify and name the parts and explain their role in sexual reproduction. They also study a range of reproduction methods that plants and fungi use, including studying fungal spores (which will need to be done in a subsequent lesson, or alternatively the spores could be prepared in advance by the technicians to use in this lesson).

Review and practice

Fruit and vegetables (10 mins)

Show students a variety of images of different fruit and vegetables (e.g., peas in a pod, potatoes, root ginger, strawberries, apples, courgettes). Ask them which are produced by sexual reproduction and which are produced by asexual reproduction.

Then complete the Interactive: *Sexual and asexual reproduction*. Students sort the advantages and disadvantages of sexual and asexual reproduction into the correct groups.

B13.4 DNA and the genome

AQA GCSE Biology specification points

4.6.1.4 DNA and the genome

Smart Curriculum Narrative

Genetics and inheritance

Theme/subtheme

Genetics and inheritance/Inheritance and DNA structure

Disciplinary knowledge

- WS 1.1, 1.4

Activity

Why does the genome matter?

	Learning outcomes		
Learning objectives	**Developing** **Aiming for 3**	**Secure** **Aiming for 5**	**Extending** **Aiming for 8**
Describe the basic structure of DNA	State that the genetic material in the nucleus of a cell is composed of DNA	**Describe the basic structure of DNA**	Suggest why the structure of DNA is so important to its function
Describe the relationship between DNA, genes, and chromosomes	Define a gene and a chromosome	**Describe the relationship between DNA, genes, and chromosomes**	Explain simply how the DNA codes for an entire organism
Explain the difference between the chromosomes in the nucleus and the genome of an organism	Define a genome	**Explain the difference between the chromosomes in the nucleus and the genome of an organism**	Explain why the size of the genome of an organism does not necessarily reflect the complexity of the organism
Describe some of the benefits of understanding the human genome	State the scientific breakthrough made by the Human Genome Project	**Describe some of the benefits of understanding the human genome**	Explain why it is important to sequence the genomes of many thousands of people

Tier 2 vocabulary	Tier 3 vocabulary
amino acid, chromosome, DNA, enzyme, gene, inheritance, mitochondrion, nucleus, polymer, protein, vaccine, virus	base pair, base sequence, double helix, genome, human evolution, human genome, Human Genome Project, inherited disorder, migration pattern, mitochondrial DNA

Lesson plan

Reactivate knowledge

1 What is the main advantage of sexual reproduction for an organism?
2 Give two ways that fungi reproduce and when they use each way.
3 Malaria parasites use two different host organisms during their lifecycle. What are they?

Common misconception

Students might not realise that your genome is the same in every cell, and may instead think different specialised cells contain different DNA.

Answers: 1 Variation in the offspring that gives a survival advantage if environmental conditions change *2* Asexual spores – normal conditions. Sexual reproduction with hyphae joining when conditions are bad *3* Mosquitoes and humans

Trigger interest

What do I know about DNA? (10 mins)

Ask students to work in pairs and write the word DNA in the middle of a sheet of paper, or on their mini-whiteboards. They should then write as much as they know about DNA around it. Ask each pair for one piece of information and collate the information from the whole class.

To support students, provide them with the following key words to help with their discussions: nucleus, genes, chromosomes.

If there is time, ask students to link their ideas to form a concept map at the end.

Exposition

Order them (10 mins)

Ask students to place the following in size order: gene, DNA, organism, nucleus, chromosome. Ask students to present their order to the class and allow others to comment.

Then ask students to suggest definitions for DNA, gene, chromosome, and genome. If students need help doing this, provide them with definitions and ask them to match up each definition to the correct key word.

Main activity

Why does the genome matter? (30 mins)

Discuss with the class what the Human Genome Project is and that it took 12 years and cost around $3 billion. Some people may view this as a waste of time and money. Ask students to use the information in the Student Book to complete the activity.

Ask students to work in small groups and discuss whether they would have their genome sequenced and analysed. Tell them that this information could help doctors to work out their risk of developing certain diseases (e.g., cancer or heart disease) and possibly prescribe personalised medicines.

However, some people are concerned that in the future, they could be asked to reveal this information in other situations (e.g., when applying for a job). Furthermore, there is a limit to how useful the information is at the moment.

Ask students to think about the advantages and disadvantages of genome sequencing, then discuss these ideas with their partner. Discuss briefly as a class before asking students to write a debate or an extended response question on it.

Support groups of students by providing them with extra information on the pros and cons of genome sequencing.

Review and practice

The Human Genome Project (5 mins)

Explain that it now only takes a few days to sequence a genome and the technology is getting quicker and cheaper. Students then complete the

Interactive: *The Human Genome Project.* Complete the interactive to consider the uses of the human genome project and identify true and false statements about the project.

B13.5 DNA structure and protein synthesis

AQA GCSE Biology specification points

4.6.1.5 DNA structure

Smart Curriculum Narrative

Genetics and inheritance

Theme/subtheme

Genetics and inheritance/Protein synthesis

Disciplinary knowledge

- WS 1.2

Practical

Extracting DNA from kiwi fruits

	Learning outcomes		
Learning objectives	**Developing Aiming for 3**	**Secure Aiming for 5**	**Extending Aiming for 8**
Describe the structure of the DNA polymer, including the repeating nucleotide units and the bases A, C, G, and T	Describe the basic structure of DNA	**Describe the structure of the DNA polymer, including the repeating nucleotide units and the bases A, C, G, and T**	(H) Describe the binding of C–G and A–T on complementary strands of DNA and explain its importance
(H) Describe simply how the DNA code is translated into specific proteins on the ribosomes	State that a sequence of three DNA bases codes for a particular amino acid	**(H) Describe simply how the DNA code is translated into specific proteins on the ribosomes**	(H) Describe the main stages of protein synthesis

Tier 2 vocabulary	Tier 3 vocabulary
amino acid, cytoplasm, DNA, enzyme, gene, hormone, nucleus, phosphate, protein, ribosome	*base*, (H) carrier molecule, (H) collagen, (H) complementary strand, double helix, mutation, nucleotide, protein synthesis, (H) template

Lesson plan

Reactivate knowledge

1 What is the main molecule of inheritance?

2 What is a genome?

3 Why is sequencing the human genome important?

Answers: 1 DNA *2* Protein synthesis *3* Helps understand inherited disorders, human evolution, and identifies genes linked to increased risk of specific diseases

Common misconception

Students may think the entire process of protein synthesis occurs on the ribosome.

Trigger interest

What do you already know? (10 mins)

Show students a model of DNA. Ask them to think and write on their mini-whiteboards what would happen to an organism if one base was changed or removed. They should consider how this would affect the animal (students are allowed to make up the gene and the organism that it belongs to).

Exposition

DNA structure (15 mins)

Show students diagrams of DNA with the complementary base pairs visible, for example, Figures 1 and 2 in the Student Book.

Ask students to copy the diagram into their notes, pointing out that the bases C and G are complementary and will always align opposite each other, and the same for A and T.

Read out the following statements and ask students to annotate their diagrams based on each statement.

- A nucleotide consists of a common sugar, phosphate group with one of four different bases attached to the sugar.
- A sequence of three bases is the code for a particular amino acid.
- The order of bases controls the order in which amino acids are assembled to produce a particular protein.
- The DNA polymer is made up of repeating nucleotide units.

Following on from this, ask students to write out the complementary DNA strand to: TAC GGT CCA CGT (Answer: ATG CCA GGT GCA)

Main activity

Extracting DNA from kiwi fruits (25 mins)

Students complete the practical and answer the questions in the activity. They will extract genetic material from kiwi fruit. This could be done as a demonstration with students helping at each stage. Alternatively, students could complete the DNA exam questions activity to test and apply their knowledge of this topic.

(H) The structure of DNA (20 mins, optional)

Introduce students to the process of protein synthesis, talking through each stage with images to help (e.g., Figure 3 from the Student Book).

Students then complete the Activity: *The structure of DNA* to describe the main stages in protein synthesis.

Review and practice

Rosalind Franklin (5 mins)

Show students an image of 'Photograph 51'. Touch on the contributions of Rosalind Franklin and how she discovered the double helix structure of DNA by taking high energy X-rays of DNA molecules.

Ⓗ B13.6 Gene expression and mutation

AQA GCSE Biology specification points

Ⓗ 4.6.1.5 DNA structure

Smart Curriculum Narrative

Genetics and inheritance

Theme/subtheme

Genetics and inheritance/Inheritance and DNA structure

Disciplinary knowledge

- WS 1.2

Practical

Make your own protein

	Learning outcomes		
Learning objectives	**Developing Aiming for 3**	**Secure Aiming for 5**	**Extending Aiming for 8**
Ⓗ Explain how the arrangement of the DNA bases determines the order of the amino acids in a protein chain, which in turn folds to form a unique shape	Describe a gene as a small section of DNA on a chromosome which codes for a particular combination of amino acids to make a specific protein	Ⓗ **Explain how the arrangement of the DNA bases determines the order of the amino acids in a protein chain, which in turn folds to form a unique shape**	Ⓗ Explain why the correct folding of a completed protein is important to its function
Ⓗ Describe mutations and how they may change the function of an enzyme or the strength of a structural protein	Define a mutation	Ⓗ **Describe mutations and how they may change the function of an enzyme or the strength of a structural protein**	Ⓗ Explain how a mutation in non-coding DNA may impact gene expression
Ⓗ Describe some of the roles of non-coding DNA	State that not all parts of the DNA code for proteins	Ⓗ **Describe some of the roles of non-coding DNA**	Ⓗ Explain the fact that humans have around 21 000 genes but there are up to 400 000 human proteins in terms of gene expression

Tier 2 vocabulary	Tier 3 vocabulary
active site, amino acid, enzyme, gene, *substrate*	*base*, mutation, non-coding DNA, phenotype

Lesson plan

Reactivate knowledge

1 Which process in the cell is controlled by the DNA in the nucleus?

2 What is a nucleotide?

3 A sequence of how many nucleotides codes for a single amino acid?

Common misconception

Students might not realise that genome is the same in every cell and instead think different specialised cells contain different DNA.

Answers: 1 Protein synthesis *2* A unit of DNA molecule containing a sugar, a phosphate group and one of four bases (A/C/G/T) *3* 3

Trigger interest

Enzyme function (10 mins)

Tell students that human DNA contains 3 billion base pairs, but only around 45 million are coding. They should work out the percentage of DNA that is non-coding (Answer: 0.015). Ask students to give their answers in standard form (Answer: 1.5×10^{-2}).

Inform students that the non-coding parts of the DNA still have a role. One function of the non-coding parts is that they can switch genes on and off, meaning the gene will or will not result in the protein being synthesised.

Exposition

What is a mutation? (10 mins)

Give students the definition of a mutation. Ask them to think about what may be a good mutation and what would be a bad mutation.

Discuss with students what the term means and how mutations can result in differently-shaped proteins being made, which may be a disadvantage, an advantage, or have no effect on survival. Point out that sometimes mutations happen but the same-shaped protein is still created, and therefore the mutation goes unnoticed.

Main activity

Make your own protein (30 mins)

Provide students with a diagram of a strand of DNA with around 30 base pairs. Also provide a table that shows the colours of amino acids, represented by beads, coded for by different codons (three bases that code for one amino acid). These can be found on the activity's teacher and technician notes. Students then use their DNA strand and the instructions on the activity sheet to build an amino acid chain. Tell them which amino acids form bonds with each other so students can fold the chain of beads into the correct shape.

Now, make one change to a base in the DNA and ask them to build a new protein. They should explain whether it has affected the protein. Students should then write an explanation of how the arrangement of the DNA bases determines the order of the amino acids in a protein chain, and how this affects the final unique shape of a protein. Students should then consider how a mutation may change the function of an enzyme or the strength of a structural protein and describe this in the notes.

Inform students that some DNA is non-coding, and therefore does not code for amino acids, but it has a role on gene expression. In pairs, ask students to consider what impact a mutation in a non-coding region of DNA may have.

To extend, tell students that humans have around 21 000 genes but there are up to 400 000 human proteins. Ask them how they think this may be possible (hint that it links to non-coding DNA and gene expression).

Review and practice

Enzymes and mutations (5 mins)

Ask students to consider why it could be harmful if a mutation occurred in the gene coding for an enzyme.

To support, prompt students to consider why the shape of an enzyme matters. Ask them which part of the enzyme needs to have a particular shape, and what could happen to the enzyme's function if this shape changed.

B13.7 Inheritance in action

AQA GCSE Biology specification points

4.6.1.6 Genetic inheritance

Smart Curriculum Narrative

Genetics and inheritance

Theme/subtheme

Genetics and inheritance/Inheritance and DNA structure

Disciplinary knowledge

- WS 1.2
- MS 1c, 2c, 2e, 3a, 4a

Activity

Inheritance card game

Punnett squares

	Learning outcomes		
Learning objectives	**Developing Aiming for 3**	**Secure Aiming for 5**	**Extending Aiming for 8**
Define and use the terms allele, dominant, recessive, homozygous, and heterozygous correctly	Recognise examples of inherited traits	**Define and use the terms allele, dominant, recessive, homozygous, and heterozygous correctly**	Explain the difference between a gene and an allele and discuss how dominant and recessive alleles interact to control the characteristics of an organism
Describe a phenotype when given the genotype	Define the terms genotype and phenotype	**Describe a phenotype when given the genotype**	Explain how the genotype affects the phenotype at a molecular level
Describe most characteristics as controlled by multiple genes interacting, and give examples	State that some characteristics are controlled by a single gene and give examples	**Describe most characteristics as controlled by multiple genes interacting, and give examples**	Discuss the use of single gene characteristics in genetic crosses even though most characteristics are the result of multiple genes interacting
Describe genetic crosses as predicting the probability of particular genotypes and phenotypes resulting from a given cross	State that a genetic cross is when you consider the potential offspring from a cross between two known parents	**Describe genetic crosses as predicting the probability of particular genotypes and phenotypes resulting from a given cross**	H Demonstrate understanding of the concept of probability in predicting the results of a single gene cross and explain why Punnett squares cannot be used to work out possible genotypes in offspring for the majority of human traits
H Complete a Punnett square diagram to predict the outcome of a monohybrid cross in terms of genotypes and phenotype ratios	Use a simple diagram to state how offspring have inherited traits	**H Complete a Punnett square diagram to predict the outcome of a monohybrid cross in terms of genotypes and phenotype ratios**	H Construct Punnett squares diagrams to model a genetic cross and explain why they are of limited use for the majority of human traits

Tier 2 vocabulary	Tier 3 vocabulary
chromosome, DNA, gamete, gene, protein	allele, *dominant*, genetic cross, genotype, heterozygote, homozygote, phenotype, Punnett square, recessive

Lesson plan

Reactivate knowledge

Common misconception

Students may think that eye colour is determined by one gene, with the alleles B for brown and b for blue, rather than it being a polygenic trait.

1 What happens to the chromosomes when an egg is fertilised by a sperm?

2 What special shape is the DNA molecule?

3 Which letters represent the four bases that determine the structure of the DNA molecule?

Answers: 1 One set from male parent and one set from female parent combine to form new cell with full set of chromosomes *2* Double helix *3* A/C/G/T

Trigger interest

Is it inherited? (5 mins)

Students use the interactive to sort a series of human traits according to whether they are inherited traits or not.

Exposition

Explaining inheritance (20 mins)

Tell the class that two black mice have a litter of offspring. Some are brown. Tell students that alleles are different versions of the same gene, and mice have two alleles for fur colour. Assign two alleles to the fur colour gene – dominant black (B) and recessive brown (b). Inform students that dominant alleles will always be expressed if they are present, but recessive alleles are only expressed if no dominant alleles are present.

Draw a simple genetic diagram to show the possible genotypes and phenotypes of offspring from two heterozygous mice (Bb).

Ask students to write any new terminology down, and to write definitions using the Student Book. Set another example with parent mice of different genotypes for them to practise.

Discuss that the majority of human traits are the result of multiple genes interacting, and some phenotypes are also impacted by the environment (e.g., human height or mass).

Main activity

Punnett squares (5 mins)

Ask students to use a Punnett square to determine the probability of two mice with black fur who are heterozygous having an offspring with brown fur. Inform students that brown fur is caused by a recessive allele, and black fur is caused by a dominant allele.

Inheritance card game (20 mins)

Students play a card game using dominant and recessive cards for different snake characteristics – length, skin colour, skin pattern, eye shape. They pick cards at random to form offspring and draw the outcome.

Review and practice

Inheritance terminology (5 mins)

Read out descriptions of the key vocabulary. Students hold up their mini-whiteboards stating the key terms.

B13.8 More about genetics

AQA GCSE Biology specification points

4.6.1.8 Sex determination

Smart Curriculum Narrative

Genetics and inheritance

Theme/subtheme

Genetics and inheritance/Inheritance and DNA structure

Disciplinary knowledge

- MS 1c, 3a

Activity

Sex determination

Learning objectives	Learning outcomes: Developing – Aiming for 3	Learning outcomes: Secure – Aiming for 5	Learning outcomes: Extending – Aiming for 8
Describe sex inheritance in humans and carry out a genetic cross using a Punnett square diagram to show the probability of a baby being female or male	State that in females the sex chromosomes are XX and in males they are XY	**Describe sex inheritance in humans and carry out a genetic cross using a Punnett square diagram to show the probability of a baby being female or male**	Explain why you only get the expected ratios in a genetic cross if there are large numbers of offspring
Use simple ratios to express the outcome of a genetic cross	Describe the outcome of a given genetic cross	**Use simple ratios to express the outcome of a genetic cross**	Use direct proportion and simple ratios to predict the outcome of a genetic cross
Extract information from a family tree to describe how people are related and how certain characteristics are inherited	Recognise a family tree and state that they are used to trace genetic characteristics through generations	**Extract information from a family tree to describe how people are related and how certain characteristics are inherited**	Interpret information from a family tree to work out the genotypes and phenotypes of different individuals

Tier 2 vocabulary	Tier 3 vocabulary
chromosome, egg cell, gamete, gene, offspring, ratio, sperm	allele, *dominant*, family tree, genetic cross, genetic diagram, genotype, heterozygous, homozygous, meiosis, phenotype, Punnett square, recessive, sex chromosome, sex inheritance

Lesson plan

Reactivate knowledge

1 What do we call different forms of the same gene?

2 An individual is homozygous for a particular characteristic. What does this mean?

3 What is the phenotype of an organism?

Answers: 1 Alleles *2* They have two identical alleles for the characteristic *3* Its physical appearance relating to a particular characteristic

Common misconception

Students may not realise that the sex chromosomes count as the 23rd pair of chromosomes in humans.

Trigger interest

Sex chromosomes (5 mins)

Henry VIII married six times as he wanted a male heir to the throne. Was it his wives or himself that determined why he had had so many daughters?

Ask students if they have heard of the terms XX chromosomes and XY chromosomes. Collate any existing knowledge students have.

Exposition

Proportion and ratio (15 mins)

Remind students about the genetic cross used in the last lesson to show inheritance in two heterozygous mice. Ask students for the ratio of black to brown offspring in this cross (3:1).

Ask them: *If the mice had a litter of four offspring, would it mean that they would definitely have three black and one brown offspring?*

Discuss why this proportion becomes more apparent the more offspring the mice have. Use this as a springboard to discuss the idea of using proportion and ratio to express the outcome of genetic crosses.

Main activity

Sex determination (30 mins)

Consider use of language in this lesson. When describing the sex of an individual, use the term 'biologically male' or 'biologically female' to clarify that the XX and XY chromosomes determine biology, not gender.

Show the class an image of human chromosomes from a biological male, in their pairs. Ask them to spot the pair that is different. Introduce this pair as the sex chromosomes. Prepare sets of sperm cards with either an X or a Y on the back, and egg cards all with an X on the back.

Working in pairs, students choose pairs of cards (one sperm card and one egg card) at random and turn them over. They record the genotype and phenotype each time and record these in a tally table. They do this for five minutes. Discuss what their evidence shows about the ratio, and how it would change if they repeated this activity 1000 times.

Students should complete question 4 in the Student Book linked to the tiger family tree.

Review and practice

Checking understanding (5 mins)

Students use the interactive to sort a series of statements about sex inheritance according to whether they are true or false. Students should then correct the false statements.

B13.9 Inherited disorders

AQA GCSE Biology specification points

4.6.1.7 Inherited disorders

4.6.2.4 Genetic engineering

Smart Curriculum Narrative

Genetics and inheritance

Theme/subtheme

Genetics and inheritance/Inheritance and DNA structure

Disciplinary knowledge

- WS 1.3

Activity

Disorder research

	Learning outcomes		
Learning objectives	**Developing Aiming for 3**	**Secure Aiming for 5**	**Extending Aiming for 8**
Describe polydactyly as an inherited disorder caused by a dominant allele and complete a Punnett square to predict the outcome of different potential crosses	Describe what is meant by an inherited disorder and recognise examples	**Describe polydactyly as an inherited disorder caused by a dominant allele and complete a Punnett square to predict the outcome of different potential crosses**	Carry out genetic crosses showing the inheritance of polydactyly and other inherited disorders caused by a dominant allele, using direct proportion, simple ratios, and probability
Describe cystic fibrosis as an inherited disorder caused by a recessive allele and complete a Punnett square to predict the outcome of different potential crosses	State that some genetic disorders are caused by dominant alleles and some are the result of recessive alleles	**Describe cystic fibrosis as an inherited disorder caused by a recessive allele and complete a Punnett square to predict the outcome of different potential crosses**	Carry out genetic crosses showing the inheritance of cystic fibrosis and other inherited disorders caused by a recessive allele, using direct proportion, simple ratios, and probability

Tier 2 vocabulary	Tier 3 vocabulary
gamete, gene, genetic engineering	allele, *carrier*, cystic fibrosis, *dominant*, family tree, genetic diagram, genetic disorder, genotype, heterozygous, homozygous, Huntington's disease, inherited disorder, phenotype, polydactyly, recessive

Lesson plan

Reactivate knowledge

1 A particular characteristic is recessive. What does this tell us?

2 What is a sex chromosome?

3 In humans, which combination of sex chromosomes results in (a) a biological female (b) a biological male.

Common misconception

Students might not realise that the genome is the same in every cell, and may instead think different specialised cells contain different DNA. They may need to be reminded that if you have the allele coding for polydactyly, this is present in every cell and not just the cells in the fingers and toes.

Answers: 1 It only appears in the phenotype/appearance if the individual is homozygous for the recessive allele *2* The chromosomes that determine your biological sex/the only pair of chromosomes which may not be the same size *3* (a) XX (b) XY

Trigger interest

Categorising disorders (5 mins)

Provide students with a list of disorders, some inherited and some not. Ask them to discuss in pairs and categorise the disorders as inherited or not inherited. For example, Inherited: colour blindness, cystic fibrosis, sickle cell disease. Take account the sensitive nature of this activity and lesson, making sure to encourage a respective and inclusive atmosphere.

Not inherited: anaemia, malaria, measles.

Exposition

Inherited disorders (5 mins)

Inform students that some disorders are the result of a change in the bases or coding of our genes and can be passed on from a biological parent to child. These types of diseases are known as inherited disorders.

Main activity

Disorder research (30 mins)

Split the class into pairs and ask each pair to research online information about the inherited disorders cystic fibrosis and polydactyly. They could also present their findings to the class at the end of the lesson. Give students a list of points that they need to find out about (e.g., how the condition is inherited, the symptoms, possible treatments, etc.).
Allow them to use the internet and/or books to find the answers and present them as a poster, presentation, or leaflet. Before they begin creating their presentations, confirm the answers to the research to ensure accuracy.

Genetic engineering (10 mins)

Set pairs of students the question: *How could genetic engineering help to cure inherited disorders in the future?* Ask them to use information from the Student Book to find the answer.

Review and practice

Test your understanding (5 mins)

Ask the class to complete the interactive and answer a series of quick questions on inherited disorders to check their understanding.

Be mindful of any student who has, or has experience of, either condition.

B13.10 Screening for genetic disorders

AQA GCSE Biology specification points

4.6.1.7 Inherited disorders

Smart Curriculum Narrative

Genetics and inheritance

Theme/subtheme

Genetics and inheritance/Inheritance and DNA structure

Disciplinary knowledge

- WS 1.3

Activity

Making a decision

	Learning outcomes		
Learning objectives	**Developing Aiming for 3**	**Secure Aiming for 5**	**Extending Aiming for 8**
Outline the methods used to screen embryos for possible inherited disorders	State that currently most genetic disorders cannot be cured	**Outline the methods used to screen embryos for possible inherited disorders**	Explain how screening may indicate whether an embryo has a genetic disorder and indicate limitations of the process
List advantages and disadvantages of embryo screening, including some social, ethical, and economic issues	Describe one concern about embryo screening	**List advantages and disadvantages of embryo screening, including some social, ethical, and economic issues**	Make an informed judgement about embryo screening and gene therapy, recognising that this may alleviate suffering but showing awareness of ethical, social, and economic issues

Tier 2 vocabulary	Tier 3 vocabulary
abortion, DNA, economic issue, embryo, ethical issue, false negative, false positive, foetus, in vitro fertilisation (IVF), miscarriage, placenta, uterus	allele, amniocentesis, chorionic villus sampling, genetic disorder, genetic screening, genetic test

Lesson plan

Reactivate knowledge

1 What is a genetic disorder?

2 What is a Punnett square?

3 What is the term for an individual who has one recessive allele for a genetic disorder but has no symptoms themselves?

Common misconception

Students might not realise that embryos divide by mitosis and therefore all the cells contain the same DNA.

Answers: 1 A disorder which results from a change in the genes and may be passed from parents to their offspring *2* A diagram that allows us to model a genetic cross *3* A carrier

Trigger interest

Genetic screening for all? (5 mins)

Tell the class that it is possible to test embryos to see if they have an inherited disorder before they are born. Ask students if they think all parents should be offered this test. They can vote by moving to different places in the classroom or by a show of hands. Ask some students to explain their opinion and be mindful of any student who has, or has experience of, an inherited disorder.

Exposition

Genetic disorders and screening (20 mins)

Tell students to put themselves in the role of parents who have a child with a serious genetic disorder and who want to have another child. Ask them to discuss in pairs why this is a big decision for the parents. Be aware that some students may have direct experience of serious genetic disorders within their family.

Screening techniques

Talk students through the two embryo screening techniques, amniocentesis and chorionic villus sampling, plus IVF with embryo screening. You can use the information from the Student Book to help.

What would you do?

Ask students to put themselves in the role of the parents from the earlier discussion. In their pairs, they should now discuss the considerations the parents would have to make to come to a decision on screening. Take account the sensitive nature of this activity, making sure to encourage a respective and inclusive atmosphere.

Main activity

Making a decision (25 mins)

Students use information from the discussions and the Student Book, if needed, to complete the activity. Discuss the answers to this activity in the lesson so students can self-mark their work, or add additional information.

Review and practice

Screening true or false? (5 mins)

Students use the interactive to answer a series of true or false questions about screening for genetic disorders.

B13 Reproduction and genetic inheritance

Introduction to checkpoint intervention

The checkpoint intervention provides suggestions for a lesson to follow up the *B13 Reproduction and genetic inheritance* assessment. Depending on the outcomes of the assessment, these suggestions can be used to move a class towards secure, to target specific learning outcomes, or to consolidate knowledge for students achieving secure outcomes. Use students' outcomes from the checkpoint assessment to address general misconceptions from the content covered. Be prepared to re-cover content as required; three key concepts from the chapter are covered in more detail below.

Checkpoint secure learning outcomes

- Describe the key elements in sexual and asexual reproduction, including the role of meiosis and mitosis.
- Explain how meiosis halves the number of chromosomes in gametes and fertilisation restores the full number.
- Describe some of the benefits of studying the human genome.
- (H) Describe mutations and how they may change the function of an enzyme or the strength of a structural protein.
- Describe sex inheritance in humans and carry out a genetic cross using a Punnett Square diagram to show the probability of a baby being female or male.
- Describe polydactyly as an inherited disorder caused by a dominant allele and complete a Punnett square to predict the outcome of different potential crosses.

Student reflection

Encourage students to reflect on whether there were any checkpoint questions they found difficult or straightforward as well as their preparation for the checkpoint assessment, including how they revised, what they revised, the time spent on revision, and what areas they could improve upon. For more strategies, see the Metacognition spread.

Intervention activity: Describe the key elements in sexual and asexual reproduction, including the role of meiosis and mitosis.

Students should write an answer to the following five-mark question:

Compare asexual and sexual reproduction.

To support students in scaffolding this answer, ask them to refer to their summary comparison tables from lesson 13.1.

Students should include the following marks in their answer:

- Sexual reproduction involves two parents, whereas asexual reproduction involves one.
- Sexual reproduction involves gametes, whereas asexual reproduction does not.
- Sexual reproduction involves the fusion of gametes to make a zygote, whereas asexual reproduction does not.
- The gametes for sexual reproduction are formed by meiosis whereas asexual reproduction only involves mitosis.
- The offspring that result from sexual reproduction have genetic variation, whereas in asexual reproduction clones are made.

Intervention activity: Describe sex inheritance in humans and carry out a genetic cross using a Punnett Square diagram to show the probability of a baby being female or male.

Ask students to state the chromosomes that would determine whether a human is biologically male or biologically female.

Next, ask students to state the potential sex chromosomes in the gametes that a biological male and a biological female could produce.

Now that the students have the gametes for two potential parents, ask them to draw a Punnett square to determine the genotypes and phenotypes of the offspring. If students require support, provide them with the empty Punnett square to complete.

Finally, ask students to look at the offspring and state what the probability is of the parents having a biological female.

Intervention activity: Describe polydactyly as an inherited disorder caused by a dominant allele and complete a Punnett square to predict the outcome of different potential crosses.

Students should recall that polydactyly is an inherited disease caused by a dominant allele. Remind students that conditions caused by dominant alleles will always be expressed (seen in the phenotype).

Students will now be taken through the stages of creating a Punnett square for two parents who are heterozygous for the polydactyly gene. If students need support, inform them that this means each parents has one dominant and one recessive allele.

Ask the students to write down the genotype of the two parents, using ‘P’ to represent the dominant allele for polydactyly and ‘p’ for the recessive allele for 5 digits on each hand.

Next, ask them to state what the potential gametes are that both parents could produce.

Now that the students have the gametes for the two parents, ask them to draw a Punnett square to determine the potential genotypes and phenotypes of any possible offspring. If students require support, provide them with the empty Punnett square to complete.

When their Punnett squares are complete, ask students to look at the possible offspring genotypes and state which would result in offspring who have polydactyly. Finally, ask students to state what the probability is of the parents having a child with polydactyly as a percentage or fraction.

B14 Variation and selection

B14.1 Variation

AQA GCSE Biology specification points
4.6.2.1 Variation

***Smart Curriculum* Narrative**
Variation and evolution

Theme/subtheme
Variation and evolution/Variation

Activity
Twin studies

Learning objectives	Learning outcomes: Developing Aiming for 3	Secure Aiming for 5	Extending Aiming for 8
Describe how variation between individuals may be due to genetic causes, environmental causes, or a combination of both	State that differences in characteristics of individuals in a population is called variation	**Describe how variation between individuals may be due to genetic causes, environmental causes, or a combination of both**	Explain with examples how the genome and the way it interacts with the environment influence the development of the phenotype of an organism
Describe examples of characteristics that show variation due to genetic causes, environmental causes, and a combination in humans, other animals, and plants	List examples of characteristics that show variation in humans, other animals, and plants	**Describe examples of characteristics that show variation due to genetic causes, environmental causes, and a combination in humans, other animals, and plants**	Explain with examples how some forms of variation are due only to genetic causes and some are due only to environmental causes
Give reasons why identical twins will start to show variation as they get older	Explain why identical twins share the same genes	**Give reasons why identical twins will start to show variation as they get older**	Discuss some of the issues scientists face when conducting twin studies
Use data to explain why studying identical twins helps scientists investigate the interaction of genetic causes and environmental causes in human variation	State why identical twins are so useful to scientists studying human variation	**Use data to explain why studying identical twins helps scientists investigate the interaction of genetic causes and environmental causes in human variation**	Explain some of the difficulties in getting valid results from twin studies

Tier 2 vocabulary	Tier 3 vocabulary
environment, gene, identical twin, variation	environmental variation, genetic variation

Lesson plan

Reactivate knowledge

Common misconception

Students may think that identical twins will be completely identical in looks and personality.

1 What is variation within a species?
2 Name two types of variation in an individual.
3 Give an example of a characteristic that shows discontinuous variation.

Answers: 1 Differences in characteristics between individuals *2* Inherited and environmental variation *3* E.g., blood group

Trigger interest

Twins (10 mins)

Show the class images of identical twin babies. Ask students to write down a scientific explanation of why they look identical. Discuss the fact that identical twins are genetically identical to each other and were formed when a fertilised egg split in half.

Ask students to explain why the twins will not look exactly identical as they get older (due to environmental changes). To support students, supply key words: egg, sperm, fertilisation, split, genes, traits, environment.

Students should then explain why non-identical twins often look similar but not identical.

Exposition

What causes variation? (15 mins)

Inform students that variation can be caused by differences in our DNA (genetic variation) or due to differences acquired throughout life due to environmental conditions we are exposed to (environmental variation). Ask the students to write down a list of variations in humans. Allow pairs or small groups to discuss their answers so that all students have a list of at least ten examples. If needed, start by giving the class some examples (e.g., race, hair colour, height, blood group).

In pairs, students then discuss environmental and inherited causes of variation to categorise the causes of each example in their lists.

Then ask students to write some examples of variation in plants. Write some of these examples on the board and then ask students to sort them into those caused by the environment, by genetics, or an interaction of both.

Discuss the students' lists and the causes they opted for as a class.

Main activity

Twin studies (25 mins)

Tell students to imagine that identical twins are each adopted by different families. Ask students to discuss in small groups how similar they think the twins would be after 25 years apart and why, giving specific examples of traits. They should work alone to study Table 1 in the Student Book and write conclusions from the data.

Groups should consider why these types of studies are rare (very few twins are raised apart). Ask students to list reasons why the most valid data results from studies where the twins were separated at birth.

Review and practice

Effects of the environment (5 mins)

Discuss environments can develop genetic traits over large periods of time. For example, people living in hotter climates have higher levels of melanin than people living in cooler climates to protect skin cells from being damaged.

B14.2 Evolution by natural selection

AQA GCSE Biology specification points

4.6.2.2 Evolution

Smart Curriculum Narrative

Variation and evolution

Theme/subtheme

Variation and evolution/Variation

Disciplinary knowledge

- WS 1.2

Practical

Modelling natural selection

	Learning outcomes		
Learning objectives	**Developing Aiming for 3**	**Secure Aiming for 5**	**Extending Aiming for 8**
Explain how a mutation may lead to a new phenotype and recall that all the variants in a population arise from mutations	Describe a mutation as a change in the DNA code	**Explain how a mutation may lead to a new phenotype and recall that all the variants in a population arise from mutations**	Explain why it is rare for a mutation to lead to a new phenotype
Describe how some variants may be better suited to a particular environment, or better adapted after an environmental change, and these organisms are most likely to survive and breed successfully	State that individual organisms within a species show a wide range of genetic and phenotypic variation	**Describe how some variants may be better suited to a particular environment, or better adapted after an environmental change, and these organisms are most likely to survive and breed successfully**	Explain how the alleles that deliver an advantageous phenotype are passed on to the next generation
Describe the steps that take place during evolution by natural selection and describe the point at which a new species has evolved	Describe the theory of evolution by natural selection as a process by which living things have evolved from simple life forms	**Describe the steps that take place during evolution by natural selection and describe the point at which a new species has evolved**	Explain why a change in the environment often triggers evolutionary change

Tier 2 vocabulary	Tier 3 vocabulary
allele, DNA, gene, mutation, phenotype, species	evolution, natural selection, resistant, variant

Lesson plan

Reactivate knowledge

Common misconception

Students may think that a mutation always results in a change, and that the change always causes a negative impact.

1 How do changes in the environment affect individuals and even whole species?

2 What type of studies are widely used to find out whether a characteristic is affected by the environment or not?

3 Are most differences between individuals the result of their genetics, their environment, or both?

Answers: 1 Means some adapted to compete more successfully, and some less *2* Twin studies *3* Both

Trigger interest

Albinism (5 mins)

Show the class images of a range of albino animals. Tell them that albinism is due to a mutation.

Remind students that a mutation is a change in the DNA code. Ask students to suggest whether this mutation is an advantage or a disadvantage, and to explain why they think this.

Exposition

Real-life examples (20 mins)

Ask students to write true or false on their mini-whiteboards in response to each of the following statements:

- Mutations always result in a change to the phenotype.
- All mutations are harmful.

Discuss how some mutations cause a change in the DNA, but still code for the same protein and therefore cause no change to the phenotype. Discuss that some mutations can be harmful, such as those resulting in the allele coding for cystic fibrosis or albinism. Other mutations may provide an organism with an advantage that helps them to survive in their environment, such as antibiotic resistance in bacteria. Use this to initiate a discussion of natural selection and evolution.

Introduce evolution as a process by which living things change over time, and natural selection as a theory that explains how this happens.

Talk through the stages of natural selection. Use the Student Book so students have a written version of this. Students can then use this to create an explanation of how a bacterium could evolve to have antibiotic resistance.

Main activity

Modelling natural selection (25 mins)

Students play a game in the practical to model natural selection. After the game, students complete the questions on the practical sheet. Discuss their answers as a class. Include in this a discussion of why there is variation in a population, and why this means some organisms are better adapted than others. What happens to the organisms that are poorly adapted? Why does the number of the best-adapted organisms tend to increase?

Review and practice

Describing natural selection (5 mins)

Students use the interactive to complete a description of the process of natural selection by filling in the gaps.

B14.3 Selective breeding

AQA GCSE Biology specification points

4.6.2.3 Selective breeding

Smart Curriculum Narrative

Genetics and inheritance

Theme/subtheme

Genetics and inheritance/Genetic manipulation

Disciplinary knowledge

- WS 1.3, 1.4

Activity

Developing dog breeds

	Learning outcomes		
Learning objectives	**Developing Aiming for 3**	**Secure Aiming for 5**	**Extending Aiming for 8**
Describe the process of selective breeding, giving examples of the results of selective breeding in animals and plants	Describe selective breeding as a process where humans choose which plants or animals to breed together for particular genetic characteristics	**Describe the process of selective breeding, giving examples of the results of selective breeding in animals and plants**	Compare and contrast natural and artificial selection
Explain why humans have used selective breeding over time	Give one example where selective breeding has been used in animals and one in plants	**Explain why humans have used selective breeding over time**	Explain how the variation of alleles in a population is reduced through selective breeding
Explain what inbreeding is, and give examples of the problems it causes in dog breeding and in agricultural crops	State that over-use of selective breeding may lead to inbreeding and this is clearly seen in some dog breeds	**Explain what inbreeding is, and give examples of the problems it causes in dog breeding and in agricultural crops**	Explain in detail why the reduction of variation in a population through selective breeding is a problem

Tier 2 vocabulary	Tier 3 vocabulary
allele, disease resistance, variation	evolution, inbreeding, population, selective breeding

Lesson plan

Reactivate knowledge

1 How do new variants arise?

2 Evolution by natural selection is often referred to as 'survival of the ....'

3 What are the main features of natural selection?

Common misconception

Students may think that selective breeding only has benefits, as advantageous traits are selected for.

Answers: *1* Mutation *2* Fittest *3* Offspring best suited to their environment most likely to survive and breed to pass on the advantageous alleles

Trigger interest

Dog breeds (5 mins)

Show students an image of different dog breeds.

Explain that all of these dogs are one species, because they can breed with each other to produce fertile offspring. Choose examples of dog breeds and ask students to suggest how the features of that breed make it useful for specific tasks; for example, a greyhound is fast so it is good for racing, a mastiff is large so it makes a good guard dog, etc.

Exposition

Selective breeding (10 mins)

Ask students what they think the term 'selective breeding' means. Discuss their answers and use this to lead to a definition they can write down.

Ask students to work in pairs to list the advantages and disadvantages of using selective breeding in animals and plants. Discuss their answers as a class.

Main activity

Developing dog breeds (35 mins)

Tell students that all modern dogs are descended from one species – the wolf.

Students work in groups to research and develop an item for a short presentation about the selective breeding of dogs.

They should explain to a lay audience how people used selective breeding to produce all the modern dog breeds, and also discuss the problems inbreeding has caused in some breeds. They can choose which breeds to focus on and produce an engaging verbal or computer presentation to present to the rest of the class.

This would work best with access to computers, but students could use the Student Book as an alternative to source the information if necessary.

Review and practice

Artificial selection (5 mins)

Tell the class that selective breeding is also called artificial selection. Ask them to suggest why on their mini-whiteboards.

B14.4 Genetic engineering

AQA GCSE Biology specification points

4.6.2.4 Genetic engineering

Smart Curriculum Narrative

Genetics and inheritance

Theme/subtheme

Genetics and inheritance/Genetic manipulation

Disciplinary knowledge

- WS 1.3, 1.4

Activity

Genetic engineering

	Learning outcomes		
Learning objectives	**Developing Aiming for 3**	**Secure Aiming for 5**	**Extending Aiming for 8**
Describe the principle of genetic engineering to produce GM organisms	Describe genetic engineering as a process that involves modifying (changing) the genome of an organism, e.g., by introducing a gene from another organism, to give a desired characteristic	**Describe the principle of genetic engineering to produce GM organisms**	(H) Describe the main steps in the process of genetic engineering using scientific vocabulary (e.g., plasmid, vector, restriction enzymes, marker genes, recombinant DNA)
Describe a number of different types of organisms that have been genetically modified, including microorganisms, animals, and crop plants, and explain the advantages the genetic modification brings	Give an example of a GM crop plant and describe its advantages over the non-engineered crop	**Describe a number of different types of organisms that have been genetically modified, including microorganisms, animals, and crop plants, and explain the advantages the genetic modification brings**	(H) Explain why gene modification takes place at the early stage of development of animals and plants

Tier 2 vocabulary	Tier 3 vocabulary
bacterial plasmid, chromosome, *culture*, DNA, enzyme, food security, gene, insulin, resistant, virus, yield	clone, CRISPR, gene editing, genetic engineering, GM crop, *vector*

Lesson plan

Reactivate knowledge

Common misconception

Students may not realise that to edit the DNA in every cell in an organism, genetic engineering must occur in the embryo.

1 How have most domestic animals and crop plants developed?
2 How does this process work?
3 How might GM plants help solve climate change?

Answers: 1 Selective breeding *2* Select only animals or plants with desired characteristics to breed, then select again from the offspring until all offspring have the desired characteristics *3* Increase uptake of carbon dioxide from the atmosphere

Trigger interest

Glow-in-the-dark mouse (5 mins)

Show an image of the genetically modified fluorescent mouse. Ask students to discuss in small groups how they think the mouse could have been produced. Explain that the mouse's genome now contains the gene for a glowing protein from a jellyfish.

Exposition

What is GM? (10 mins)

Ask students if they have heard of GM issues and explain what is meant by genetic engineering. Ask the students to use information from the Student Book, or a class discussion, to write a list of examples of genetically modified (GM) organisms that have been produced by genetic engineering.

Discuss as a class why scientists have produced GM organisms. Ask students to consider the advantages to a farmer of using GM crops.

Main activity

Genetic engineering (30 mins)

Provide students with data in the activity sheet that show how crop yields of GM corn and normal corn compare. Ask them to analyse the data and write a letter to a farmer explaining what the data shows, and whether they would recommend growing GM corn.

Introduce another function of genetic engineering, such as making GM bacteria, that produce human insulin. Provide students with an unlabelled diagram showing the principles of genetic engineering and ask them to go through the diagram in pairs, working out how the process is carried out. Allow them to use a labelled diagram to check their understanding, before writing down the stages of how GM corn is produced.

Review and practice

Genetic engineering for people (10 mins)

Remind students that genetic engineering could be used to help cure inherited disorders such as cystic fibrosis. Ask students to use their knowledge from the lesson to suggest and research online about how this could be done. For example, using CRISPR technology. As support, students can read the information from the Student Book before discussing. Students with higher prior attainment should also discuss the limitations of this technique.

B14.5 Cloning

AQA GCSE Biology specification points

4.6.2.5 Cloning

Smart Curriculum Narrative

Genetics and inheritance

Theme/subtheme

Genetics and inheritance/Genetic manipulation

Disciplinary knowledge

- WS 1.3, 1.4

Practical

Stem and leaf cuttings

	Learning outcomes		
Learning objectives	Developing Aiming for 3	Secure Aiming for 5	Extending Aiming for 8
Describe how to clone plants using cuttings and tissue culture	Define the term clone, and state plants are cloned by taking cuttings and by tissue culture	**Describe how to clone plants using cuttings and tissue culture**	Compare and contrast the two main methods of plant cloning
Describe how embryo transplants are produced, and how embryo splitting produces offspring which are clones	State that animals are cloned by embryo splitting, embryo transplants, and adult cell cloning	**Describe how embryo transplants are produced, and how embryo splitting produces offspring which are clones**	Discuss the benefits of embryo transplantation and embryo splitting to farmers
Describe the potential benefits and risks of cloning plants and animals	List one potential benefit of cloning animals and cloning plants	**Describe the potential benefits and risks of cloning plants and animals**	Explain the benefits of embryo transplants and embryo splitting over standard sexual reproduction for farmers

Tier 2 vocabulary	Tier 3 vocabulary
agar jelly, asexual reproduction, disease resistance, embryo, hormone, plant hormone, plantlet, yield	callus, clone, cloning, embryo splitting, embryo transplant, genetically modified, plant cutting, tissue culture

Lesson plan

Reactivate knowledge

1 What is genetic engineering?

2 Name one human hormone made by genetically engineered bacteria.

3 Give one way in which genetically modified plants might help overcome the problems of global warming/climate change.

Answers: 1 Changing the genetic material of an organism *2* Insulin *3* Taking up more carbon dioxide in photosynthesis/being able to withstand more extreme weather, for example, flooding/drought.

Common misconception

Students may think that a clone would be created the same age as the individual they were cloned from. It should be emphasised that clones still go through the same developmental growth stages, but with identical DNA to the individual they are a clone of.

Trigger interest

Using cloning (10 mins)

Briefly remind students what cloning is. Ask students to suggest some possible uses of cloning. Encourage them to think about uses of both plant and animal cloning.

Exposition

Embryo transplants (10 mins)

Tell the class to imagine that they are horse breeders. They have an excellent racehorse that has won them lots of money, and it would be ideal if they could produce more horses like this one.

Ask students to discuss in pairs what they would do. Listen to their suggestions.

Students should then read through the information from the Student Book about cloning animals. Ask them to describe how they could use each technique (embryo transplants and embryo splitting) with their racehorse, which technique they would choose and why. Ask them why each of the embryos produced by embryo splitting will be a clone.

Main activity

Stem and leaf cuttings (25 mins)

Introduce cuttings as a way of cloning plants using asexual reproduction. Students take leaf and stem cuttings from plants. They can take photographs of the cuttings before they are placed in the soil, and then again after a period of 4–6 weeks to show any root development. Students answer the questions based on the experiment within the practical sheet, testing understanding of the purpose of each step and the advantages of growing plants using cuttings compared with seeds.

Inform students about tissue culture as another method of cloning plants (detailed in the Student Book). Ask students to look at the pros and cons of each method and compare them.

Review and practice

Key meanings (10 mins)

Students use the interactive to match definitions and key words from the topic (e.g., tissue culture, cuttings, embryo transplants, cloning).

Students should then list one advantage and one disadvantage of cloning plants on their mini-whiteboards. Next, ask students to explain why it is an advantage or disadvantage.

Students with higher prior attainment should be able to explain the benefits of embryo transplants over sexual reproduction for farmers.

B14.6 Adult cell cloning

AQA GCSE Biology specification points

4.6.2.5 Cloning

Smart Curriculum Narrative

Genetics and inheritance

Theme/subtheme

Genetics and inheritance/Genetic manipulation

Disciplinary knowledge

- WS 1.3, 1.4

Activity

Future uses?

Learning objectives	Learning outcomes: Developing – Aiming for 3	Learning outcomes: Secure – Aiming for 5	Learning outcomes: Extending – Aiming for 8
Describe the main steps in adult cell cloning	Define adult cell cloning as producing a new animal from the cell of another adult animal	**Describe the main steps in adult cell cloning**	Explain why the organism produced by adult cell cloning will not be identical to the animal which supplied the adult cell that was cloned
Design a flow chart to describe the process of adult cell cloning	Describe the process of adult cell cloning using a diagram	**Design a flow chart to describe the process of adult cell cloning**	Use advanced terminology to explain the process of adult cell cloning
Explain some of the potential benefits and risks of adult cell cloning	State one potential benefit of adult cell cloning	**Explain some of the potential benefits and risks of adult cell cloning**	Compare and contrast the processes of adult cell and embryo cloning

Tier 2 vocabulary	Tier 3 vocabulary
asexual reproduction, egg cell, embryo, extinction, gene, nucleus, population, variety	adult cell cloning, clone, cloning

Lesson plan

Reactivate knowledge

1 What is a clone?

2 Name two ways of cloning plants.

3 Give an advantage of cloning both animals and plants.

Common misconception

Students may think that a clone would be created the same age as the individual they were cloned from. It should be emphasised that clones still go through the same developmental growth stages, but with identical DNA to the individual they are a clone of.

Answers: 1 An individual produced asexually which is genetically identical to its parent *2* Taking cuttings/tissue culture *3* To produce lots of copies of an organism/genetically identical organisms for research

Trigger interest

Dolly (5 mins)

Ask if any students have heard of Dolly the sheep. Share a video or news article from when Dolly's birth was announced to the public, on 22nd February 1997.

Exposition

How to clone an animal (15 mins)

Give students a diagram showing the process of adult cell cloning (e.g., Figure 2 in the Student Book). Ask them to study the diagram and then explain to a partner how the process is carried out. Allow students to use information from the Student Book to design a flow chart to show how adult cell clones are produced.

They should then explain why the baby born is a clone.

Challenge students and ask them to use the terms somatic cell, surrogate mother, and enucleation in their descriptions. Define these words for them.

Main activity

Future uses? (30 mins)

Split the class into groups and give each group one possible future use of adult cell cloning. For example, bringing back extinct animals, cloning cows/sheep that produce useful substances in their milk, cloning dead pets, or cloning humans.

Provide students with information on each potential use (or direct them to suitable webpages) and ask them to discuss whether they think it is a good idea or not.

Ask groups to present their thoughts to the rest of the class. Then, after each presentation, ask the class to stand in continuum lines to show how they feel about each use of cloning (with one end of the line being 'not acceptable' and the other end 'good idea').

Review and practice

Cloning words (5 mins)

Write a list of words associated with cloning on the board, but leave out the vowels. For example,

- CLNNG (cloning)
- NCLS (nucleus)
- GNS (genes)
- SRRGT (surrogate)
- MBRY (embryo)

Ask students to write out the words.

To support students, indicate where the vowels should go by using dashes.

Students should then write a definition of each word and explain how it is linked to adult cell cloning.

B14.7 Ethics of genetic technologies

AQA GCSE Biology specification points

4.6.2.4 Genetic engineering

Smart Curriculum Narrative

Genetics and inheritance

Theme/subtheme

Genetics and inheritance/Genetic manipulation

Disciplinary knowledge

- WS 1.3, 1.4

Activity

Benefits and risks of genetic engineering

	Learning outcomes		
Learning objectives	**Developing** **Aiming for 3**	**Secure** **Aiming for 5**	**Extending** **Aiming for 8**
Summarise the potential benefits and risks of genetic engineering and compare them to those of cloning in plants	State one possible benefit and one potential risk of modern genetic techniques in plants	**Summarise the potential benefits and risks of genetic engineering and compare them to those of cloning in plants**	Interpret data and evaluate the potential benefits and risks of genetic engineering and cloning in plants
Summarise the potential benefits and risks of genetic engineering and compare them to those of cloning in animals	State one possible benefit and one potential risk of modern genetic techniques in animals	**Summarise the potential benefits and risks of genetic engineering and compare them to those of cloning in animals**	Interpret data and evaluate the potential benefits and risks of genetic engineering and cloning in animals
Describe the current state of gene therapy research and use in medicine	Define gene therapy	**Describe the current state of gene therapy research and use in medicine**	Discuss, using scientific knowledge, the concerns some people express about gene therapy
Describe some economic and ethical issues that people may have about modern genomic technologies	State that many people support the developments in gene therapy, gene editing, and cloning but some express concerns	**Describe some economic and ethical issues that people may have about modern genomic technologies**	Use scientific knowledge to make informed judgements on the significance of recent developments in our ability to edit genomes and clone organisms

Tier 2 vocabulary	Tier 3 vocabulary
bone marrow, cystic fibrosis, diabetes, embryo, ethical issue, gene, growth hormone, herbicide, inherited disorder, insulin, pesticide, sickle cell anaemia, thalassemia, vitamin A, yield	cloning, gene therapy, genetic engineering, genetic modification, genetic technology, GM crop, human engineering

Lesson plan

Reactivate knowledge

Common misconception

Students may think that cloning and genetic engineering only have benefits.

1 Give one reason why genetic engineering is more effective than selective breeding.
2 What do the letters GM mean in GM food or GM organism?
3 Which types of organisms are used to edit genetic material to make human medicines?

Answers: 1 Any from: get results faster; avoids inbreeding; can remove or insert very specific genes; can introduce genes from different organisms *2* Genetically modified *3* Bacteria/microorganisms

Trigger interest

Definitions (5 mins)

Ask students to work in pairs. One student should define genetic engineering and the other should define cloning. Ask them to read out their definitions to each other and suggest changes until they are happy with their definitions.

As a prompt, provide pairs with several definitions of the terms and ask them to pick out the ones that they think are best.

Exposition

Cloning concerns (10 mins)

Present the class with a range of numerical facts about cloning:

- The process that produced Dolly in 1996 used 277 fertilised eggs, which formed 29 viable embryos, and produced three lambs at birth, one of which lived.
- Dolly died young (at six years old – most sheep live until they are 12). She had lung problems and arthritis.
- In 2001 it took 188 attempts to make Cc, the first cloned cat, producing 87 cloned embryos, only one of which resulted in a kitten.
- You can clone your dog for £60 000.

Ask students to use these figures to present data supporting the argument against the use of adult cell cloning.

Students should consider both economic and ethical issues. For example, it costs money to clone animals, and the success rate is low. This could be seen as wasting money that could be used for other pursuits. Also, many embryos will die during the process.

Main activity

Benefits and risks of genetic engineering (30 mins)

Provide students with resources that outline arguments for and against genetic engineering in agriculture and medicine.

Ask them to write a balanced article for a newspaper or news website on genetic engineering.

Review and practice

Peer assessment (10 mins)

Ask students to peer assess each other's articles from the previous activity, giving one positive comment about the work and one suggestion for how it could be improved.

B14 Variation and selection

Introduction to checkpoint intervention

The checkpoint intervention provides suggestions for a lesson to follow up the *B14 Variation and selection* assessment. Depending on the outcomes of the assessment, these suggestions can be used to move a class towards secure, to target specific learning outcomes, or to consolidate knowledge for students achieving secure outcomes. Use students' outcomes from the checkpoint assessment to address general misconceptions from the content covered. Be prepared to re-cover content as required; three key concepts from the chapter are covered in more detail below.

Checkpoint secure learning outcomes

- Describe how variation between individuals may be due to genetic causes, environmental causes or a combination of both.
- Explain how a mutation may lead to a new phenotype and recall that all the variants in a population arise from mutations.
- Explain what inbreeding is, and give examples of the problems it causes in dog breeding and in agricultural crops.
- Describe a number of different types of organisms that have been genetically modified, including microorganisms, animals and crop plants and explain the advantages the genetic modification brings.
- Summarise the potential benefits and risks of genetic engineering and compare them to those of cloning in animals.
- Describe how embryo transplants are produced and explain why the offspring are clones.

Student reflection

Encourage students to reflect on whether there were any checkpoint questions they found difficult or straightforward as well as their preparation for the checkpoint assessment, including how they revised, what they revised, the time spent on revision, and what areas they could improve upon. For more strategies, see the Metacognition spread.

Intervention activity: Describe how variation between individuals may be due to genetic causes, environmental causes or a combination of both.

Ask students to list five examples of variation in plants. For students who require more support, you could provide them with the following list of five:

- stem height
- flower colour
- leaf colour
- time of year flowers bloom
- direction of plant growth (bending of plant towards light).

Students then draw out the table below and categorise each type of variation according to whether it was caused by genetics, the environment, or an interaction of both.

Genetic cause	Environmental cause	Interaction of both

Intervention activity: Explain what inbreeding is, and why it is a problem in dog breeding.

Students write a blog post to inform the public about what inbreeding is, and the issues this has resulted in for farmers and crops.

The blog post must include:

- an explanation of what is meant by inbreeding
- an explanation of why inbreeding has been occurring in crops
- a specific explanation of the challenges farmers have due to inbreeding.

Students can use the Student Book or research online to help them write their blog post.

Intervention activity: Describe how embryo transplants are produced and explain why the offspring are clones.

Provide students with Figure 2 about embryo splitting from B14.5 of the Student Book, but without the labels. Ask students to write a description to explain what is happening in each stage of the flow diagram.

Ask students to use information from the flow diagram to explain why all the calves that are born are clones of each other.

Biology only

B15 Evolution and classification

B15.1 The history of genetics

AQA GCSE Biology specification points

4.6.3.3 The understanding of genetics

Smart Curriculum Narrative

Genetics and inheritance

Theme/subtheme

Genetics and inheritance/Inheritance and DNA structure

Disciplinary knowledge

- WS 1.1

Activity

Gene theory timeline

	Learning outcomes		
Learning objectives	**Developing Aiming for 3**	**Secure Aiming for 5**	**Extending Aiming for 8**
Describe the key elements of Mendel's work that made it so unusual for its time	Describe the role of Gregor Mendel in the development of our understanding of genetics	**Describe the key elements of Mendel's work that made it so unusual for its time**	Suggest why Mendel's work on genetics was not recognised in his lifetime
Describe the roles played by Franklin, Wilkins, Watson, and Crick in the discovery of the double helix structure of DNA	Correctly order important discoveries in gene theory	**Describe the roles played by Franklin, Wilkins, Watson, and Crick in the discovery of the double helix structure of DNA**	Discuss reasons why Mendel's work on genetics and gene theory was not recognised until 16 years after his death, but work on the structure of DNA was recognised almost immediately and the Nobel Prize awarded within nine years
Describe some of the characteristics Mendel studied in peas and choose an example to describe the results from Mendel's experiments	State the type of plant Mendel used most often in his breeding experiments	**Describe some of the characteristics Mendel studied in peas and choose an example to describe the results from Mendel's experiments**	Construct and use a Punnett square to draw conclusions from the results of Mendel's experiments

Tier 2 vocabulary	Tier 3 vocabulary
chromosome, DNA, gene, offspring, protein, X-ray	breeding experiment, *dominant*, double helix, epigenetics, gene theory, genetics, genetic cross, monohybrid inheritance, units of inheritance

Lesson plan

Reactivate knowledge

1 What is heredity?

2 What is the relationship between chromosomes, genes, and DNA?

3 What type of square do we use to model a genetic cross?

Common misconception

Students may think that chromosomes always appear as a double chromatid 'X' structure, whereas in fact this is what a chromosome looks like only when the DNA has replicated before cell division.

Answers: 1 The process by which genetic information is passed from one generation to another *2* Chromosomes contain DNA, the genetic material. Genes are small sections of DNA on the chromosome which code for particular characteristics *3* Punnett square

Trigger interest

Genetic definitions (5 mins)

Students use the Interactive: *Genetic definitions* to match genetics terms with their definitions. Students should be able to complete part one of the interactive using prior knowledge. If they score full marks, they can move onto part two which tests definitions for recessive, dominant, homozygous, and heterozygous.

Exposition

Pea parents (15 mins)

Tell the class that peas come in two colours – green and yellow. Ask students to suggest how two plants with yellow peas can have offspring with green peas. Ask students if they can use a Punnett square to show this. They can refer to Figure 1 in the Student Book to help.

Discuss Mendel's work and findings with the class, including why it was so unusual for the time. Lead this discussion into how the work of Rosalind Franklin, Maurice Wilkins, James Watson, and Francis Crick led to the discovery of the DNA double helix structure.

Main activity

Gene theory timeline (30 mins)

Provide students with cards from the activity showing major discoveries in gene theory, from Mendel's work to the modern day.

Ask them to work in groups to create a timeline of the development of gene theory.

As an extension, students should then annotate their timeline with further details about each stage, using the Student Book or the internet.

Review and practice

Recognition (5 mins)

Students write down possible reasons why Franklin and Mendel's work was not recognised during their lifetime, but the work of Watson and Crick was. Gather ideas as a class to create an agreed list.

B15.2 Theories of evolution

AQA GCSE Biology specification points

4.6.3.1 Theory of evolution

Smart Curriculum Narrative

Variation and evolution

Theme/subtheme

Variation and evolution/Evolution, extinction, and the fossil record

Disciplinary knowledge

- WS 1.1, 1.3

Activity

Darwin's theory

	Learning outcomes		
Learning objectives	**Developing** **Aiming for 3**	**Secure** **Aiming for 5**	**Extending** **Aiming for 8**
Describe the theory of inheritance of acquired characteristics proposed by Jean-Baptiste Lamarck	State that scientists had been trying to explain the existence of different species for many years before Darwin, and give Jean-Baptiste Lamarck's theory of inheritance of acquired characteristics as an example	**Describe the theory of inheritance of acquired characteristics proposed by Jean-Baptiste Lamarck**	Explain why Lamarck's theory is no longer accepted in the vast majority of cases
Describe the main principles of evolution through natural selection and explain how Darwin developed his ideas	State that Charles Darwin proposed the theory of evolution by natural selection, which is now widely accepted by biologists	**Describe the main principles of evolution through natural selection and explain how Darwin developed his ideas**	Explain how and why scientific models, such as the theory of evolution by natural selection, are developed over time and are based on information gathered by many different scientists

Tier 2 vocabulary	Tier 3 vocabulary
inheritance, offspring, species, variation	acquired characteristic, evolution, natural selection

Lesson plan

Reactivate knowledge

1 Who is known as the father of genetics?

2 Why was Mendel so ahead of his time?

3 Name the female scientist who took the X-ray photos that helped work out the double helix structure of DNA.

Common misconception

Students may say 'an advantageous allele is more likely to survive' instead of 'an individual with an advantageous allele is more likely to survive' when describing natural selection.

Answers: 1 Gregor Mendel *2* No one knew about chromosomes and genes so they could not find an explanation to his models *3* Rosalind Franklin

Trigger interest

How much do you already know? (10 mins)

Write the following terms on the board: evolution, natural selection, Charles Darwin. Ask students to either write down on their mini-whiteboards, or discuss in pairs, what they already know about each term. Then discuss as a class.

Exposition

Lamarck's theory (15 mins)

Review previous knowledge about evolution, then explain that over time scientists have proposed different theories for how it occurs.

Introduce Lamarck as a scientist who proposed a theory of how evolution takes place. Apply this to the example in the Student Book, showing Lamarck's ideas about how the neck of the giraffe got longer.

Ask students to suggest reasons why this theory is no longer recognised as correct in the majority of circumstances.

Main activity

Darwin's theory (25 mins)

Students should have already covered natural selection in B14.2. They should use this knowledge to complete a storyboard for an animation or film on the work of Charles Darwin. As a class, use information from the Student Book to agree on the different and most important scenes that need to be included. As well as drawing the scene (or describing it in words), students should write a narration that explains what it is showing and why this was important in the development of Darwin's theory of natural selection. Review the activity with the class and ask them to summarise similarities and differences between Darwin and Lamarck's theory on their mini-whiteboards.

As an extension, ask students to consider why theories such as evolution develop over time.

Review and practice

Natural selection (5 mins)

Students use the Interactive: *Natural selection* to put the stages of natural selection into the correct order. Students can complete this on their mini-whiteboards or it can be done as a class by *cold-calling*.

B15.3 Accepting Darwin's ideas

AQA GCSE Biology specification points

4.6.3.1 Theory of evolution

Smart Curriculum Narrative

Variation and evolution

Theme/subtheme

Variation and evolution/Evolution, extinction, and the fossil record

Disciplinary knowledge

- WS 1.1, 1.3

Practical

Darwin's finches

	Learning outcomes		
Learning objectives	**Developing Aiming for 3**	**Secure Aiming for 5**	**Extending Aiming for 8**
Explain how finches on different islands evolved different-shaped beaks by natural selection	Describe how finches have different-shaped beaks so they can eat different foods	**Explain how finches on different islands evolved different-shaped beaks by natural selection**	Explain how the finch species on the different Galapagos Islands are evidence for evolution by natural selection
Describe, using an example, how modern knowledge of genetics provides a mechanism for Darwin's theory of evolution	State that we now know that useful characteristics are passed from parents to their offspring through the alleles they inherit, and this supports Darwin's model of evolution through natural selection	**Describe, using an example, how modern knowledge of genetics provides a mechanism for Darwin's theory of evolution**	Apply modern understanding of genetics to suggest how Darwin's evidence from pigeon breeding would support his ideas on natural selection and evolution
Describe several reasons why many people did not accept Darwin's theory when it was first published	State one reason why most people did not accept Darwin's theory when it was first published	**Describe several reasons why many people did not accept Darwin's theory when it was first published**	Discuss why Darwin was conflicted over publishing his theory
Explain why it was important that Darwin collected a variety of evidence	Give one piece of evidence that support's Darwin's theory of natural selection	**Explain why it was important that Darwin collected a variety of evidence**	Explain why scientists eventually accepted Darwin's theory

Tier 2 vocabulary	Tier 3 vocabulary
chromosome, evidence, gene, inheritance, species, theory, variation	artificial selection, breeding experiment, evolution, natural selection

Lesson plan

Reactivate knowledge

1 Who had a theory of evolution based on 'the fountain of life'?

2 What was the name of the ship which carried Darwin on his epic voyage?

3 Where did Darwin make the observations about different finches and tortoises that helped shape his theory of natural selection?

Common misconception

Students may say 'an advantageous allele is more likely to survive' instead of 'an individual with an advantageous allele is more likely to survive' when describing natural selection.

Answers: 1 Jean-Baptiste Lamarck *2 The Beagle 3* The Galapagos Islands

Trigger interest

Darwin cartoon (5 mins)

Search online and show students the famous cartoon of Charles Darwin with an ape's body.

Ask students to suggest what it shows, and why it was published.

Exposition

Accepting a theory (10 mins)

Tell students that there is a theory that taller people are more intelligent than shorter people (this is based on a real study and news articles can be found online).

Discuss with the class what they think about this theory, and what evidence they would need in order to accept the theory. Link this to how people reacted when Darwin published his theory of evolution by natural selection.

Then ask students to consider reasons why Darwin's theory gradually came to be accepted by the scientific community.

Main activity

Darwin's finches (35 mins)

Show students the location of the Galapagos Islands on a map, where Charles Darwin visited on his journey.

Discuss how the different islands resulted in the finches being separated from each other and from the mainland. Students studying GCSE geography may be familiar with this.

Then, show students images of the different finch species that live on each island (e.g., Figure 1 from the Student Book). Ask students to suggest features that the finches from different islands share, and features that are different.

Students then carry out the practical activity, where they try to pick up beads using different-shaped 'beaks'. After the practical, students complete the accompanying questions and review the answers as a class.

Review and practice

Why was Darwin's theory not accepted by some people? (5 mins)

Students use the Interactive: *Why was Darwin's theory not accepted?* to sort a series of statements according to whether or not they were reasons why Darwin's theory of evolution was not immediately accepted.

Students then sort evidence according to whether it supports Darwin's theory of evolution or not.

If students are struggling with this, encourage them to work in groups and use the Student Book to help them.

B15.4 Evolution and speciation

AQA GCSE Biology specification points

4.6.3.2 Speciation

Smart Curriculum Narrative

Variation and evolution

Theme/subtheme

Variation and evolution/Evolution, extinction, and the fossil record

Disciplinary knowledge

- WS 1.1

Activity

The process of speciation

	Learning outcomes		
Learning objectives	**Developing** **Aiming for 3**	**Secure** **Aiming for 5**	**Extending** **Aiming for 8**
Describe the steps in the process of speciation	Define speciation as the point at which two populations become so different that successful interbreeding is no longer possible	**Describe the steps in the process of speciation**	Explain the role of natural selection in the evolution of new species
Explain why Wallace's work prompted Darwin to publish *The Origin of Species*	Give an example of an important discovery by Wallace	**Explain why Wallace's work prompted Darwin to publish *The Origin of Species***	Discuss how Wallace's work has informed our current ideas on speciation
Explain why the species living on islands such as Madagascar, Borneo, the Galapagos Islands, and Australia often share some similarities with species found elsewhere, but may also be unique	State that geographical isolation and environmental isolation are the two main factors that drive speciation	**Explain why the species living on islands such as Madagascar, Borneo, the Galapagos Islands, and Australia often share some similarities with species found elsewhere, but may also be unique**	Explain the relationship between the length of isolation and the number of unique species that evolve

Tier 2 vocabulary	Tier 3 vocabulary
fertile offspring, population, species	allele, environmental isolation, evolution, genetic variation, geographical isolation, interbreed, natural selection, speciation, *The Origin of Species*, warning colouration

Lesson plan

Reactivate knowledge

1 Darwin spent years investigating and experimenting to confirm his ideas. Name one species he studied.

2 What was the title of Darwin's famous book?

3 Why did some people object to Darwin's ideas?

Common misconception

Students may say 'an advantageous allele is more likely to survive' instead of 'an individual with an advantageous allele is more likely to survive' when describing natural selection.

Answers: 1 Pigeons, finches, barnacles *2 The Origin of Species 3* Challenged by religious beliefs, there was not enough evidence, and no other studies to explain chromosomes and genes

Trigger interest

Madagascar (10 mins)

Show the class images of animals that are only found in Madagascar. These could include the aye-aye, various other lemurs (e.g., bamboo, black-and-white ruffed, ring-tailed, and mouse lemur), Verreaux's sifaka, fossa, tomato frog, giant leaf-tailed gecko, satanic leaf-tailed gecko, giraffe weevil, and miniature chameleon *Brookesia micra*. Ask students to make notes on the features of these animals and on similarities and differences between these and other related animals not found in Madagascar.

Exposition

Wallace and Darwin (10 mins)

Discuss Alfred Russel Wallace's work with the class. Have them summarise the most vital information as five bullet points about who Wallace was, and his work. This can be based on discussions and/or using the Student Book.

In pairs, have students write ideas on their mini-whiteboards to suggest and explain how Wallace's work prompted Darwin to publish *The Origin of Species*.

Main activity

The process of speciation (30 mins)

Give students a couple of minutes to discuss how the stages in the process of natural selection lead to evolution. Then 'bounce' around the class, asking different students to recall the stages in order.

Describe speciation as the process by which new species evolve. Inform students that this happens when one population of a species becomes isolated, so goes through natural selection and evolves to the conditions in the new environment. Address the misconception here.

Students then complete the Activity: *Speciation*. Provide students with the diagrams from this activity to show the process of speciation. Students should put the diagrams in the correct order and explain what is happening in each one.

Students use their knowledge to complete a cloze paragraph to explain how lemurs evolved through the process of speciation.

Review and practice

How organisms change through time (5 mins)

Students use this interactive to answer a series of true or false questions about the information covered in the lesson.

B15.5 Evidence for evolution

AQA GCSE Biology specification points

4.6.3.4 Evidence for evolution

4.6.3.5 Fossils

Smart Curriculum Narrative

Variation and evolution

Theme/subtheme

Variation and evolution/Evolution, extinction, and the fossil record

Disciplinary knowledge

- WS 1.3
- MS 2c, 4a

Activity

History of the Earth timeline

	Learning outcomes		
Learning objectives	**Developing Aiming for 3**	**Secure Aiming for 5**	**Extending Aiming for 8**
Describe different ways in which fossils are formed	Describe what a fossil is and give an example	**Describe different ways in which fossils are formed**	Discuss the particular value of ice and peat fossils to scientists
Describe how fossils provide evidence for evolution by natural selection	State that fossils are evidence for evolution by natural selection	**Describe how fossils provide evidence for evolution by natural selection**	Explain why early fossils of soft-bodied organisms are so rare and yet so important to evolutionary science
Give reasons why the fossil record is not complete	State that there is little fossil record from the earliest days of life on the Earth and that this lack of evidence limits our understanding of how the earliest forms of life evolved	**Give reasons why the fossil record is not complete**	Evaluate the use of fossils as evidence for evolution by natural selection and how life first formed

Tier 2 vocabulary	Tier 3 vocabulary
bacteria, carnivore, DNA, erosion, evidence, fossil, fossil record, herbivore, mineral, oxygen, temperature, theory	cast fossil, decay, evolution, geological activity, ice fossil, mould fossil, natural selection, peat, rock fossil, rootlet trace

Lesson plan

Reactivate knowledge

1 What was the name of the scientist who also came up with the idea of natural selection?

2 What work is Wallace best remembered for?

3 Name the two ways that populations become isolated.

Answers: 1 Alfred Russel Wallace *2* Warning colouration and the role of geographical isolation in speciation *3* Geographical and environmental isolation

Common misconception

Students may think that all organisms form a fossil when they die or that all fossils are made from bone or shell.

Trigger interest

Frozen mammoth (5 mins)

Show the class an image of a mammoth preserved in ice (such as Figure 1 in Student Book). Discuss what scientists could learn from studying the mammoth.

Exposition

Fossil evidence (15 mins)

Start by assessing what students already know. Ask students to write a definition for a fossil on their mini-whiteboards and to suggest how they form.

Discuss students' answers and then have them write an agreed definition into their books, and also a description of how fossils form. They could use the Student Book to help them, or base their descriptions on your class discussion.

Discuss what can be learnt about evolution from studying fossils. Ask students to read through the section 'An incomplete record' in the Student Book. They should then write down three reasons why the fossil record is not complete.

Main activity

History of the Earth timeline (25 mins)

Provide students with the Activity: *History of the Earth timeline*, which contains some major events in the evolution of life on the Earth. Ask students to use the internet to research when these events happened and to place them into the correct order. The correct answer is:

- the Earth formed (4.6 billion years ago)
- oceans and continents start to form (4.4 billion years ago)
- first evidence of life (3.5 billion years ago)
- atmospheric oxygen forms (2.4 billion years ago)
- evolution of eukaryotic life (1.8 billion years ago)
- plants move on to land (450 million years ago)
- animals move on to land (430 million years ago)
- Permo-Triassic mass extinction (251 million years ago)
- evolution of mammals (195 million years ago)
- extinction of the dinosaurs (65 million years ago)
- evolution of humans (200 000 years ago)
- extinction of the woolly mammoth (10 000 years ago).

Some students may struggle with the maths skills required here so inform them that a billion is bigger than a million, which is bigger than a thousand.

Students with higher prior attainment should then express the timescales in standard form.

Review and practice

Fossil evaluation (10 mins)

Ask students to write down why fossils are evidence for evolution by natural selection and then share with the class.

Students should also evaluate the use of fossils as evidence for evolution and how life started on the Earth, making sure they consider the limitations of the fossil record.

B15.6 Fossils and extinction

AQA GCSE Biology specification points

4.6.3.5 Fossils

4.6.3.6 Extinction

Smart Curriculum Narrative

Variation and evolution

Theme/subtheme

Variation and evolution/Evolution, extinction, and the fossil record

Disciplinary knowledge

- WS 1.1
- MS 2c, 4a

Activity

Fossils and extinction

	Learning outcomes		
Learning objectives	**Developing** **Aiming for 3**	**Secure** **Aiming for 5**	**Extending** **Aiming for 8**
Explain why complete fossil records are rare and extract information from the evolutionary tree of the horse to demonstrate their value as evidence of evolution by natural selection	State that relatively complete fossil records show how much or how little organisms have changed over time	**Explain why complete fossil records are rare and extract information from the evolutionary tree of the horse to demonstrate their value as evidence of evolution by natural selection**	Suggest how relatively complete fossil records help reconstruct how the ecology, climate, and environment has changed over time, using the horse as an example
Describe the evolution of the modern horse based on diagrams from the available fossil record	Order fossil diagrams to show the evolution of the horse	**Describe the evolution of the modern horse based on diagrams from the available fossil record**	Discuss what can be learnt from the relatively complete fossil record of sharks
Describe change in the environment as the key cause of extinction and give examples of the types of environmental changes that may lead to a single extinction or many species becoming extinct	Define extinction	**Describe change in the environment as the key cause of extinction and give examples of the types of environmental changes that may lead to a single extinction or many species becoming extinct**	Describe the importance of extinction in the process of evolution by natural selection
Describe how other organisms can cause an animal or plant to become extinct	State that the great majority of species that have evolved have also become extinct	**Describe how other organisms can cause an animal or plant to become extinct**	Evaluate the impact of the human organism on extinction rates

Tier 2 vocabulary	Tier 3 vocabulary
extinction, fossil, fossil record, predator, prey, species	competition, evolution, fossil evidence, natural selection

Lesson plan

Reactivate knowledge

Common misconception

Students may think that all organisms form a fossil when they die or that all fossils are made from bone or shell.

1 What are fossils?

2 There are several different types of fossils. Name two of them.

3 Why is the fossil record so incomplete?

Answers: 1 Remains of organisms from millions of years ago preserved in some way *2* Any from: Ice fossils/peat rock fossils/trace fossils *3* Soft-bodied animals leave little trace/fossils destroyed/fossils still not found buried in rocks or beneath sea/conditions for fossilisation rare

Trigger interest

Extinction (10 mins)

Ask the class to work in small groups to discuss and list the ways in which organisms become extinct. Go through their ideas and produce the following list – asteroid impact, climate change, competition with other species, new diseases, new predators.

Highlight the causes associated with other organisms and tell the class that this is what they will be focusing on in this lesson.

If required, use images as visual clues for the methods of extinction.

Exposition

Fossil review (10 mins)

Students complete a mini-whiteboard quiz to review what they can remember about fossils and extinction. Ask students to write the following on their boards: the definition of a fossil; reasons why complete fossil records are rare; the definition of extinction. Then discuss their answers as a class.

Main activity

Fossils and extinction (25 mins)

Students complete the first part of the Activity: *Fossils and extinction*. They should review the set of pictures showing the evolution of the horse and then put them into the correct sequence and explain what they show. They can then use the Student Book to check their work.

Next, introduce students to the extinct bird, the dodo (a video may be useful here). Discuss what scientists can deduce about the bird using its skeleton. Tell students that the dodo became extinct when humans and their livestock settled for the first time on the island of Mauritius, where the dodos lived. Ask students to work in pairs to complete the second part of the activity to propose several hypotheses for why the dodo became extinct.

Review and practice

Endangered animals (10 mins)

Give groups of students a set of images of different endangered/threatened animals (e.g., polar bear, black rhino, orangutan). Ask the students to discuss the reasons why these animals are endangered/threatened. If students are struggling, provide them with reasons to choose from.

As an extension, students can then interpret the following data below.

As of December 2023, the populations of the following animals in the wild are: 22 000 polar bears, 22 137 black rhinos, 230 000 organgutans, 1864 giant pandas.

B15.7 More about extinction

AQA GCSE Biology specification points

4.6.3.6 Extinction

Smart Curriculum Narrative

Variation and evolution

Theme/subtheme

Variation and evolution/Evolution, extinction, and the fossil record

Activity

Extinction

Learning objectives	Learning outcomes: Developing Aiming for 3	Secure Aiming for 5	Extending Aiming for 8
Explain how environmental change can cause extinctions	State that environmental change and a catastrophic event are two possible causes of mass extinction	**Explain how environmental change can cause extinctions**	Suggest why mass extinctions are important for the evolution of life on the Earth
Suggest the effects of an asteroid, comet, or meteorite strike on the Earth	Describe one theory that explains why the dinosaurs became extinct	**Suggest the effects of an asteroid, comet, or meteorite strike on the Earth**	Link ideas to give a scientific explanation of why an asteroid could have caused the dinosaurs to become extinct
Identify strengths and weaknesses in two different theories for the mass extinction event that destroyed most of the dinosaurs	Describe what a mass extinction is	**Identify strengths and weaknesses in two different theories for the mass extinction event that destroyed most of the dinosaurs**	Evaluate two theories for the mass extinction of the dinosaurs, and explain why scientists cannot be completely sure what caused either mass extinctions such as the dinosaurs or extinctions of species such as the mammoths

Tier 2 vocabulary	Tier 3 vocabulary
asteroid, atmosphere, climate, dinosaur, environmental change, extinction, fossil, fossil record, species, volcanic eruption	catastrophic event, extinction event, geological time, Ice Age, iridium, mass extinction, natural selection

Lesson plan

Reactivate knowledge

Common misconception

Students may be unaware that there are different theories to suggest why the dinosaurs became extinct.

1 What is extinction?

2 Why is extinction important?

3 What are the four main ways in which other organisms may cause the extinction of a species?

Answers: 1 The permanent loss of all members of a species either locally or globally *2* To make space for new species better suited to the environment *3* Predation (new predators)/disease (new pathogens)/competition (new species compete more successfully)/human activities (such as hunting, fishing, habitat destruction, pollution, climate change)

Trigger interest

Human extinction (5 mins)

Ask students to discuss whether they think humans will ever become extinct and, if so, when and how.

Introduce the term mass extinctions and use Figure 2 in the Student Book to show when these have occurred in the past.

Exposition

Environmental change and mass extinction (10 mins)

Show the class a graph that shows how average temperatures on the Earth have varied. Then revisit Figure 2 in the Student Book. In pairs, they should discuss whether they think there is a link between changes in temperature and mass extinctions. As a class, discuss why there is a link.

Remind students that there is a scientific consensus that global temperatures are rising. Ask them to suggest how this could be affecting organisms such as the polar bear.

Main activity

Extinction (35 mins)

Ask students to read through the section 'What destroyed the dinosaurs?' in the Student Book, which outlines two popular theories. Students should use this information and their own research to complete the table in the Activity: *Extinction,* to evaluate the evidence and to identify strengths and weaknesses in each case.

They should then discuss their ideas in small groups, and then as a whole class. Discuss the fact that theories are ideas supported by evidence that become accepted by a general consensus, or majority, and are then often superseded over the years. Explain why scientists are not sure what caused the extinction of the dinosaurs.

To extend, students should then use their knowledge of asteroid strikes, photosynthesis, and energy flow through food chains and webs to produce a concept map linking all of these factors to the mass extinction of the dinosaurs.

Review and practice

What do you think? (5 mins)

Introduce to the class the opinion that, in the long term, mass extinction is beneficial to the development of life. Discuss students' thoughts as a class.

If students are struggling, ask them to consider some benefits that occurred due to the dinosaurs going extinct.

B15.8 Antibiotic resistant bacteria

AQA GCSE Biology specification points

4.6.3.7 Resistant bacteria

Smart Curriculum Narrative

Variation and evolution

Theme/subtheme

Variation and evolution/Evolution, extinction, and the fossil record

Activity

Public information

	Learning outcomes		
Learning objectives	**Developing** **Aiming for 3**	**Secure** **Aiming for 5**	**Extending** **Aiming for 8**
Describe how antibiotic resistant bacteria evolve	Describe what is meant by antibiotic resistant bacteria	**Describe how antibiotic resistant bacteria evolve**	Explain how a fast reproduction rate is linked to the development of antibiotic resistant strains of bacteria
Summarise the reasons why the development of new antibiotics is unlikely to keep up with the emergence of new antibiotic resistant strains of bacteria	State that the development of antibiotic resistant bacteria is evidence for evolution by natural selection	**Summarise the reasons why the development of new antibiotics is unlikely to keep up with the emergence of new antibiotic resistant strains of bacteria**	Discuss the development of antibiotic resistant bacteria as evidence for evolution
Explain how the three main ways of reducing the development of antibiotic resistant bacteria have their effect	State three ways of slowing the development of antibiotic resistant bacteria	**Explain how the three main ways of reducing the development of antibiotic resistant bacteria have their effect**	Discuss the problems of preventing the development of antibiotic resistant bacteria in a global context
Describe the measures put in place in hospitals and care homes to control the spread of MRSA and evaluate their effectiveness	Describe MRSA as a strain of antibiotic resistant bacteria which became common in UK hospitals in the early 21st century	**Describe the measures put in place in hospitals and care homes to control the spread of MRSA and evaluate their effectiveness**	Explain in detail how the likelihood of antibiotic resistance in bacteria increases when people do not complete a course of antibiotics because they feel better

Tier 2 vocabulary	Tier 3 vocabulary
antibiotic, bacteria	antibiotic resistance, antibiotic resistant strain, evolution, methicillin-resistant *Staphylococcus aureus* (MRSA), microorganism, mutation, natural selection, pathogen, virus

Lesson plan

Common misconception

Students may incorrectly use the word 'immune' instead of 'resistant' when describing antibiotic resistant bacteria.

Reactivate knowledge

1 How many major extinction events have there been in the history of life on the Earth?

2 What are the two main theories for the cause of the extinction of the dinosaurs?

3 How could an asteroid strike cause extinction of the dinosaurs?

Answers: 1 Five *2* Massive asteroid impact/global warming causing ice melt and fall in sea temperature *3* Material blasted into atmosphere made everywhere dark, plants struggled to photosynthesise, global winter, not enough food for animal species

Trigger interest

Please wash your hands (5 mins)

Show students images of antibacterial hand gel for public use in hospitals, or hand-washing stations for medical professionals. Ask them why these are important. Introduce the fact that some strains of bacteria are especially dangerous because they cannot be controlled by antibiotics.

Exposition

Resistance bacteria (20 mins)

As a class, use the fact that bacteria reproduce by binary fission to discuss why bacteria evolve much more quickly than animals.

Summarise reasons why new antibiotic development lags behind antibiotic-resistant bacteria emergence. Ask students to discuss what factors contribute to the development of antibiotic-resistant bacteria, why it is challenging to develop new antibiotics, and how the process of antibiotic discovery and development works.

Factors contributing to antibiotic resistance in bacteria include: overuse and misuse of antibiotics leading to selective pressure, incomplete courses of antibiotics allowing surviving bacteria to develop resistance.

Developing new antibiotics is challenging due to: decreasing effectiveness of existing antibiotics against resistant strains, high costs and time required for research and development, limited discovery of new antibiotic compounds with different mechanisms of action.

The process of antibiotic discovery and development typically involves: testing compounds for efficacy, preclinical and clinical trials to assess safety and efficacy in humans.

Obstacles in developing new antibiotics include: high failure rates in clinical trials due to toxicity or lack of efficacy, emergence of resistance during clinical testing, limited financial incentives for pharmaceutical companies to invest in antibiotic research.

Main activity

Public information (25 mins)

In pairs, students list the three most important ways to reduce the rate of developing antibiotic resistance. They then write an information sheet for the general public outlining important facts about antibiotic resistant bacteria.

Review and practice

Order the stages (5 mins)

Students complete the interactive to arrange development of antibiotic resistant bacteria in order.

B15.9 Classification

AQA GCSE Biology specification points

4.6.4 Classification of living organisms

Smart Curriculum Narrative

Variation and evolution

Theme/subtheme

Variation and evolution/Classification systems

Disciplinary knowledge

- WS 1.1

Activity

Classifying animals

	Learning outcomes		
Learning objectives	**Developing Aiming for 3**	**Secure Aiming for 5**	**Extending Aiming for 8**
Describe the importance of classifying organisms	Describe classification	**Describe the importance of classifying organisms**	Explain how the concept of biodiversity depends on an effective classification system
Describe the classification system developed by Carl Linnaeus, to include the order of the taxonomic groups and explain why a binomial naming system is useful	State that the classification system used today is based on a system developed by Carl Linnaeus in the 18th century	**Describe the classification system developed by Carl Linnaeus, to include the order of the taxonomic groups and explain why a binomial naming system is useful**	Explain the difference between the living world as it is known today, and when Linnaeus developed his classification system
Describe how developments in biology have impacted classification systems	State that developments in biology have changed the way organisms are classified	**Describe how developments in biology have impacted classification systems**	Explain the impact of developments in biology on the complexity of the modern classification system

Tier 2 vocabulary	Tier 3 vocabulary
animal, biodiversity, cellulose cell wall, chloroplast, fertile offspring, fungus, microscope, plant, species	archaebacteria, *class*, classification, *domain*, eubacteria, *family*, genome, genus, *kingdom*, Linnaean system, *order*, phylum, prokaryote, protist

Lesson plan

Reactivate knowledge

1 What change happens to make bacteria antibiotic resistant?

2 Suggest one way to prevent more resistant strains of bacteria appearing.

3 How can hospitals prevent the spread of antibiotic resistant bacteria?

Common misconception

Students might think that the categories in the Linnaean system are rigid and unchanging. They may not realise that as scientific knowledge advances, classifications can be updated and revised to better reflect our understanding of evolutionary relationships.

Answers: 1 Mutation in genetic material *2* Don't overuse antibiotics *3* Treat specific bacteria with specific antibiotics

Trigger interest

Classify (10 mins)

Give groups of students a range of images of different animal species. Ask students to classify the animals into groups based on their shared characteristics. Allow groups of students to share their method of classification with the class. Use this as a basis for a discussion about why it is a good idea for everyone to use the same system.

Exposition

Linnaeus (15 mins)

Introduce students to the work of Carl Linnaeus. Talk through what classification is and explain that it makes it easier to share scientific data if everyone uses the same classification system. Modern technology has allowed for organisms to be named more accurately against the classification system. The domain system is widely accepted which includes three main domains: Archaea, Eukaryota, and Bacteria.

Show students the hierarchy of taxa. Provide students with the following mnemonic: **K**ing **P**rawn **C**urry **O**r **F**ried **G**reasy **S**ausage. Challenge students to come up with their own mnemonics.

It is important to explain to students that whilst Linnaeus' classification system is widely used and has brought many advantages, his original descriptions to classify humans of different races are not scientific and are inappropriate and offensive.

Main activity

Classifying animals (20 mins)

Provide pairs of students with cards that contain the Linnaean classification for some different animal species (provided within the Activity: *Classifying animals*).

For example, for a lion there would be seven cards, stating:

- Kingdom – Animalia (animals)
- Phylum – Chordata (vertebrates)
- Class – Mammalia (mammals)
- Order – Carnivora (meat eaters)
- Family – Felidae (all cats)
- Genus – *Panthera* (great cats)
- Species – leo.

Students should correctly order the cards and suggest the common name of the animal. Provide different pairs with different animals, each pair with several animals, or have all pairs doing the same animal. Go through the correct answers and introduce the term binomial system, including reasons why it is useful to scientists.

Students then complete the questions and review the answers as a class.

Review and practice

Classifying humans (10 mins)

Students use the interactive to put the taxonomic groups in the correct order. They then put the classification of humans into the correct order.

B15.10 New systems of classification

AQA GCSE Biology specification points

4.6.4 Classification of living organisms

Smart Curriculum Narrative

Variation and evolution

Theme/subtheme

Variation and evolution/Classification systems

Disciplinary knowledge

- WS 1.1, 1.2

Activity

Evolutionary trees

	Learning outcomes		
Learning objectives	**Developing** **Aiming for 3**	**Secure** **Aiming for 5**	**Extending** **Aiming for 8**
Describe how organisms are divided in the three-domain system developed by Carl Woese	Name the three domains and six kingdoms	**Describe how organisms are divided in the three-domain system developed by Carl Woese**	Compare and contrast the Linnaean system with the three-domain system
Describe why the three-domain system was proposed, including the advances in evidence available to scientists to help classify organisms	Recognise that ideas about classification have changed over time	**Describe why the three-domain system was proposed, including the advances in evidence available to scientists to help classify organisms**	Discuss how and why ideas about classification have developed over time
Explain why evolutionary trees are built up using DNA evidence to show evolutionary relationships between living organisms but fossil evidence is used to build evolutionary relationships between extinct animals, and draw conclusions about the relationships between species from a simple evolutionary tree	State that evolutionary trees are developed to show the relationships between organisms and recognise the connections between members of an evolutionary tree	**Explain why evolutionary trees are built up using DNA evidence to show evolutionary relationships between living organisms but fossil evidence is used to build evolutionary relationships between extinct animals, and draw conclusions about the relationships between species from a simple evolutionary tree**	Use more complex evolutionary trees and draw conclusions about the evolutionary relationships of organisms shown

Tier 2 vocabulary	Tier 3 vocabulary
animal, bacteria, biochemistry, biodiversity, DNA, fossil, fungus, *nucleus*, plant, species	archaea, archaebacteria, biochemical analysis, classification, cyanobacteria, *domain*, eubacteria, eukaryota, eukaryote, evolutionary relationship, evolutionary tree, extremophile, *kingdom*, Linnaean system, microscopy, protist, ribosome, three-domain system

Lesson plan

Reactivate knowledge

1 Who developed the first systematic classification system of the living world?

2 What is the smallest group of the main classification system?

3 What is the largest group used in modern classification systems called, and how many are there?

Common misconception

Students might think that the categories in the Linnaean system are rigid and unchanging. They may not realise that as scientific knowledge advances, classifications can be updated and revised to better reflect our understanding of evolutionary relationships.

Answers: 1 Carl Linnaeus *2* Species *3* Domain, three

Trigger interest

The tree of life (10 mins)

Project or print out copies of an example of a 'tree of life' found online. Let pairs study the diagram and discuss what they think it shows before opening it up as a class discussion.

Ask students to write two key things the diagram shows on their mini-whiteboards and share these ideas with the class.

Exposition

The three-domain system (15 mins)

Show students images of different dog breeds (e.g., labrador, poodle, Yorkshire terrier, etc.) and ask if these are all the same species? The answer is yes. Then ask students why you might think they are not the same species. Students could consider that it is because physically they look very different. Inform students that this is an example of why classification has advanced to examine similarities in DNA and RNA base sequences, and this resulted in Carl Woese updating the classification system to create the three-domain system.

Show the students Table 1 in the Student Book, which outlines Carl Woese's three-domain system. Students can copy this table, showing the different classifications.

As an extension, students should compare the three-domain system to the Linnaean system.

Main activity

Evolutionary trees (20 mins)

Students apply their knowledge of what evolutionary trees show to answer the questions in the Activity: *Evolutionary trees*. Review the answers with the class for students to self-mark.

Review and practice

Which kingdom? (10 mins)

Students use the Interactive: *Which kingdom?* to classify a list of organisms as archaebacteria, eubacteria, protista, fungi, plants, or animals.

B15 Evolution and classification

Introduction to checkpoint intervention

The checkpoint intervention provides suggestions for a lesson to follow up the *B15 Evolution and classification* assessment. Depending on the outcomes of the assessment, these suggestions can be used to move a class towards secure, to target specific learning outcomes, or to consolidate knowledge for students achieving secure outcomes. Use students' outcomes from the checkpoint assessment to address general misconceptions from the content covered. Be prepared to re-cover content as required; three key concepts from the chapter are covered in more detail below.

Checkpoint secure learning outcomes

- Describe the key elements of Mendel's work that made it so unusual for its time.
- Describe the roles played by Franklin, Wilkins, Watson and Crick in the discovery of the double helix structure of DNA.
- Describe how fossils are evidence for evolution by natural selection.
- Describe the classification system developed by Carl Linnaeus, to include the order of the taxonomic groups.
- Explain why a binomial naming system is useful.
- Explain why evolutionary trees are built up using DNA evidence to show evolutionary relationships between living organisms but fossil evidence is used to build evolutionary relationships between extinct animals, and draw conclusions about the relationships between species from a simple evolutionary tree.

Student reflection

Encourage students to reflect on whether there were any checkpoint questions they found difficult or straightforward as well as their preparation for the checkpoint assessment, including how they revised, what they revised, the time spent on revision, and what areas they could improve upon. For more strategies, see the Metacognition spread.

Intervention activity: Describe the classification system developed by Carl Linnaeus, to include the order of the taxonomic groups.

Ask students to use the table to answer the questions below.

Taxon	Human
Kingdom	Animalia
	Chordata
Order	Mammalia
	Primates
Family	Hominidae
	Homo
	sapiens

1. Complete the missing taxa in the table for the Linnaeus classification system.
2. What is the binomial name for a human?
3. Complete the table below for the domestic cat.

Taxon	Domestic cat
	Animalia
	Chordata
Order	Mammalia
	Carnivora
	Felidae
Genus	*Felis*
Species	*catus*

Intervention activity: Explain why a binomial naming system is useful.

Ask students to select the statements below that correctly explain why a binomial system is useful.

- A binomial system can be used to indicate how closely related different species are. (true)
- A binomial system is used universally, so it means scientists across the world can share data, without any confusing use of different names from different languages. (true)
- A binomial system enables scientists to identify if an organism is a plant or an animal. (false)

Intervention activity: Explain why evolutionary trees are built up using DNA evidence to show evolutionary relationships between living organisms but fossil evidence is used to build evolutionary relationships between extinct animals, and draw conclusions about the relationships between species from a simple evolutionary tree.

Show students Figure 2 from B15.10 of the Student Book.

Ask students to use the evolutionary tree to determine:

- which species 'other bears' are most closely related to
- which family weasels are most closely related to
- what every branch point in an evolutionary tree represents.

5 Ecology

Introduction to unit

This unit will look at how organisms are adapted to their environment and how they compete for mates and resources. It will explore how ecosystems are organised, how organisms play their role within the ecosystem, and how scientists monitor the population and spread of organisms. Students will also examine the impact of the human population on the planet and measures being taken to reduce the harm caused.

Assessment

Suggested minimum guided learning hours: 24 hours

Exam: Paper 2

Prerequisite knowledge from KS3 (chapter references from *Oxford Smart* Activate)

B2.3.3 Ecosystems

- A population is a group of a species in a habitat at a particular time.
- A community is composed of all the species in a habitat at one time.
- Changes to a population have an effect on other organisms in the community.
- Organisms are affected by factors that are living and non-living.

B2.3.4 Competition and adaptation

- Individuals will compete for resources.
- Adaptations help an organism survive in its habitat.

B3.4.4 Apply: Investigating competition

- The different types of average – mode, mean, median.
- How to add scales to axes on graphs and plot data using different formats.

B3.4.6 Reach: Maintaining biodiversity

- Biodiversity is the variety of all the different species of organisms.

B2.3.1 Food chains and webs

- Food chains show the flow of energy from one organism to another.
- Food webs are made of multiple food chains.

B2.3.2 Disruption to food chains and webs

- Humans are destroying habitats for land for building or agriculture.

AQA GCSE Biology specification links

4.7 Ecology

- 4.7.1 Adaptations, interdependence, and competition
- 4.7.2 Organisation of an ecosystem
- 4.7.3 Biodiversity and the effect of human interaction on ecosystems
- 4.7.4 Trophic levels in an ecosystem (biology only)

Identity, awe, and wonder

Feeding the growing human population is a worldwide issue and sustainable methods need to be found to prevent more species becoming extinct. Ecologists provide expertise to policymakers to inform environmental policies and regulations. They use new methods of monitoring organisms to see where and when they move about in their habitat.

Required practicals

AQA required practicals	Practical skills	Topic
Required practical 9: Field investigations Measure the population size of a common species in a habitat. Use sampling techniques to investigate the effect of a factor on the distribution of this species.	AT 1 – use appropriate apparatus to record length and area. AT 3 – use transect lines and quadrats to measure distribution of a species. AT 4 – safe and ethical use of organisms and response to a factor in the environment. AT 6 – application of appropriate sampling techniques to investigate the distribution and abundance of organisms in an ecosystem via direct use in the field. AT 8 – use of appropriate techniques in more complex contexts including continuous sampling in an investigation.	B16.3
Required practical 10: Decay Investigate the effect of temperature on the rate of decay of fresh milk by measuring pH change.	AT 1 – use appropriate apparatus to record temperature and pH. AT 3 – the use of appropriate apparatus to measure anaerobic decay. AT 4 – safe use of microorganisms. AT 5 – measurement of rate of decay by pH change.	B17.4

Working scientifically

There is high relevancy between the topics covered in this unit and a student's everyday life. This includes the impacts of human activity resulting in climate change and leading to extreme weather changes. This will be a good chance for students to practice their science communication skills and learn to present scientific data to a range of different audiences.

Students should also be able to monitor the distribution of plant species and organisms through a series of practical experiments by recording their observations and measurements using a range of apparatus and methods such as a quadrat or transect line.

Maths skills

In this unit, students will develop their numerical skills including being able to plot two variables on a graph from data they have collected and processed. Students should be able to choose a graph most suitable for the data (bar charts, scatter graphs, histograms) and handle data by calculating a mean, median, or mode.

Literacy links

There are a number of activities that focus on developing literacy skills in this unit. These include: B16.6 (WebQuest), B16.7, B18.6, and B18.7. Any of the practical investigations could be used to develop scientific and instructional writing skills such using a step-by-step method in B18.3.

B16 Adaptations, interdependence and competition

B16.1 The importance of communities

AQA GCSE Biology specification points

4.7.1.1 Communities

Smart Curriculum Narrative

Ecosystems and environment

Theme/subtheme

Ecosystems and environment/Ecosystems

Disciplinary knowledge

- WS 2.6
- MS 2c, 4a

Activity

A country garden

	Learning outcomes		
Learning objectives	**Developing Aiming for 3**	**Secure Aiming for 5**	**Extending Aiming for 8**
Define the terms habitat, abiotic factor, and biotic factor	Define the terms ecosystem, population, and community	**Define the terms habitat, abiotic factor, and biotic factor**	Explain the importance of photosynthesis as an energy source in most communities
Describe a range of examples of how different species may interact within a community	List some resources that living things need	**Describe a range of examples of how different species may interact within a community**	Explain why it is hard to replace a stable community if it is lost
Describe the importance of stable communities	Describe a stable community and give an example	**Describe the importance of stable communities**	Suggest the impact of removing a population from a stable community

Tier 2 vocabulary	Tier 3 vocabulary
biodiversity, *community*, competition, ecosystem, habitat, microorganism, pollination, species	abiotic factor, biotic factor, interdependence, stable community

Lesson plan

Reactivate knowledge

1 What is interdependence?

2 What is an ecosystem?

3 What is a community?

Common misconception

Students often mix up key terms such as population, community, and ecosystem. Make sure they are given the correct definitions and keep revisiting them until the class is confident.

Answers: 1 The way in which living organisms depend on each other to survive, grow, and reproduce 2 A particular area and the animals and plants found there 3 The organisms in an ecosystem

Trigger interest

Ecological words (10 mins)

Ask students to organise the following key terms from the largest to the smallest: population, ecosystem, community (correct answer: ecosystem, community, population).

Then, students use the Interactive: *Ecological Words* to match key words (e.g., community, population, habitat, ecosystem, distribution, abiotic factor, biotic factor) with their definitions.

Exposition

Food webs and interdependence (10 mins)

Show students a slide containing a food web.

Then ask them to answer the following questions about it on their mini-whiteboards:

- How many plant species are in this food web?
- How do plants enable energy to enter the food web?

Ask the students other questions about what might happen to the population of one species when the population of another changes. Ask them to explain their answers.

From this, lead on to a discussion of what is meant by a stable community and interdependence.

Main activity

A country garden (25 mins)

Assign each student a biotic or abiotic factor found in a garden. Biotic factors could include, for example: rose, ladybird, bee, goldfish, hedgehog, algae, grass, or soil bacteria. Abiotic factors could include, for example: water, nutrients, temperature sunlight, air, soil, or wind. Ask students to think about their own role in the ecosystem and list their ideas.

Then, put students into groups of four and work out how each biotic factor in their group is dependent on all the other factors, and how they work together to create a stable community. For example, the bee is dependent on the soil because it provides plants with water and minerals to grow flowers, which provide nectar for the bee. As a class, talk through some of the examples of interdependence that were discussed.

Review and practice

Concept mapping (10 mins)

Give students the key words used in the interactive at the start of the lesson. Ask them to link the words together to create a concept map (a spider diagram with words/sentences linking the boxes) to outline what they learnt in the lesson. This should give a good idea of what the class has understood.

B16.2 Organisms in their environment

AQA GCSE Biology specification points

4.7.1.2 Abiotic factors

4.7.1.3 Biotic factors

Smart Curriculum Narrative

Ecosystems and environment

Theme/subtheme

Ecosystems and environment/Ecosystems

Disciplinary knowledge

- WS 1.2
- MS 2c, 4a

Activity

Understanding abiotic and biotic factors

	Learning outcomes		
Learning objectives	**Developing Aiming for 3**	**Secure Aiming for 5**	**Extending Aiming for 8**
Describe how a range of abiotic factors might affect communities	Identify environmental factors as biotic or abiotic	**Describe how a range of abiotic factors might affect communities**	Explain how the impact of abiotic factors on different populations might impact the community in an ecosystem
Describe how a range of biotic factors might affect communities	Name one biotic factor that might affect an animal population and one biotic factor that would affect a plant population	**Describe how a range of biotic factors might affect communities**	Explain how competition within a species and between species might affect populations of organisms

Tier 2 vocabulary	Tier 3 vocabulary
adaptation, carnivore, *community*, competition, decay, ecosystem, herbivore, mineral ion, moisture level, nitrate ion, pH, population, predator, species, temperature	abiotic factor, biotic factor, distribution, native species, outcompete, parasite, pathogen, *resistance*, transpiration

Lesson plan

Reactivate knowledge

1 What is a stable community?

2 Name two things that organisms are dependent on each other for in a community.

3 What type of organism is all the other organisms in the community dependent on?

Answers: 1 A community where the balance of organisms remains relatively constant 2 Any from: food, shelter, pollination, seed dispersal 3 Plants/producers

Common misconception

Students may think 'food' is an abiotic factor, whereas it is actually a biotic factor because the plants and animals that make up the food are living organisms.

Trigger interest

What will be different? (10 mins)

Show students images of two highly contrasting environments, such as rainforest and tundra, or desert and savannah. Ask the students to work in pairs to compare the two environments in terms of their abiotic factors, and to suggest biotic factors for each environment.

Exposition

Abiotic and biotic factors (20 mins)

Ask students to draw a line splitting their mini-whiteboards in half. They should label one half 'abiotic' and the other half 'biotic'. Students should then list all the abiotic and biotic factors that they can think of in the correct place on their boards.

Next, ask students to consider a local ecosystem, such as a local woodland or park. Students should work through the abiotic and biotic factors given in the Student Book, and create a table that explains what effect each factor may have on the community of the chosen ecosystem. They should give specific examples where relevant.

Main activity

Understanding abiotic and biotic factors (20 mins)

Ask students to read the list of factors that can affect a community. They should use this list to help them answer the questions within the activity.

Discuss the answers as a class. Students can then self-mark or peer-mark the work.

Review and practice

Sorting factors (5 mins)

Students use the interactive to classify a list of factors as abiotic factors or biotic factors.

As an extension, ask students to work in pairs to discuss how each of the factors could affect a population of native red squirrels.

B16.3 Distribution and abundance

AQA GCSE Biology specification points

4.7.2.1 Levels of organisation

Smart Curriculum Narrative

Ecosystems and environment

Theme/subtheme

Biodiversity and human interactions with the environment/Studying ecosystems

Disciplinary knowledge

- WS 1.2, 2.1, 2.2, 2.3
- MS 1d, 2b, 2d, 2f, 3a, 4a, 4c

Required Practical 9

Field investigations

	Learning outcomes		
Learning objectives	**Developing Aiming for 3**	**Secure Aiming for 5**	**Extending Aiming for 8**
Explain why scientists measure the abundance and distribution of organisms in an ecosystem	Define the terms abundance and distribution of an organism	**Explain why scientists measure the abundance and distribution of organisms in an ecosystem**	Suggest ways in which measurements of the abundance and distribution of organisms in an ecosystem might be used
Explain how a quadrat and a transect are used to estimate population sizes and distributions	Describe a quadrat and a transect	**Explain how a quadrat and a transect are used to estimate population sizes and distributions**	Discuss what factors determine the size of the quadrat used
Calculate the range, mean, median, and mode of a data set to calculate the population of an organism in a given area	Calculate the mean of a set of results	**Calculate the range, mean, median, and mode of a data set to calculate the population of an organism in a given area**	Compare and contrast the use of quadrats and transects in measuring the abundance and distribution of organisms in a habitat

Tier 2 vocabulary	Tier 3 vocabulary
community, ecosystem, habitat, *mean*, median, *mode*, population, *range*, species	abiotic factor, abundance, biotic factor, distribution, line transect, quadrat, quantitative sampling, random sampling, sample area, sample size, sampling, transect

Lesson plan

Reactivate knowledge

1 What is an abiotic factor in an ecosystem?

2 Give three examples of abiotic factors that might affect an ecosystem.

3 Name four biotic factors that might affect a community.

Common misconception

Students may confuse the difference between mean, median, and mode as measures of average. An easy way to remember this is that median is the 'middle' and MOde is the 'MOst commonly occuring'.

Answers: 1 A non-living factor 2 Any from: light intensity, temperature, moisture levels, soil pH, soil mineral content, wind intensity and direction, availability of oxygen (water), availability of carbon dioxide 3 Availability of food, pathogens and parasites, predators, competition

Trigger interest

How many daisies? (5 mins)

Show students an image of a field full of daisies (this can be sourced by searching online). Ask students to suggest how they could estimate how many daisies are in the field.

If students are struggling to answer the question, tell them to consider what equipment they might need, how they might sample the field, and whether repeats should be taken.

Extend students' understanding by asking them to consider how they can make sure their estimate is non-biased and representative.

Exposition

Why do we sample? (10 mins)

Following on from the previous activity, ask students why biologists would sample the field to obtain an estimate of the number of daisies, rather than counting every individual daisy.

Discuss their ideas and then inform students that if we are going to sample an area, we must make sure the sample accurately represents the whole population. Talk through the importance of large sample sizes and random sampling to avoid bias.

Extend by asking students if they think randomisation and sampling is only used in biology. Tell students that there are lots of careers which require the use of sampling such as marketing and finance.

Main activity

Field investigations (30 mins)

Choose a suitable location (e.g., the school field or local park). Let students study the area and choose an organism to investigate. Discuss how to use a quadrat, how to choose random coordinates, why this is important, and how many samples to take. Students should then work in pairs to write their plan, which should include the equipment needed, method used, and a suitable results table. Go through with the class how to calculate range, mean, median, and mode, and how these could be used to analyse their results. Students will need another lesson to collect their data and analyse their results.

Review and practice

Quadrat calculation (10 mins)

Give the class the following situation:

Students are estimating the number of dandelions on a field that is 20 m long by 20 m wide. They are using 50 cm × 50 cm square quadrats. Go through the questions to work out the answers as a class or ask students to answer the questions individually.

Questions:

1. How many quadrats would cover the whole area? (Answer: 1600)
2. If they wanted to cover 2% of the field, how many samples would they need to take? (Answer: 32)
3. The mean number of dandelions in a quadrat is 2.2. Estimate the number in the whole field. (Answer: 3520)

B16.4 Competition in animals

AQA GCSE Biology specification points

4.7.1.1 Communities

Smart Curriculum Narrative

Ecosystems and environment

Theme/subtheme

Ecosystems and environment/Ecosystems

Disciplinary knowledge

- WS 2.6
- MS 2c, 4a

Activity

Survival rivals

Learning objectives	Learning outcomes: Developing Aiming for 3	Secure Aiming for 5	Extending Aiming for 8
Use information to suggest factors that animals are competing for in a given habitat	State that animals compete with each other for resources and list the main resources they compete with each other for	**Use information to suggest factors that animals are competing for in a given habitat**	Use the terms interspecific and intraspecific competition, and give examples of each
Describe some adaptations that help an animal compete successfully for a named resource	Identify adaptations that help an animal compete for a named resource	**Describe some adaptations that help an animal compete successfully for a named resource**	Suggest how animals are adapted to compete for named resources and explain how the adaptations described give the organism a competitive advantage
Describe how the distribution of a species may change because of competition	State that only animals that compete successfully will survive and reproduce	**Describe how the distribution of a species may change because of competition**	Discuss how interactions and competition between organisms affect populations in a given ecosystem

Tier 2 vocabulary	Tier 3 vocabulary
adaptation, camouflage, carnivore, *competition*, extinction, herbivore, *mate*, predator, prey, species, territory	interspecific competition, intraspecific competition, outcompete, warning colour

Lesson plan

Reactivate knowledge

1 What is a quadrat?

2 What do we measure using a quadrat?

3 What do animals in an ecosystem compete for?

Common misconception

Students may think that when an animal is outcompeted by another, the only possible outcome is that the animal dies and the population size decreases. They may not have considered that animals may relocate to another area with similar resources.

Answers: 1 A sample area used to count the number of organisms 2 Abundance and distribution of organisms in a habitat 3 Food, water, space, mates

Trigger interest

Grass eaters (5 mins)

Give students one minute to write a list of as many animals as they can think of that eat grass.

Ask them to swap lists with another student to count how many animals they found. Gather results and announce the winner. Use this to introduce the idea of competition between animals for food.

Exposition

Animal adaptations (10 mins)

Show students an image of a cheetah (or another suitable animal).

Ask students to list:

- the resources a cheetah would need to survive
- the adaptations a cheetah has to enable it to compete successfully for these resources.

As an extension, ask students to suggest what would happen to the cheetah population in an area if they were outcompeted by another animal. This will be a good opportunity to address the misconception. Students are likely to state that the cheetahs will die out due to lack of food, and may not have considered that some animals may move to a different location. This is another way that competition can affect distribution.

Main activity

Survival rivals (30 mins)

Print out some small cards, each showing one of the following words: food, territory, water, mates. Make enough cards so that most, but not all, members of the class can have a full set. Before the lesson, place the cards around the room. Ask the students to move around and collect cards to get a full set. The rules are that they can only hold one of each type at a time, they must not take any by force, and they can't swap. Once the students have worked through the cards, those without the full set did not 'survive'.

As a class, discuss how this models real life, why each of these resources is needed, and how they could have increased their chances of survival.

Then provide students with some information about competition between red and grey squirrels in the UK. Ask the students to write down an explanation of why there are many more non-native grey squirrels now living in the UK than native red squirrels, using ideas about competition. If students require support here, set them up with some questions that lead on to an explanation.

As an extension, repeat the game with the cards but this time let students have the option of exchanging cards with other students. Discuss whether this happens in real life, and why (e.g., parents giving up their share of food for their offspring).

Review and practice

Name the resource (10 mins)

Use the interactive to show the students a range of tactics that different animals use when competing for resources (e.g., a peacock using feathers to attract a mate, a male robin fighting another male in his territory, an owl hunting at night). Students choose the resource that each tactic helps the animal to compete for.

Support students by giving them two possible resources to choose from.

As an extension, ask students how each adaptation helps the animal to survive.

B16.5 Competition in plants

AQA GCSE Biology specification points

4.7.1.1 Communities

Smart Curriculum Narrative

Ecosystems and environment

Theme/subtheme

Ecosystems and environment/Ecosystems

Disciplinary knowledge

- WS 2.6
- MS 2c, 4a

Practical

Density of sowing

	Learning outcomes		
Learning objectives	**Developing Aiming for 3**	**Secure Aiming for 5**	**Extending Aiming for 8**
Describe the resources that plants compete for in a given habitat and explain why each resource is needed by the plant	List resources that plants compete with each other for	**Describe the resources that plants compete for in a given habitat and explain why each resource is needed by the plant**	Discuss which environmental factors are most important to plants in their competition to succeed and explain your choice
Describe the methods plants use to outcompete others, or avoid competition, and explain how they work	Describe one way in which plants are adapted to avoid direct competition with each other	**Describe the methods plants use to outcompete others, or avoid competition, and explain how they work**	Suggest problems that may be caused in an ecosystem by plant species that easily outcompete others
Explain why plants disperse their seeds, and describe a range of seed dispersal methods	Define seed dispersal and give three examples of different seed dispersal methods	**Explain why plants disperse their seeds, and describe a range of seed dispersal methods**	Discuss the effectiveness of seed dispersal in terms of interspecific competition
Plan an investigation into the effect of overcrowding in plants and display given data appropriately	Describe an investigation into the effect of overcrowding in plants	**Plan an investigation into the effect of overcrowding in plants and display given data appropriately**	Display appropriately and analyse data to explain the effects of overcrowding

Tier 2 vocabulary	Tier 3 vocabulary
adaptation, chlorophyll, *competition*, germination, mineral ion, nutrient, photosynthesis, seed dispersal	outcompete

Lesson plan

Reactivate knowledge

1 What do herbivores compete for?
2 Why do animals compete for territories?
3 Where is the competition fiercest – between different species or between members of the same species?

Common misconception

Students may think that plants only need carbon dioxide (for photosynthesis), and forget that plants also need oxygen for respiration.

Answers: 1 Plant food 2 For food/to reproduce and rear young 3 Between members of the same species

Trigger interest

What do plants compete for? (10 mins)

Show the students Relatability: could this be the image of a local park or forest? I'm not a specialist but I think this is something relatively easy to spot with plants and trees in everyday life - could this be encouraged here?

More extreme examples could then be shown using images of far flung jungles and rainforests, perhaps? (Lots of cool examples in the Amazon!) On mini-whiteboards, ask students to:

- list the resources the plants would be competing for in this forest
- explain why the plant needs each of these resources to survive.

Ask students to suggest which of the resources is the most important for the survival of a plant and explain why.

Exposition

Spreading the seeds (15 mins)

Show the class a series of images of familiar plants that are well-adapted for competition, for example, dandelion (spreads seeds a long way), ivy (climbs towards the light), clover (makes nitrates), and snowdrop (grows in winter). Ask students to suggest how each is adapted in terms of competition. If necessary, prompt them with hints, such as, clover can turn nitrogen from the air into nitrates in the soil.

Now focus on the dandelion spreading its seeds, and ask students what methods of seed dispersal they can remember from KS3. (You could show them some images as prompts if necessary.) Ask students to use their understanding of competition to explain why plants use seed dispersal methods.

Main activity

Density of sowing (25 mins)

Let students study some seed packets for information on how to sow seeds. Ask them why the packet gives information about how far apart to sow the seeds, and why this is different for each species of plant.

Allow students to study seedlings that have been set up in advance to show the effects of density of sowing on the seedlings.

Split the class into pairs. Each pair should make observations and measure the length of shoot, length of root, and mass for one seedling from each set of trays. They can then share results as a class in order to analyse the data and draw conclusions.

Review and practice

Weed removal (5 mins)

Ask students to explain why farmers want to remove weeds from their crop fields, using ideas about competition.

Discuss as a class, providing prompts for students if they are struggling. This can include considering abiotic factors that affect the survival of plants.

B16.6 Adapt and survive

AQA GCSE Biology specification points

4.7.1.4 Adaptations

***Smart Curriculum* Narrative**

Ecosystems and environment

Theme/subtheme

Ecosystems and environment/Adaptation

Activity

Making a living organism

Learning objectives	Learning outcomes Developing Aiming for 3	Secure Aiming for 5	Extending Aiming for 8
Describe some common adaptations that enable plants and animals to compete successfully in their habitat	State that organisms have adaptations (features) that enable them to survive in the conditions where they normally live	**Describe some common adaptations that enable plants and animals to compete successfully in their habitat**	Explain how common adaptations help organisms survive in a particular habitat
Explain how adaptations allow an organism to survive in its habitat	Give one example of an animal adaptation to a named habitat and one plant adaptation to the same or a different habitat	**Explain how adaptations allow an organism to survive in its habitat**	Explain how an organism might evolve the adaptations it needs to survive in a changed or extreme habitat
Describe three extreme habitats and some of the adaptations that enables organisms to survive in each one	Define an extremophile and give an example	**Describe three extreme habitats and some of the adaptations that enables organisms to survive in each one**	Explain why finding extremophiles living at very high temperatures and in very salty environments was unexpected

Tier 2 vocabulary	Tier 3 vocabulary
adaptation, bacteria, carnivore, denature, enzyme, herbivore, microorganism, mineral ion, osmosis, pH, photosynthesis, pollination	antifreeze, archaea, epiphyte, extremophile, salt gland, thermophile

Lesson plan

Reactivate knowledge

Common misconception

Students may think that no organisms can survive at very high temperatures.

1 What do plants compete for?

2 Why do most plants disperse their seeds as widely as possible?

3 Give two different ways in which plant seeds are spread successfully.

Answers: 1 Light, water, soil nutrients, space to grow 2 To avoid competition with the parent plant 3 Any two from: air, wind, explosive pods, animals, water

Trigger interest

What do microorganisms need? (5 mins)

Ask students to discuss in pairs what resources they think microorganisms compete for. Share their ideas as a class, and discuss how microorganisms need a range of resources. Include the idea that some microorganisms are similar to plants and animals, while others don't need oxygen or light to survive.

Exposition

Extremophiles (20 mins)

Give students some examples of extremophiles (organisms that live in extreme conditions) – such as the red flat bark beetle (*Cucujus clavipes*), Sahara desert ant (*Cataglyphis bicolor*), Himalayan jumping spider (*Euophrys omnisuperstes*), Pompeii worm (*Alvinella pompejana*), *Aquifex* genus of bacteria, the microorganism *Halobacterium salinarum*, or the Saguaro cactus (*Carnegiea gigantea*)

Ask students to use the internet to research one of these organisms. They should find out where it lives, why this is an extreme environment, how it is adapted, and how this helps it to survive. Students could present their findings to the rest of the class.

Main activity

Making a living organism (20 mins)

Ask students to name some extreme environments where it would be difficult for organisms to live (e.g., inside a volcano, on the top of Everest, in deep ocean trenches, or on the surface of Mars). Ask students to explain why it would be difficult to live there. Then students should pick one location and design an organism that has adaptations to help it to survive in their chosen location. They draw the organism, label the adaptations, and explain how they help it to survive.

To support students, choose one suitable location together as a class.

As an extension, students should explain in detail how their organisms evolved to have these adaptations.

Review and practice

The life of blister beetles (10 mins)

Show the class a video of how blister beetle larvae get food. Ask the students to describe why the beetle's habitat is extreme, why getting food is difficult, and how the larvae are adapted to find food and survive.

B16.7 Adaptation in animals

AQA GCSE Biology specification points

4.7.1.4 Adaptations

Smart Curriculum Narrative

Ecosystems and environment

Theme/subtheme

Ecosystems and environment/Adaptation

Activity

Insulation

	Learning outcomes		
Learning objectives	**Developing Aiming for 3**	**Secure Aiming for 5**	**Extending Aiming for 8**
Explain why it is important that most animals maintain the correct body temperature	State that it is important for most animals to maintain the correct body temperature	**Explain why it is important that most animals maintain the correct body temperature**	Suggest differences you would expect to see in the required body temperature of mammals and birds compared with reptiles or insects and explain how this might impact on the adaptations they evolve to cope with cold and heat
Describe structural, behavioural, and functional adaptations and give examples of each	State that adaptations that enable organisms to survive in the conditions where they normally live may be structural, behavioural, or functional	**Describe structural, behavioural, and functional adaptations and give examples of each**	Discuss the differences between structural, functional, and behavioural adaptations
Describe how fur or feathers may keep an animal warm, camouflage it, or keep it cool	State that fur or feathers may be adaptations to keep an animal warm, camouflage it, or keep it cool	**Describe how fur or feathers may keep an animal warm, camouflage it, or keep it cool**	Describe the challenges to an animal of living in a dry climate and explain how a range of adaptations help overcome these challenges
Describe ways in which animals are adapted to live in hot, dry, and cold habitats	State that different animals are adapted to live in hot, dry, and cold habitats	**Describe ways in which animals are adapted to live in hot, dry, and cold habitats**	Explain and illustrate how surface area to volume ratio is linked to adaptations that maintain the correct body temperature

Tier 2 vocabulary	Tier 3 vocabulary
adaptation, camouflage, metabolism, migration, surface area to volume ratio	antifreeze, behavioural adaptation, functional adaptation, structural adaptation

Lesson plan

Reactivate knowledge

1 What is an adaptation?

2 Give an example of an animal adaptation and a plant adaptation – and what the adaptation is for.

3 What is an extremophile?

Common misconception

Students may think that animals only have adaptations that are visible from the outside, and may not realise that they can have functional and behavioural adaptations too.

Answers: 1 A specific feature that enables an organism to survive in a particular environment 2 Any suitable example e.g., herbivores have flat teeth for grinding up plant material/water plants may have large air spaces in the leaves so they float 3 An organism that lives and reproduces in extreme conditions

Trigger interest

Camouflage (5 mins)

Show the class some examples of camouflaged animals in their habitat (there are many suitable images available online), and ask students to spot them. Use one example and ask the students to write down how this adaptation aids the animal's survival.

Exposition

Temperature regulation (15 mins)

Discuss why it is important that animals maintain a constant body temperature.

Discuss how mammals and birds can regulate their body temperature, but other animals cannot.

Review what is meant by surface area to volume ratio and how to calculate it (students could use the Maths skills activity from B4.5 to recap this).

Ask students to work in pairs to suggest how surface area to volume ratio and insulation are important for temperature regulation in animals.

Main activity

Insulation (25 mins)

Provide students with the Working scientifically worksheet. Ask half of the class to research animals that live in cold conditions in polar regions, and the other half to research animals that live in hot, dry deserts. Students should use books and/or the internet to list some animals that live in this habitat, the problems they face, and their adaptations.

Invite some students from each group to present one thing they found out. Allow all students to make notes on adaptations of animals that live in these extremes of temperature. Students can supplement their notes with information from the Student Book.

To support, ask students to either research the polar bear or the camel.

As an extension, or if time permits, encourage students to find examples of unusual animals.

Review and practice

Polar problem (10 mins)

Discuss the fact that the Arctic is getting warmer. Ask students to suggest why this is affecting the population of polar bears.

Extend this discussion to how climate change is affecting the distribution of other animals, and why.

B16.8 Adaptations in plants

AQA GCSE Biology specification points

4.7.1.4 Adaptations

***Smart Curriculum* Narrative**

Ecosystems and environment

Theme/subtheme

Ecosystems and environment/Adaptation

Activity

Analysing transpiration data

Learning objectives	Learning outcomes: Developing – Aiming for 3	Learning outcomes: Secure – Aiming for 5	Learning outcomes: Extending – Aiming for 8
Describe conditions where a plant needs to reduce water loss by transpiration	Describe why plants need a constant supply of water	**Describe conditions where a plant needs to reduce water loss by transpiration**	Explain how the rate of transpiration is linked to the structure of the leaf
Explain how different plant adaptations to reduce water loss enable the plants to survive in hot and/or dry environments	Describe one example of a plant adaptation to prevent water loss	**Explain how different plant adaptations to reduce water loss enable the plants to survive in hot and/or dry environments**	Examine an unfamiliar plant and suggest its habitat, explaining your suggestions in terms of the adaptations it shows
Describe a range of plant adaptations to hot, dry climates and explain how they are effective	Describe one plant adaptation for collecting water from the soil, and one for storing water	**Describe a range of plant adaptations to hot, dry climates and explain how they are effective**	Suggest and explain why a cactus would not survive in a cold climate

Tier 2 vocabulary	Tier 3 vocabulary
adaptation, evaporation, photosynthesis, respiration, stomata, surface area, surface area to volume ratio, water vapour, xylem	*cuticle*, functional adaptation, structural adaptation, transpiration

Lesson plan

Reactivate knowledge

1 What are the three main types of adaptations in animals?

2 Give two animal adaptions for cold climates that involve their surface area to volume ratio.

3 Give one example of a functional animal adaptation against the cold, and one against heat and lack of water.

Common misconception

Students may only be aware of plants having adaptations to survive in dry climates (e.g, cacti), and not realise that plants like pond lilies have adaptations to survive in wet habitats.

Answers: 1 Structural, functional, behavioural 2 Being large and the surface area of the thin-skinned areas of their bodies, such as their ears, is small 3 Any suitable example e.g. cold: antifreeze in cells, lack of heat and water: kidneys adapted to make very concentrated urine

Trigger interest

The need for water (10 mins)

Ask students to list reasons why water is important for plants. Pairs of students can then compare lists and feed back their ideas to the class. Discuss as a class how plants are adapted to take up water efficiently.

Exposition

Transpiration (5 mins)

Review students' knowledge about transpiration by asking them to work in pairs to complete a spider diagram showing everything they already know. Use the students' responses on the spider diagrams to describe and explain adaptations plants have to reduce transpiration (e.g., spines on cacti instead of leaves to reduce surface area, fewer stomata to reduce surface area, and curled leaves to increase the humidity around the stomata).

Discuss how plants lose water by transpiration, and the adaptations that some plants have to reduce this.

Main activity

Analysing transpiration data (35 mins)

Provide students with a method and raw data from an investigation into the rate of transpiration using shoots from different species of plant. Ask students to analyse the data using methods of their choosing. Show the students images of each plant and ask them to explain why the rate was different based on the structure of the leaves. Students should explain why these plants might have these adaptations.

To support, guide students through how to analyse the data (e.g., what type of graph to use).

Review and practice

Why wouldn't you find...? (5 mins)

Ask students to explain why you wouldn't find a cactus in the Arctic.

B16 Adaptations, interdependence, and competition

Introduction to checkpoint intervention

The checkpoint intervention provides suggestions for a lesson to follow up the *B16 Adaptations, interdependence, and competition* assessment. Depending on the outcomes of the assessment, these suggestions can be used to move a class towards secure, to target specific learning outcomes, or to consolidate knowledge for students achieving secure outcomes. Use students' outcomes from the checkpoint assessment to address general misconceptions from the content covered. Be prepared to re-cover content as required; three key concepts from the chapter are covered in more detail below.

Checkpoint secure learning outcomes

- Describe the importance of stable communities and explain why it is hard to replace them if they are lost.
- Describe how a range of abiotic factors influence different populations.
- Design a method to estimate a population using a sampling technique.
- Describe the methods plants use to outcompete others, or avoid competition, and explain how they work.
- Describe ways in which animals are adapted to live in hot, dry, and cold habitats.
- Explain how adaptations allow an organism to survive in its habitat.

Student reflection

Encourage students to reflect on whether there were any checkpoint questions they found difficult or straightforward as well as their preparation for the checkpoint assessment, including how they revised, what they revised, the time spent on revision, and what areas they could improve upon. For more strategies, see the Metacognition spread.

Intervention activity: Describe how a range of abiotic factors influence different populations.

Show students a photo of a rainforest (this can be sourced online).

Ask the students to list three to five abiotic factors in the rainforest that would affect the survival of the organisms there.

The students should then explain how each of the abiotic factors they listed would affect the survival of the plants and animals there.

Intervention activity: Explain how adaptations allow an organism to survive in its habitat.

Show the students an image of a holly bush (or another plant or animal of your choice). Ask students to list all the visible adaptations the holly bush has to help it to survive.

For each adaptation the students list, they should then explain how it helps the plant to survive.

Intervention activity: Design a method to estimate a population using a sampling technique.

Below is a method a student wrote to describe how to use sampling to estimate how many daisies are on the school field. Ask students to correct/improve each step of this method to enable accurate data to be collected.

Method:

1. Place two tape measures in the field.
   Improvement: place tape measures at right angles.
2. Place the quadrat within this area.
   Improvement: students should describe a method to make this placement of the quadrat random.
3. Count the daisies.
   Improvement: the instruction should be to count how many daisies are within that quadrat.
4. Repeat.
   Improvement: students should state at least 20 repeats.
5. Work out the area of the field and the area of the quadrat, and calculate how many times the quadrat would fit into the field. Multiply the number of daisies in one quadrat by how many times the quadrat fits into the field to obtain your final estimate.
   Improvement: the instruction should be to calculate a mean number of daisies per quadrat and use this to calculate the final estimate.

B17 Organising an ecosystem

B17.1 Feeding relationships

AQA GCSE Biology specification points

4.7.2.1 Levels of organisation

Smart Curriculum Narrative

Ecosystems and environment

Theme/subtheme

Ecosystems and environment/Feeding relationships

Disciplinary knowledge

- WS 1.2
- MS 4a

Activity

Predator–prey relationship

	Learning outcomes		
Learning objectives	**Developing Aiming for 3**	**Secure Aiming for 5**	**Extending Aiming for 8**
Describe how producers synthesise glucose molecules by photosynthesis	State that green plants and algae (photosynthetic organisms) are the producers of biomass for life on Earth and that feeding relationships within a community can be represented by food chains	**Describe how producers synthesise glucose molecules by photosynthesis**	Explain how all biomass originates with producers
Identify producers, primary consumers, secondary consumers, and tertiary consumers in a food chain	Define the terms producer, primary consumer, secondary consumer, and tertiary consumer and give examples of each	**Identify producers, primary consumers, secondary consumers, and tertiary consumers in a food chain**	Evaluate food chains as models of feeding relationships
Describe the relationships between predators and prey in a stable community	Define the terms predator and prey and identify both in a food chain and food web	**Describe the relationships between predators and prey in a stable community**	Explain how the observed rises and falls of predators and prey are nevertheless regarded as part of a stable community
Describe and interpret a graph showing how the numbers of predators and prey change over time	Describe a graph showing how the numbers of predators and prey change over time	**Describe and interpret a graph showing how the numbers of predators and prey change over time**	Make predictions based on data on a predator–prey relationship

Tier 2 vocabulary	Tier 3 vocabulary
feeding relationship, food chain, predator, prey, primary consumer, *producer*, secondary consumer, tertiary consumer	biomass, gut microorganism, phytoplankton, stable community, zooplankton

Lesson plan

Reactivate knowledge

1 What is a producer?
2 What is a prey organism?
3 What is a predator?

Common misconception

Students may draw arrows the wrong way in food chains, thinking they represent which organism consumes which, rather than representing the direction of energy flow.

Answers: 1 Organism that makes its own food by photosynthesis 2 An animal that is eaten by other animals 3 An animal that eats another animal

Trigger interest

Food chains (10 mins)

Provide students with the following organisms: Grass, cow, human.

On mini-whiteboards, ask students to:

- draw the food chain for these organisms
- state what the direction of the arrows represents
- label the producer, primary consumer, and secondary consumer
- label a predator
- label a prey animal.

Exposition

Food webs (10 mins)

Provide students with an example of a food web, for example, one sourced online. Set the class a series of questions where they must categorise the different organisms (e.g., producer, consumer, etc.) and about how a change in the population of one organism could affect another.

This could be completed on mini-whiteboards or written out in their books.

Main activity

Predator–prey relationship (30 mins)

Students are provided with data showing changes in the populations of a prey animal and a predator in the Working scientifically worksheet. Ask students to plot the data on a suitable graph (both populations should be shown on the same set of axes). Then ask them to explain the shape of the graph. They can use the explanation of the predator–prey cycle in the Student Book to assess their own answer.

Review and practice

Predator and prey (5 mins)

Ask students to suggest what would happen to a predator–prey graph if the following variables changed:

- prey birthrate
- predator effectiveness.

They should answer the question in their books as if it were an exam question, followed by a justification of their answers.

B17.2 Materials cycling

AQA GCSE Biology specification points

4.7.2.2 How materials are cycled

Smart Curriculum Narrative

Earth and environment

Theme/subtheme

Ecosystems and environment/Material cycling in ecosystems

Disciplinary knowledge

- WS 1.2

Activity

The decay cycle

	Learning outcomes		
Learning objectives	**Developing Aiming for 3**	**Secure Aiming for 5**	**Extending Aiming for 8**
Explain why decomposers are important to a stable ecosystem	Describe what a decomposer is and give examples	**Explain why decomposers are important to a stable ecosystem**	Explain how detritivores increase the rate of decay using ideas about digestion and surface area
Explain the importance of recycling substances and describe the events of the decay cycle	Name some substances that are recycled in the living world	**Explain the importance of recycling substances and describe the events of the decay cycle**	Describe the key positive role of microorganisms in the decay cycle and comment on the limitations of a simple model of decay
Describe the events in the water cycle, including the continual evaporation and precipitation of water, and explain its importance to life on the Earth	State that the water cycle provides fresh water for plants and animals on land before draining into rivers and seas, and identify the main stages of the water cycle	**Describe the events in the water cycle, including the continual evaporation and precipitation of water, and explain its importance to life on the Earth**	Discuss the concept that every drop of water drunk has passed through many living organisms previously

Tier 2 vocabulary	Tier 3 vocabulary
bacteria, carbon dioxide, condensation, evaporation, microorganism, mineral ion, *precipitation*, respiration, water cycle	abiotic, biotic, decay, decay cycle, decomposer, detritivore, detritus feeder, percolation, transpiration

Lesson plan

Reactivate knowledge

1 What is a primary consumer?
2 What is a secondary consumer?
3 Give an example of a primary, secondary, and tertiary consumer.

Common misconception

Students may think that small organisms, like worms, are decomposers rather than understanding that decomposers are all microorganisms.

Answers: 1 An animal that eats producers 2 An animal that eats primary consumers 3 Any suitable examples e.g., fish, seal, killer whale

Trigger interest

A world without decay (10 mins)

Ask students to discuss in pairs what the world would be like if there was no decay. Ask each pair to feed back their ideas and draw out the idea that decay is useful for recycling organic matter.

Exposition

Cycles (15 mins)

Ask students to work in pairs to write a suitable narration that explains what is happening in Figure 1 of the Student Book. Their narration should include which organisms are carrying out the process of decay and why decay is vital for the survival of other organisms in the ecosystem. They can use the information in the Student Book to help them.

As an extension, students should give scientific explanations for the change in appearance of the decaying organism.

Next, consider the water cycle. Go a round the class and ask each student to state one thing that would happen if there was no water cycle. Discuss how important the water cycle is to life on the Earth.

Main activity

The decay cycle (20 mins)

Using the activity, provide students with cards that describe the parts of the decay cycle (Figure 2 in the Student Book). Ask students to arrange the cards to create a cycle that shows how carbon atoms and mineral ions are recycled. They can check their answers against the diagram in the Student Book.

For students with lower prior attainment, provide a cycle with boxes on an A3 piece of paper. Students can place the cards in the correct boxes.

Review and practice

Modelling decay (10 mins)

Provide students with simple models made out of interlocking plastic blocks. Then ask them to rearrange the blocks to build something else, explaining how this models the decay cycle.

B17.3 The carbon cycle

AQA GCSE Biology specification points

4.7.2.2 How materials are cycled

Smart Curriculum Narrative

Earth and environment

Theme/subtheme

Ecosystems and environment/Material cycling in ecosystems

Disciplinary knowledge

- WS 1.2

Activity

Processes in the carbon cycle

	Learning outcomes		
Learning objectives	**Developing Aiming for 3**	**Secure Aiming for 5**	**Extending Aiming for 8**
Describe the events in the carbon cycle and explain why it is important to life on Earth	State that the carbon cycle returns carbon from organisms to the atmosphere as carbon dioxide to be used by plants in photosynthesis	**Describe the events in the carbon cycle and explain why it is important to life on Earth**	Explain why the concentration of carbon dioxide in the atmosphere is rising, and why this is an issue
Draw and label a diagram of the carbon cycle and write word equations for photosynthesis, respiration, and combustion	Complete a diagram of the carbon cycle and use it to describe the main processes involved	**Draw and label a diagram of the carbon cycle and write word equations for photosynthesis, respiration, and combustion**	Explain the links between photosynthesis, respiration, and combustion in the carbon cycle
Describe a number of carbon sinks	Define a carbon sink	**Describe a number of carbon sinks**	Predict the impact on the carbon cycle of the reduction of carbon sinks, e.g., deforestation, burning fossil fuels

Tier 2 vocabulary	Tier 3 vocabulary
algae, biomass, carbon cycle, carbon dioxide, combustion, *community*, *consumer*, glucose, microorganism, oxygen, photosynthesis, protein, respiration	carbonate rock, carbon sink, decay, decomposer, organic compound

Lesson plan

Reactivate knowledge

1 What is a decomposer?
2 Why are decomposers so important in the soil?
3 What are the main stages of the water cycle?

Common misconception

Students may think carbon only cycles around in the form of carbon dioxide, rather than being incorporated into molecules like carbohydrates, proteins, and lipids.

Answers: 1 Organisms that break down dead animals and plants and animal waste 2 They return mineral ions to the soil from the plant and animal material they break down 3 Evaporation/condensation/precipitation/percolation/respiration/transpiration

Trigger interest

Study the cycle (10 mins)

Split the class into groups of four to five students, and give each group a piece of A4 paper. One student from each group comes to the front and has 30 seconds to study an image of the carbon cycle. They then return to their group and describe what they saw, with the rest of the group working together to create a diagram of the carbon cycle based on this description. The next student from each group then comes up and gets 30 seconds to study the image, before returning to their group to report back so they can add to their image. This continues until each member of the group has studied the image and reported back. Invite each group to share their diagram with the rest of the class. The class can then judge which group recreated the carbon cycle most accurately.

Exposition

Pass the carbon (10 mins)

Provide the same groups with a soft ball to represent a carbon atom. Give students roles as parts of the carbon cycle (the atmosphere, a plant, an animal, a fossil fuel, a decomposer). Ask the group to pass the ball around to model how carbon is recycled, and to name each process that they are showing.

As an extension, students should use their model to explain why the concentration of carbon dioxide in the atmosphere is rising, and why this is an issue.

Main activity

Processes in the carbon cycle (30 mins)

Split the class into groups and give each group a process in the carbon cycle to study in more detail (photosynthesis, respiration, combustion, decomposition, feeding). Ask each group to research how their process moves carbon from one place to another, name the places involved, write down any word equations for the process, and draw a picture to represent it. Then, ask the groups to present their findings to the rest of the class, and use this to compile a class diagram of the carbon cycle. Show the students the Animation: *Carbon Cycle* to check and reinforce students' understanding.

As an extension, students should discuss what could happen if the processes of photosynthesis, respiration, or combustion were removed.

Review and practice

Protecting carbon sinks (5 mins)

Show students a short video clip about efforts to preserve carbon sinks in places like the Amazon rainforest. Relate this to ways of tackling climate change, explaining that deforestation in the Amazon was reduced by 50% in 2023.

B17.4 Rates of decomposition

AQA GCSE Biology specification points

4.7.2.3 Decomposition

Smart Curriculum Narrative

Earth and environment

Theme/subtheme

Ecosystems and environment/Material cycling in ecosystems

Disciplinary knowledge

- WS 2.1, 2.4, 2.6, 2.7
- MS 1c, 4a, 4c

Required Practical 10

Decay

Learning objectives	Learning outcomes Developing Aiming for 3	Learning outcomes Secure Aiming for 5	Learning outcomes Extending Aiming for 8
Explain why different factors speed up or slow down the rate of decomposition	List the conditions that affect the rate of decay	**Explain why different factors speed up or slow down the rate of decomposition**	Explain why a rise in temperature increases the rate of compost production up to around 70°C, after which decomposition stops
Explain the importance of decomposition in recycling kitchen and garden waste into compost	Describe the main stages in the process of making compost	**Explain the importance of decomposition in recycling kitchen and garden waste into compost**	Apply factors that affect the rate of decay to real life situations (e.g., compost making, preserving food)
Describe how biogas generators are used to produce methane gas as a fuel	State that anaerobic decay produces methane gas	**Describe how biogas generators are used to produce methane gas as a fuel**	Explain the differences in using biogas generators in warm climates and cooler ones

Tier 2 vocabulary	Tier 3 vocabulary
aerobic, anaerobic, bacteria, denature, ecosystem, enzyme, exothermic, methane, microorganism, oxygen, pH, temperature	biogas generator, compost, decay, decomposer, decomposition, fertiliser, organic matter, scavenger

Lesson plan

Reactivate knowledge

1 What are the main carbon sinks on the Earth?

2 What is the carbon cycle?

3 What are the main processes involved in the carbon cycle?

Answers: 1 Fossil fuels/carbonate rocks/forests and marine algae/carbon dioxide in the atmosphere/carbon dioxide dissolved in water 2 The constant cycling of carbon between living things and the environment 3 Photosynthesis/respiration/combustion

Common misconception

Students may think that small organisms, like worms, are decomposers rather than understanding that decomposers are all microorganisms.

Trigger interest

What affects the rate of decay? (10 mins)

Students complete the interactive to sort conditions (warm temperature, cold temperature, lots of oxygen, no oxygen, moist conditions, dry conditions, no carbon dioxide, lots of carbon dioxide) into groups of what will speed up, slow down, or have no effect on decay.

Challenge students by asking them to explain why each condition has this effect. These can be written in their books.

Exposition

Applications of decomposition (15 mins)

Following on from the previous activity, inform students that knowledge of factors that speed up decay has been used by humans to their advantage. Talk students through the production of compost by gardeners and of methane in biogas generators.

Main activity

Decay (25 mins)

Students should spend 25 minutes in this lesson planning the required practical. The practical itself can then be completed in a subsequent lesson, with additional time given to analyse data.

Ask students to work through the questions in the required practical box in the Student Book to help them plan their investigation. Ideally, students should have the equipment they will need at the front of the classroom to help them consider how to use it for the experiment they are planning.

Students should produce an equipment list and a method. Discuss how they will collect quantitative data. Students need to measure the rate of decay by measuring pH change caused by the milk becoming more acidic as it decays. Check methods and allow students to set up their experiment.

Students should use time in subsequent lessons to collect data. This should be represented as a line graph with more than one line plotted (e.g., for the different temperatures they used), with time on the *x*-axis and decay (measured using change in pH) on the *y*-axis.

Students with higher prior attainment should use their line graph to calculate the rate of decay.

Review and practice

Ice mammoth (5 mins)

Show students an image of a mammoth preserved in ice. Ask them to explain why the mammoth hasn't decayed, using ideas about what affects the rate of decay.

B17 Organising an ecosystem

Introduction to checkpoint intervention

The checkpoint intervention provides suggestions for a lesson to follow up the *B17 Organising an ecosystem* assessment. Depending on the outcomes of the assessment, these suggestions can be used to move a class towards secure, to target specific learning outcomes, or to consolidate knowledge for students achieving secure outcomes. Use students' outcomes from the checkpoint assessment to address general misconceptions from the content covered. Be prepared to re-cover content as required; three key concepts from the chapter are covered in more detail below.

Checkpoint secure learning outcomes

- Identify producers, primary consumers, secondary consumers, tertiary consumers, predators, and prey in a food web.
- Describe and interpret a graph showing how the numbers of predators and prey change over time.
- Explain why decomposers are important to a stable ecosystem.
- Explain the importance of recycling substances and describe the events of the decay cycle.
- Explain why the carbon cycle is vital to life on the Earth and describe the current threats to the balance of the carbon cycle.
- Plan and carry out an investigation into the effect of temperature on the rate of decay of fresh milk by measuring pH change.

Student reflection

Encourage students to reflect on whether there were any checkpoint questions they found difficult or straightforward as well as their preparation for the checkpoint assessment, including how they revised, what they revised, the time spent on revision, and what areas they could improve upon. For more strategies, see the Metacognition spread.

Intervention activity: Identify producers, primary consumers, secondary consumers, tertiary consumers, predators, and prey in a food web.

Ask students to match up the key term to the definition:

Key terms

- producer
- primary consumer
- secondary consumer
- tertiary consumer
- predator
- prey

Definitions

- An organism at the start of a food web that produces glucose by photosynthesis.
- An organism that eats producers. It will always be an herbivore.
- An organism that eats primary consumers. It will be an omnivore or carnivore.
- An organism that eats secondary consumers. It will be an omnivore or carnivore.
- An animal that hunts and eats other animals.
- An animal that is hunted and eaten by another animal.

Students should then use these key terms to create a food web containing organisms of their choosing. They should label the organisms using the key terms.

Intervention activity: Explain why decomposers are important to a stable ecosystem.

Ask students to complete the paragraph below by filling in the gaps using the following words:

active, waste, deficient, mineral, roots, proteins

Decomposers break down dead bodies and ______________ material. In doing this, decomposers return ______________ ions, such as nitrates, to the soil. This is important, because plants can then absorb the nitrates by ______________ transport through their ______________. The nitrates are used by the plant to make ______________ and other chemicals in their cells. Without decomposers, soils would become nutrient ______________.

Answers: waste, mineral, active, roots, proteins, deficient

Intervention activity: Plan and carry out an investigation into the effect of temperature on the rate of decay of fresh milk by measuring pH change.

Give students the following information and ask them to answer the questions that follow.

A student wants to investigate the effect of temperature on how quickly milk decays. They have decided to test five different temperatures and to measure the pH of the milk after it has been left at the specified temperature for five days. The more acidic the milk, the more it has decayed.

1. What is the dependent variable in this investigation? (Answer: pH of milk)
2. What is the independent variable in this investigation? (Answer: temperature)
3. Suggest five temperatures the student could use in this investigation. (Answer: students could pick any temperature between 0 °C and 65 °C (inform them above 65 °C could be potentially hazardous and enzymes would already be denatured, so temperatures above this are not as appropriate)
4. Using the information above, state one variable the student has controlled. (Answer: the duration (5 days) the milk was left for)
5. Identify one potential hazard and risk in this method, and suggest how the student can keep themself safe from this. (Answer: microbes within the milk are a hazard, the risk is that if they are accidentally ingested it could make you sick. Prevent this by washing hands after any contact with decayed milk)

B18 Biodiversity and ecosystems

B18.1 The human population explosion

AQA GCSE Biology specification points

4.7.3.1 Biodiversity

4.7.3.2 Waste management

4.7.3.3 Land use

Smart Curriculum Narrative

Earth and environment

Theme/subtheme

Biodiversity and human interactions with the environment/Biodiversity

Disciplinary knowledge

- WS 1.4

Activity

The problem with the human population

Learning objectives	Learning outcomes: Developing Aiming for 3	Learning outcomes: Secure Aiming for 5	Learning outcomes: Extending Aiming for 8
Describe the importance of a good level of biodiversity to the stability of an ecosystem	Define diversity as the variety of all the different species of organisms on Earth, or within an ecosystem	**Describe the importance of a good level of biodiversity to the stability of an ecosystem**	Explain in detail why a high level of biodiversity is important to the stability of ecosystems
Give reasons for the recent growth in human population	State that the human population is over 8 billion and growing fast	**Give reasons for the recent growth in human population**	Explain why human population change differs from population change of other animals
Analyse and interpret data and information concerning human population growth	Extract information from graphs on human population growth	**Analyse and interpret data and information concerning human population growth**	Suggest and evaluate solutions to the problems caused by human population growth
Describe some of the effects of human population growth on land use and other resources	List some resources that humans are using up	**Describe some of the effects of human population growth on land use and other resources**	Explain in detail how human activities are affecting the land and resources

Tier 2 vocabulary	Tier 3 vocabulary
ecosystem, energy resource, extinction, fertiliser, fossil fuel, habitat, human population, metal ore, pollution, predator, resource, species	biodiversity, industrial waste, sewage

Lesson plan

Reactivate knowledge

1 What is biodiversity?
2 What is interdependence?
3 Why is classification important to scientists studying biodiversity?

Common misconception

Students may only link the increase in human population size to the challenge of a lack of space, and not consider the impact on pollution and destruction of natural resources.

Answers: 1 The range of organisms in a particular area 2 When one organism or type of organism depends on another 3 Allows them to measure biodiversity

Trigger interest

Estimating human population (5 mins)

Ask students to write on a piece of paper their estimation of the total number of people in the world. Ask them to stand in a line so their estimates are in numerical order. Reveal whose estimate is the closest. You can show students a population counter (found online) that shows how fast the current population is changing.

Students can use standard form to express their population number. They could also calculate the range, median, and mean of the estimates given.

Exposition

Biodiversity (15 mins)

Tell the class that a rainforest has a high level of biodiversity, and ask students to suggest what this means. Ask pairs of students to list possible reasons why a high level of biodiversity is important to:

- other plants and animals in the same ecosystem
- the entire human population of the world.

If students are struggling to think of ideas, they can use the information in the Student Book to help them.

Main activity

The problem with the human population (25 mins)

Ask students for examples of resources that humans need, such as air, clean water, and food.

Discuss where we get each of these things. Divide the class into small groups and give each group a factor – air, water, land, natural resources (e.g., minerals/fossil fuels), or biodiversity. Ask them to prepare a short presentation on how the growth in the human population is impacting negatively on their factor and why. Students can use information from the Student Book, other books, or conduct research online.

Review and practice

Solutions (10 mins)

For each of the problems presented in the previous activity, ask students to suggest solutions. For example, the depletion of fossil fuels could be reduced by using more renewable energy resources.

Students should then choose one solution and evaluate the suggestion in detail.

B18.2 Land and water pollution

AQA GCSE Biology specification points

4.7.3.2 Waste management

Smart Curriculum Narrative

Earth and environment

Theme/subtheme

Biodiversity and human interactions with the environment/Pollution

Practical

Eutrophication

	Learning outcomes		
Learning objectives	**Developing** **Aiming for 3**	**Secure** **Aiming for 5**	**Extending** **Aiming for 8**
Describe the process of eutrophication, and explain how it affects biodiversity	List some substances that pollute water and land	**Describe the process of eutrophication, and explain how it affects biodiversity**	Explain how oxygen levels in water are depleted during the process of eutrophication
Describe how pesticides, and herbicides pollute the land and water and reduce biodiversity	Describe some effects of rubbish, pesticides, and sewage on land and water	**Describe how pesticides, and herbicides pollute the land and water and reduce biodiversity**	Compare and contrast biomagnification and bioaccumulation and explain how they interact to reduce biodiversity

Tier 2 vocabulary	Tier 3 vocabulary
algae, bioaccumulation, *competition*, fertiliser, food chain, habitat, human population, microorganism, mineral ion, nitrate, oxygen, pollution, population, primary consumer, radioactivity, secondary consumer, top carnivore	biodiversity, bioindicator, biomagnification, data logger, decomposer, eutrophication, herbicide, industrial waste, landfill site, pesticide, sewage

Lesson plan

Reactivate knowledge

1 What is biodiversity on a global scale?

2 Approximately how many people are alive on the Earth at the moment?

3 How does the human population use land?

Common misconception

Students may only link the increase in human population size to the challenge of a lack of space and not consider the impact on pollution and destruction of natural resources.

Answers: 1 The variety of all the different species of organisms on the Earth 2 Over 8 billion 3 Building (houses/shops/industrial plants/landfills etc.), farming, quarrying

Trigger interest

Waste (5 mins)

Ask students to list the things that get thrown away in their home every day. Discuss the fact that these waste items include paper, metals, plastics, and food waste, as well as sewage and waste water from household appliances such as washing machines and dishwashers.

Exposition

Impact of waste (15 mins)

Ask students to consider where the waste from people's homes ends up. Talk through with them how this waste can cause both water and land pollution. They can use the information from the Student Book to provide ideas to support this discussion.

Then present students with the following food chain:

phytoplankton → zooplankton → small fish → large fish

Inform students that eutrophication can lead to the growth of harmful algae, which produce toxic chemicals called cyanotoxins. Cyanotoxins are taken in by the zooplankton in this food chain, but have little effect on them.

Ask students to suggest why this situation can be very harmful for the large fish in the food chain. Have students share their answers and inform them that the term bioaccumulation is used to describe this process.

Main activity

Eutrophication (25 mins)

Set up this practical three weeks in advance as a demonstration to show the effect of extra nitrates and phosphates in the water supply on the growth of plants. Talk about the issue of lakes and streams becoming covered in weeds (i.e., eutrophication). Students follow the instructions on the Practical sheet to count the leaves and record the results in the table. Ask them to link this to the problem of eutrophication, using the information in the Student Book.

To support students, provide them with cards (from the activity) that show the events in eutrophication. Students can then arrange the cards into the correct order.

Review and practice

Effects of farming (10 mins)

Give small groups of students a substance that is used or produced by farms (e.g., fertilisers, herbicides, pesticides, sewage, etc.).

Ask the students to discuss in their groups how this substance can damage ecosystems.

B18.3 Air pollution

AQA GCSE Biology specification points

4.7.3.2 Waste management

Smart Curriculum Narrative

Earth and environment

Theme/subtheme

Chemistry of the Earth/Climate change

Practical

The effect of acid rain

	Learning outcomes		
Learning objectives	**Developing Aiming for 3**	**Secure Aiming for 5**	**Extending Aiming for 8**
Describe how acid rain is formed	State that acid rain is caused when acidic gases, produced when fossil fuels are burned, dissolve in rain	**Describe how acid rain is formed**	Explain why acid rain does not always fall on the area that has produced the acidic gas air pollution
Describe the impact of acid rain on forests and lakes and state ways that the risk of acid rain can be reduced	List some effects of acid rain on plants and animals	**Describe the impact of acid rain on forests and lakes and state ways that the risk of acid rain can be reduced**	Explain how preventative measures to reduce acid rain work
Describe the causes of global dimming and smog, and describe their effects	State that acid rain, smoke, and smog are forms of air pollution	**Describe the causes of global dimming and smog, and describe their effects**	Explain the complexity of smog as a combination of smoke and compounds such as sulfur dioxide and nitrogen oxides
Describe and interpret data on air pollution and acid rain	Interpret observations and data on air pollution and acid rain, with guidance	**Describe and interpret data on air pollution and acid rain**	Analyse in detail data showing air pollution over time and suggest reasons for any trends observed

Tier 2 vocabulary	Tier 3 vocabulary
air pollution, carbon dioxide, ecosystem, fossil fuel, nitric acid, pH, pollution, sulfuric acid	acid rain, emission, global dimming, particulate, smog, smoke pollution

Lesson plan

Reactivate knowledge

1 Why is pollution a problem?
2 Give two ways that people pollute the land.
3 What is bioaccumulation?

Common misconception

Students may not realise that rain is naturally slightly acidic because carbon dioxide reacts with rainwater to form an acid. However, acid rain refers to higher levels of acidity in rainwater caused by nitrogen oxides and sulfur dioxide.

Answers: 1 Kills plants and animals/reduces biodiversity/threatens human health 2 Any from: sewage, waste, industrial processes, chemicals used in farming 3 The build up of toxins in an organism over its lifetime

Trigger interest

Production of acidic gases (10 mins)

Carry out the following demonstration to show that the burning of a fuel produces acidic gases. Make sure eye protection is worn and carry out in a fume cupboard. Using a Bunsen burner, set light to some coal in a deflagrating spoon and place it in a gas jar contain universal indicator. Shake to observe the colour change.

Discuss where coal is burnt, and ask students to suggest why burning coal has an impact on the environment. Link this to smoke and smog being forms of air pollution.

As a follow-up, ask students to construct word and symbol equations to show how the carbon and/or sulfur in the fuel produces acidic oxides.

Exposition

Acid rain formation (15 mins)

Supply students with Figure 1 from B18.3 of the Student Book, but with the labels blanked out. Ask them to complete the diagram using information from the Student Book. Discuss that rain is naturally slightly acidic because carbon dioxide reacts with rainwater to form an acid. However, the addition of nitrogen oxides and sulfur dioxide from burning fuels makes rainwater more acidic and produces acid rain.

Students with higher prior attainment can include word and symbol equations on their diagrams.

Main activity

The effect of acid rain (25 mins)

Ask students to use what they have learnt in biology and chemistry about the effects acid rain to list the effects of it on living organisms. Students could use information from the Student Book for support with this.

Students then plan an investigation into the effect of acid rain on the germination of cress seeds. Ask them to write a method and prediction, then allow them to carry out the investigation. Time will be needed in subsequent lessons to gather the results and analyse them. Students can calculate the percentage germination and display the results in a suitable chart or graph.

Students with prior lower attainment can be provided with a suitable method to follow.

Review and practice

Acid rain (5 mins)

Using the interactive, students order the stages in the production of acid rain.

B18.4 Deforestation and peat destruction

AQA GCSE Biology specification points

4.7.3.3 Land use

4.7.3.4 Deforestation

Smart Curriculum Narrative

Earth and environment

Theme/subtheme

Biodiversity and human interactions with the environment/Land use, deforestation, and food

Disciplinary knowledge

- WS 1.4, 1.5

Activity

Peat problems

	Learning outcomes		
Learning objectives	**Developing** **Aiming for 3**	**Secure** **Aiming for 5**	**Extending** **Aiming for 8**
Describe the main reasons for large-scale deforestation globally	Define deforestation	**Describe the main reasons for large-scale deforestation globally**	Analyse data on deforestation rates, and give an explanation for any trends observed
Describe the effects of deforestation and peat removal in terms of atmospheric carbon dioxide concentrations and biodiversity	Describe an effect of deforestation	**Describe the effects of deforestation and peat removal in terms of atmospheric carbon dioxide concentrations and biodiversity**	Evaluate the environmental implications of deforestation
Describe reasons for and against the use of peat for food production	Give a use for peat	**Describe reasons for and against the use of peat for food production**	Discuss the conflict between using peat to increase food production and the need to conserve peat bogs

Tier 2 vocabulary	Tier 3 vocabulary
carbon dioxide, combustion, deforestation, ecosystem, ethanol, extinction, habitat, microorganism, photosynthesis, species	biodiversity, biofuel, carbon sink, carnivorous plant, compost, conservation, decay, decomposition, endangered species, monoculture, palm oil, peat, peat bog, peatland, 'slash-and-burn' clearance, staple food

Lesson plan

Reactivate knowledge

1 What is the main source of acid rain?

2 Name two types of gases that dissolve in water to cause acid rain.

3 Give one measure that will reduce acid rain.

Common misconception

Some students may assume that deforestation only happens to provide wood. They may not understand that trees are often cut down to provide space and then burnt or left to rot, releasing carbon dioxide.

Answers: 1 Burning fossil fuels (for transport or power generation) 2 Sulfur dioxide/nitrogen oxides 3 Low-sulfur fuels for vehicles and power generation/catalytic converters/electric vehicles/using nuclear fuels/biofuels

Trigger interest

Looking at deforestation (5 mins)

Show the class an image of deforestation. Ask them to list on their mini-whiteboards reasons why deforestation is considered a bad thing. Then ask if there are any benefits of deforestation.

To support, prompt students to think of disadvantages by asking them to consider carbon dioxide levels in the atmosphere, and habitats for organisms.

As an extension, students should consider who benefits from deforestation.

Exposition

Deforestation summary (20 mins)

Students use the worksheet to help them organise the information on deforestation. Let students use information from the Student Book or other resources to study the reasons for, and effects of, deforestation. Ask students to organise this information as a table with the headings 'Reasons for deforestation', 'Impacts', and 'Solutions'.

Students should then categorise each of their reasons and impacts as environmental, social, economic, and/or political, and show how they are linked together.

Main activity

Peat problems (25 mins)

Let students study (or demonstrate) samples of peat and peat-free compost. Students should wear gloves if handling the samples.

Ask students to write down their observations about the differences between the two types of compost. Ask them to suggest why gardeners and food producers have preferred to use peat in their compost. (It improves the properties of the soil by increasing mineral content and helping with water retention).

Give the class a series of true or false statements on peat – what it is, where it comes from, what it is used for, and problems with its use.

1. Peat is a type of soil made up of partially decayed plant material (T)
2. Peat forms in oxygen-rich environments (F)
3. Peat is primarily found in arid desert regions (F)
4. Peat can be used as a fuel (T)
5. Peat extraction involves draining and drying out wetlands (T)
6. Peat is used in gardening as compost (T)
7. The extraction of peat has no impact on carbon dioxide emissions (F)
8. Peatlands are important habitats for biodiversity (T)
9. The use of peat as a fuel contributes to greenhouse gas emissions (T)

Ask students to use information from the Student Book to help them decide which statements are true and which are false. Students should copy the true statements and write corrected versions of the false ones.

Discuss the conflict between using peat to increase food production and the need to conserve peat bogs.

Review and practice

Deforestation rates (5 mins)

Use the interactive, in which students look at a chart of deforestation over time and answer a multiple choice question about possible reasons for increased deforestation rates.

B18.5 Global warming

AQA GCSE Biology specification points

4.7.3.5 Global warming

Smart Curriculum Narrative

Earth and environment

Theme/subtheme

Chemistry of the Earth/Climate change

Disciplinary knowledge

- WS 1.3, 1.6

Activity

Changing conditions

Learning objectives	Learning outcomes: Developing Aiming for 3	Secure Aiming for 5	Extending Aiming for 8
Describe the greenhouse effect and explain its importance to life on Earth	Use the terms greenhouse effect, global warming, and climate change correctly	**Describe the greenhouse effect and explain its importance to life on Earth**	Discuss the misuse of the term greenhouse effect as a damaging process
Describe how increased levels of carbon dioxide and methane in the atmosphere are causing global warming and climate change	State that increased levels of carbon dioxide and methane in the atmosphere are adding to the greenhouse effect and causing global warming	**Describe how increased levels of carbon dioxide and methane in the atmosphere are causing global warming and climate change**	Explain in detail the causes and effects of rising carbon dioxide and methane levels in the atmosphere
Describe in detail the biological consequences of global warming and climate change	Give one biological consequence of global warming	**Describe in detail the biological consequences of global warming and climate change**	Discuss the implications of global warming for biodiversity

Tier 2 vocabulary	Tier 3 vocabulary
atmosphere, carbon dioxide, climate change, deforestation, extinction, fossil fuel, global warming, greenhouse effect, greenhouse gas, habitat, methane, migration, photosynthesis, respiration	biodiversity, carbon sink, distribution

Lesson plan

Reactivate knowledge

Common misconception

Students may assume that global warming only causes an increase in temperature, whereas in fact the effects on climate are much more varied.

1. What are the three main reasons for deforestation?
2. What term describes cutting down rainforests and burning them to clear land for farming?
3. Forests are a major carbon sink. Name another which is widely used for fuel and for growing plants.

Answers: 1 Crops for food/cattle for beef/crops for biofuels 2 Slash-and-burn 3 Peat bogs

Trigger interest

Carbon cycle review (10 mins)

Ask students to help compile a diagram of the carbon cycle on the board by thinking about ways in which carbon dioxide is removed from the atmosphere, and ways in which it is added. Discuss the idea that for thousands of years, these processes were in balance and the amount of carbon dioxide in the air remained relatively constant.

Exposition

Global warming (20 mins)

Show the students a diagram of the greenhouse effect (Figure 2 in the Student Book), to show how greenhouse gases (methane and carbon dioxide) in the atmosphere help maintain the Earth at a temperature suitable for life. Students should have already learnt about the greenhouse effect in physics. Ask students to use their knowledge from physics, and from the diagram, to write out a description of the greenhouse effect. Ask them to include the following key terms in their answers: reflect, transmit, atmosphere, infrared, Sun, source.

Discuss how human activity has increased the volume of greenhouse gases in the air. The Activity: *Global warming* will be useful in this.

As an extension, encourage students to give their opinions about climate change. Discuss whether they think reducing climate change should be a priority for all governments.

Main activity

Changing conditions (20 mins)

Tell the class about the two main consequences of global warming – rising sea levels and climate change.

Use a map to show predictions for flooding in areas of the UK, or possibly the local area, as sea levels rise.

Ask students to divide a sheet of paper in two and write 'Rising sea levels' in the middle of one side of the page and 'Climate change' in the middle of the other. They can then use information from the Student Book to draw a spider diagram showing how each of these consequences will affect living organisms.

Review and practice

Effect of climate change on people (5 mins)

Tell the class the following statement and ask them to discuss in pairs: *The effects of climate change are global and their projected impacts concern every area in the world, however scientists suggest that climate risks disproportionately affect the poorest countries and people, who are more exposed, and more vulnerable to their impacts.*

Students can discuss to what extend they agree or disagree as well as make suggestions to how people living in poorer countries are more affected by climate change.

Ⓗ B18.6 The impact of change

AQA GCSE Biology specification points

Ⓗ 4.7.2.4 Impact of environmental change

Smart Curriculum Narrative

Earth and environment

Theme/subtheme

Ecosystems and environment/Ecosystems

Disciplinary knowledge

- WS 1.4

Activity

Changing birds of Britain

	Learning outcomes		
Learning objectives	**Developing** **Aiming for 3**	**Secure** **Aiming for 5**	**Extending** **Aiming for 8**
Categorise environmental changes as due to seasonal changes, geographical changes, human interaction, or a combination of them all	State some examples of environmental changes that affect the distribution of species in an ecosystem	**Categorise environmental changes as due to seasonal changes, geographical changes, human interaction, or a combination of them all**	Explain why environmental changes might change the distribution of a species
Explain how humans may cause environmental changes which have positive or negative impacts on the distribution of species in an ecosystem	Describe an example of how environmental change has affected the distribution of a species	**Explain how humans may cause environmental changes which have positive or negative impacts on the distribution of species in an ecosystem**	Predict and explain how an environmental change will affect the distribution of an organism
Analyse and interpret data linking environmental change and species distributions	Describe graphs showing environmental changes and species distribution	**Analyse and interpret data linking environmental change and species distributions**	Manipulate and interpret data to predict future environmental changes and the impacts they may have on species distribution

Tier 2 vocabulary	Tier 3 vocabulary
adaptation, altitude, climate change, deforestation, environmental change, extinction, global warming, habitat, migration, pH, predator, species	acid rain, biodiversity, distribution, geographical change, pathogen, seasonal change, temperate area, tropical area

Lesson plan

Reactivate knowledge

1 Give two examples of greenhouse gases.
2 Why is the greenhouse effect so important for life on the Earth?
3 What potential biological consequences may result if global warming continues at its current rate?

Common misconception

Students may assume that global warming only causes an increase in temperature, whereas in fact the effects on climate are much more varied.

Answers: 1 Carbon dioxide and methane 2 Keeps the temperature at the surface warm enough for organisms to survive 3 Any from: loss of habitat, changes in distribution, changes in migration patterns, reduced biodiversity

Trigger interest

What lives where? (10 mins)

Split the class into small groups. Give each group an image of a different environment, e.g., polar region, African savannah, rainforest, UK woodland. Ask students to describe the abiotic factors in each environment and suggest organisms that live in each one. As a class, discuss why different organisms are found in different environments.

As an extension, ask students to predict what would happen to the organisms in the environment if the abiotic factors were to change.

Exposition

What would happen? (10 mins)

Give students examples of organisms in their habitats, e.g., rabbit in a UK field.

Present possible environmental changes, e.g., increase in rainfall over the summer or a very cold winter.

Ask students to discuss in pairs their predictions for how the organisms' distribution might change, with reasons.

Main activity

Changing birds of Britain (30 mins)

Students read the newspaper article on the Activity worksheet and then summarise the information.

Students should include:

- abiotic factors affecting the birds' distribution
- how this is changing
- why they are changing
- the effect on bird population
- what is being done to prevent further disruption to distribution.

Alternatively, use the example of the Dartford warbler from the corresponding section in the Student Book.

Review and practice

Types of environmental change (5 mins)

Using the interactive, students categorise environmental changes into whether they are due to seasonal changes, geographical changes, or human interaction.

B18.7 Maintaining biodiversity

AQA GCSE Biology specification points

4.7.3.6 Maintaining biodiversity

Smart Curriculum Narrative

Earth and environment

Theme/subtheme

Biodiversity and human interactions with the environment/Conservation

Disciplinary knowledge

- WS 1.4, 1.5

Activity

Heathland restoration

	Learning outcomes		
Learning objectives	**Developing Aiming for 3**	**Secure Aiming for 5**	**Extending Aiming for 8**
Describe how people are working to maintain biodiversity directly	List some ways in which people can help maintain biodiversity	**Describe how people are working to maintain biodiversity directly**	Evaluate the conflicting pressures on maintaining biodiversity in some habitats
Describe projects designed to prevent more environmental damage and so maintain biodiversity	Describe the reasons why some habitats are at risk	**Describe projects designed to prevent more environmental damage and so maintain biodiversity**	Discuss some of the issues arising from attempts to prevent further environmental damage or to restore biodiversity directly
Use information and data to explain the conflicting pressures on maintaining biodiversity	Draw simple conclusions from given data	**Use information and data to explain the conflicting pressures on maintaining biodiversity**	Interpret and evaluate data on different methods of tackling human impacts on the environment

Tier 2 vocabulary	Tier 3 vocabulary
carbon dioxide, deforestation, ecosystem, global warming, habitat, pH, pollution, population, recycling, species	biodiversity, breeding programme, compost, coral reef, heathland, hedgerow, landfill, mangrove, methane generator, salinity, soil erosion

Lesson plan

Reactivate knowledge

1 What is biodiversity?

2 How will rising sea levels affect biodiversity?

3 State one way in which global warming and climate change will affect biodiversity.

Answers: 1 The variety of species of animals and plants in an ecosystem or globally 2 Flood low-lying areas/remove habitat and so reduce biodiversity 3 Increase rate of extinctions as organisms cannot adapt to changes fast enough

Common misconception

Students may only consider the harmful impacts of large animals going extinct, and may not think about the implications of plant and insect species going extinct.

Trigger interest

Odd one out (5 mins)

Ask students to choose which is the odd one out from: orangutan, polar bear, blue whale, and bottlenose dolphin. The bottlenose dolphin is the only animal in the list that is not currently an endangered species.

Exposition

Protecting endangered species (20 mins)

As a class, discuss reasons why some species are at risk of extinction, and why this has an impact on other organisms, including humans.

Ask students to work in groups of four and give them some sticky notes. Ask them to use information from the Student Book to write down the six ways to protect biodiversity (breeding programmes, protection of rare habitats, reintroduction of hedgerows, reduction of deforestation, reductions of carbon dioxide, and recycling), one on each note, and stick them to the desk.

Then, ask students to each choose one method and work alone. They should find one example of an organism that is being protected in this way, state one place in the world this method is being used, and summarise how the method can help maintain biodiversity.

They should write each piece of information on a sticky note and place it with the protection method.

Each student can then explain their protection method, with the example, to the rest of their group.

Consider photocopying the information in the Student Book and allowing students to highlight relevant parts.

Main activity

Heathland restoration (20 mins)

Students use the map of the UK showing areas of heathland in this activity to locate their nearest area of heathland. Students should research what a heathland is, which species live in heathlands, the reasons why the habitat is under threat, problems this can bring, and methods that are being used to maintain these areas (access to computers will be needed).

Students can present what they find out as a written report, presentation, or leaflet, which could be completed for homework.

As an extension, students should consider conflicting interests, e.g., the need for housing for a growing population versus the need to protect unique habitats.

Review and practice

Ban zoos? (10 mins)

Remind the class that zoos carry out breeding programmes. Divide the class into two. Ask one half to argue that keeping animals captive in zoos should not be allowed, and the other half to argue the opposite side.

To support, ask students to work in pairs and give them statements for and against zoos. They can sort the statements into those for and those against zoos.

B18.8 Trophic levels and biomass

AQA GCSE Biology specification points

4.7.4.1 Trophic levels

4.7.4.2 Pyramids of biomass

Smart Curriculum Narrative

Ecosystems and environment

Theme/subtheme

Ecosystems and environment/Feeding relationships

Disciplinary knowledge

- WS 1.2
- MS 2c

Practical

Leaf litter investigation

Learning objectives	Learning outcomes: Developing – Aiming for 3	Learning outcomes: Secure – Aiming for 5	Learning outcomes: Extending – Aiming for 8
Number and describe the trophic levels in a food chain, food web, and pyramid of biomass	Define the term trophic levels	**Number and describe the trophic levels in a food chain, food web, and pyramid of biomass**	Explain how trophic levels fit into a pyramid of biomass
Explain the shape of pyramids of biomass	Identify a producer, primary consumer, and secondary consumer from a food chain and pyramid of biomass	**Explain the shape of pyramids of biomass**	Explain the role of the decomposers in the ecosystem
Use appropriate data to draw a pyramid of biomass and explain what it shows	Describe a pyramid of biomass	**Use appropriate data to draw a pyramid of biomass and explain what it shows**	Explain the value of a pyramid of biomass compared to a pyramid of numbers

Tier 2 vocabulary	Tier 3 vocabulary
algae, carnivore, concentration gradient, enzyme, food chain, food web, herbivore, microorganism, photosynthesis, predator, primary consumer, *producer*, secondary consumer, tertiary consumer	apex predator, biomass, decomposer, dry biomass, feeding relationship, pyramid of biomass, trophic level, wet biomass

Lesson plan

Reactivate knowledge

1 What are the main problems with breeding programmes for endangered species?

2 How are people working to reduce the carbon emissions of global transport?

3 What is recycling?

Answers: 1 May not breed easily or fast/habitat may have been lost/need to avoid inbreeding 2 Any from: cleaner fuels, more efficient engines, more electric vehicles, using renewable sources to generate electricity 3 Reusing as many materials as possible such as glass, plastics, metal, and paper

Common misconception

Students may think that the arrows in a food chain represent what eats what, rather than the transfer of energy.

Trigger interest

Pyramids of number (10 mins)

Show the class the food chain:

oak tree → aphids → ladybirds. Ask students to draw a pyramid of numbers. Reveal the answer (see Figure 4 in the Student Book).

If the class are struggling, do this activity together.

Once students have completed their pyramids, ask them to discuss why pyramids of number are not always in a pyramidal shape.

Exposition

Trophic levels (10 mins)

Ask students to write down a food chain with four organisms. Then ask them to label it using the terms producer, primary consumer, secondary consumer, and tertiary consumer.

Use a food chain chosen by a student as an example to introduce the term 'trophic levels'. Ask students to label each of their trophic levels with a number, starting from the bottom.

Students should then draw a pyramid of biomass for their trophic levels (not to scale, but following the concept that because energy and biomass are lost at each trophic level, each bar going up the pyramid becomes narrower).

Main activity

Leaf litter investigation (30 mins)

Students can either identify the organisms in leaf litter and measure their mass to construct pyramids of biomass (see the practical instructions), or alternatively provide students with a method and results from a similar investigation.

They can use squared paper to construct their pyramids of biomass using a suitable scale. They should explain what the pyramid shows about the habitat. Students may need another lesson for this analysis.

Review and practice

Spot the pyramid (5 mins)

Students complete the interactive to select the correct pyramid of biomass for each food chain. This can either be done individually or *cold-calls* from the class.

B18.9 Biomass transfers

AQA GCSE Biology specification points

4.7.4.3 Transfer of biomass

Smart Curriculum Narrative

Ecosystems and environment

Theme/subtheme

Ecosystems and environment/Feeding relationships

Disciplinary knowledge

- MS 1c

Activity

Loss of biomass and energy transfers

Learning objectives	Learning outcomes: Developing Aiming for 3	Secure Aiming for 5	Extending Aiming for 8
Describe the energy transfer that occurs during photosynthesis	State that biomass is lost between different trophic levels	**Describe the energy transfer that occurs during photosynthesis**	Suggest reasons why only 1% of the incident energy from light is used for photosynthesis
Describe how biomass is lost between trophic levels	List the ways biomass is lost between trophic levels	**Describe how biomass is lost between trophic levels**	Explain the difference between biomass lost in faeces and biomass lost in waste
Plan an investigation to demonstrate that biomass is lost as organisms respire, transferring energy to their surroundings	Describe a simple investigation to show that the surroundings warm up as energy is transferred to the surroundings during respiration in organisms	**Plan an investigation to demonstrate that biomass is lost as organisms respire, transferring energy to their surroundings**	Make a detailed plan of an investigation which shows both that biomass is lost as organisms respire, and that energy is transferred to the surroundings, warming them up

Tier 2 vocabulary	Tier 3 vocabulary
algae, amino acid, ammonia, carnivore, cellulose, food chain, herbivore, photosynthesis, prey, producer, protein, respiration, urea	biomass, decomposer, dry biomass, pyramid of biomass, trophic level, wet biomass

Lesson plan

Reactivate knowledge

1 Which organisms make up trophic level one?

2 At which trophic levels would you expect to find carnivores?

3 What do we call the biological material produced by living organisms?

Answers: 1 The producers – plants and algae 2 Trophic level three, trophic level four 3 Biomass

Common misconception

Students may think that the arrows in a food chain represent what eats what, rather than the transfer of energy.

Trigger interest

Photosynthesis (10 mins)

Ask students to write out the word equation for photosynthesis on their mini-whiteboards. Use this to remind students that plants use photosynthesis to produce glucose from carbon dioxide and water, and that light is needed.

Ask students to write down the energy transfer taking place (light energy transferred to the chemical energy store). Discuss the fact that only around 1% of the light falling on producers (plants and algae) is transferred.

As an extension, ask students to suggest what happens to the rest of the light (absorbed by other parts of the leaf rather than the chloroplasts, reflected, passes through the leaf, etc).

Exposition

Biomass efficiency (15 mins)

Using the worksheet Calculation: *Efficiency of energy transfer*, show the class a pyramid of biomass and ask students to calculate the percentage of biomass passed between each trophic level. Then they can calculate the efficiency of transfer for each.

Ask students to use information from the Student Book to summarise the reasons why biomass decreases as you go up a food chain.

To support, show students how to calculate the percentage efficiency of energy transfer from producer to primary consumer, and then allow students to calculate this for the transfer between primary and secondary consumer.

Main activity

Loss of biomass and energy transfers (25 mins)

Ask students to plan an investigation to explore biomass loss and energy transfer during respiration in plant seeds. They can work in pairs to develop a detailed investigation plan. The main objective is for students to consider variables such as:

- the type of seeds (germinating vs. non-germinating)
- the duration of the experiment
- what measurements they will take
- the control of environmental factors like temperature and humidity.

As an extension, students can develop more intricate investigation plans, considering both measurements to take to measure loss of biomass and energy transfer. They can then evaluate their current or other models used to discuss its limitations.

Review and practice

Marine food chains (5 mins)

Students use the interactive to answer a multiple choice question about why marine food chains can be a lot longer than those on land.

Biology only

B18.10 Factors affecting food security

AQA GCSE Biology specification points

4.7.5.1 Factors affecting food security

Smart Curriculum Narrative

Ecosystems and environment

Theme/subtheme

Biodiversity and human interactions with the environment/Land use, deforestation, and food production

Disciplinary knowledge

- WS 1.4

Activity

Food security convention

Learning objectives	Learning outcomes		
	Developing Aiming for 3	Secure Aiming for 5	Extending Aiming for 8
Define sustainable food production and explain how it could help increase food security	Define food security	**Define sustainable food production and explain how it could help increase food security**	Discuss whether malnutrition is just a problem in developing countries
Describe how biological factors affect food security	List birth rate, diet changes, and new pests and pathogens as examples of biological factors that reduce food security	**Describe how biological factors affect food security**	Suggest how biological factors might impact sustainable food production
Describe how environmental factors affect food security	List droughts and floods, the costs of agricultural inputs, and human conflicts as environmental factors that reduce food security	**Describe how environmental factors affect food security**	Suggest how environmental factors might impact sustainable food production
Interpret population and food production statistics to evaluate food security	Extract information from population and food production statistics with support	**Interpret population and food production statistics to evaluate food security**	Present information on food security clearly with supporting evidence and suggest ways of developing sustainable food production

Tier 2 vocabulary	Tier 3 vocabulary
balanced diet, climate change, fertiliser, genetic engineering, global warming, pesticide, yield	agricultural input, biological factor, birth rate, fish stock, food security, irrigation system, malnutrition, pathogen, pest, sustainable food production

Lesson plan

Reactivate knowledge

1 How is biomass lost in faeces?

2 What are the waste products of respiration and excess protein breakdown?

3 Which group of organisms probably contains the most biomass?

Common misconception

Students may think malnutrition refers only to a lack of food (in terms of quantity), rather than considering a lack of particular food groups in someone's diet. Students may not realise obesity is classed as a type of malnutrition.

Answers: 1 Some animals cannot digest the cellulose in the plant material they eat/carnivores cannot digest all of the animals they eat/undigested biomass passed out in faeces so not passed on 2 Carbon dioxide and urea 3 Decomposers

Trigger interest

Malnutrition (5 mins)

Ask students to discuss the idea of malnutrition in pairs and come up with a definition for it.

Ask students to discuss whether malnutrition is just a problem in developing countries.

Exposition

Defining food security (10 mins)

Introduce the term food security. Ask students to use the Student Book to find the three main aspects of food security (biological factors, the environmental changes, and sustainable solutions). Split them into three groups and assign them one aspect to draw a range of images to illustrate. A selection of the images drawn for each category could be shown to the class using a visualiser.

Main activity

Food security convention (30 mins)

Split the class into seven groups. Six of the groups contain delegates from countries where food security is an issue:

- Uganda – increasing birth rate
- Bolivia – changing diet (quinoa being exported)
- Cuba – new pathogen (banana disease called Black Sigatoka)
- Pakistan – environmental changes (floods)
- Kenya – high costs of agricultural inputs
- Ukraine – conflict.

Ask students to research the problems they face in their given country using the Student Book and online, if available. Students should find out about the problem, how it affects food security, and what they need from other countries to help them solve it. The seventh group contains food security experts. They should research how sustainable solutions can help each particular issue. After completing their research, each group will present the challenge their country faces. The food security experts can present possible solutions.

This activity can be altered for a class of prior lower attainment by holding a class debate instead. The teacher can act as the food security expert.

Review and practice

Refugee crisis (10 mins)

Discuss how war and conflicts can impact food security both nationally and globally. Be mindful of any students who have personal experience of this.

To support, ask students to consider the impacts this may have on exports, imports, transport links, land destruction, and money available for the country.

As an extension or for homework, students can read news stories linked to food security and summarise their findings.

B18.11 Making food production efficient

AQA GCSE Biology specification points

4.7.5.2 Farming techniques

Smart Curriculum Narrative

Ecosystems and environment

Theme/subtheme

Biodiversity and human interactions with the environment/Land use, deforestation, and food production

Disciplinary knowledge

- WS 1.3, 1.4

Activity

Chicken speech

Learning objectives	Learning outcomes: Developing – Aiming for 3	Learning outcomes: Secure – Aiming for 5	Learning outcomes: Extending – Aiming for 8
Describe three elements of artificially managed food production and explain how each helps increase the rate of growth of food animals	State that the efficiency of food production can be improved by limiting the movement of food animals and controlling the temperature of their surroundings	**Describe three elements of artificially managed food production and explain how each helps increase the rate of growth of food animals**	Explain in detail why, in terms of food production efficiency, it is a good idea to reduce meat in the diet
Explain why there are ethical objections to some intensive farming techniques	List some advantages and disadvantages of free-range and intensive farming of animals	**Explain why there are ethical objections to some intensive farming techniques**	Evaluate the advantages and disadvantages of modern farming techniques
Compare fish farming in the sea with intensive farming of chickens or cows on the land	State that fish are farmed intensively in fish farms	**Compare fish farming in the sea with intensive farming of chickens or cows on the land**	Discuss the attitudes of people to fish farming compared to intensive farming on the land

Tier 2 vocabulary	Tier 3 vocabulary
efficiency, energy transfer, food chain, respiration, urea	biomass, fish stock, pyramid of biomass

Lesson plan

Reactivate knowledge

1 What is food security?

2 State three threats to global food security.

3 How might we overcome some of the problems of food security?

Common misconception

Students may think it is more energy efficient to eat animals further up the food chain than those at lower trophic levels.

Answers: 1 Having enough food to feed a population 2 Any three from: increasing birth rate, changing diets in developed countries, new pests and pathogens, environmental changes 3 Maintain soil quality/find more efficient ways to produce food/take care of fish stocks

Trigger interest

Meat-free eating (10 mins)

Tell the class that eating less meat could mean more food for everyone. Ask students to discuss in pairs why this is, in terms of biomass loss. To help with this, present them with food chains for plants and meat, for example:
wheat → person; wheat → chicken → person.

Exposition

Increasing efficiency (20 mins)

Remind students that biomass is lost as animals use it for respiration to release energy. This energy is needed for movement and for controlling their body temperature.

Ask the class how they feel about eating insects. Ask them to suggest why, in terms of food production efficiency, this could be a good idea. Information about eating insects can be found on online. You could even consider having samples of insects for students to try.

Next, ask the class to consider how caged hens are kept, and explain why this increases the rate of growth. Students can use information from the Student Book for further guidance. Inform students about techniques used to supply human diets with fish, which have developed to meet the demands globally (e.g., trawling, farming). Talk through the damage this causes to biodiversity. In pairs, ask students to identify any common advantages and disadvantages both intensive chicken farming and fish farming share.

Main activity

Chicken speech (20 mins)

Students write a short speech in support of factory farming of chickens, and a second speech against it. They should include the advantages and disadvantages of this method, and use scientific facts and data as evidence. Then ask students to draw lots as to which side to present during a class debate. You could open up the discussion to debate views on the raising of calves in pens to produce veal.

As an extension, students could present viewpoints from different viewpoints, e.g., battery hen farmer, parent on a low income, etc.

Review and practice

Fish farming (5 mins)

In this interactive, students are presented with ways that growth is increased in farmed fish (e.g., small cages and feeding with high-protein food) and must match the techniques to explanations of how they work.

B18.12 Sustainable food production

AQA GCSE Biology specification points

4.7.5.3 Sustainable fisheries

4.7.5.4 Role of biotechnology

Smart Curriculum Narrative

Ecosystems and environment

Theme/subtheme

Biodiversity and human interactions with the environment/Land use, deforestation, and food production

Disciplinary knowledge

- WS 1.4

Activity

The role of biotechnology

Learning objectives	Learning outcomes		
	Developing Aiming for 3	Secure Aiming for 5	Extending Aiming for 8
Describe the reasons why fish stocks in the oceans are decreasing	Give one reason why fish stocks in the oceans are decreasing	**Describe the reasons why fish stocks in the oceans are decreasing**	Explain the negative impacts of fishing restrictions on communities
Describe the techniques used to conserve fish stocks	Describe one reason why a reduction in fish stocks is a problem	**Describe the techniques used to conserve fish stocks**	Interpret data on the impact of different conservation techniques on the recovery of fish stocks
Describe what mycoprotein is and how it is produced	State that biotechnology techniques enable large quantities of microorganisms to be cultured for food and describe mycoprotein as a novel food based on fungi	**Describe what mycoprotein is and how it is produced**	Discuss the advantages of eating mycoprotein or other foods produced by biotechnology for the individual and for society as a whole

Tier 2 vocabulary	Tier 3 vocabulary
conservation, extinction, fungus, microorganism, population, vitamin A, yield	aerobic, biotechnology, culture, *Fusarium*, mycoprotein, overfishing, quota

Lesson plan

Reactivate knowledge

Common misconception

Students may confuse the concept of genetic engineering with cloning.

1 Why do farmers often feed high protein food to animals?

2 What effect does rearing animals in temperature-controlled indoor spaces have on the efficiency of meat production?

3 Why is good animal welfare important in intensive farming?

Answers: 1 To increase their growth rate 2 Less movement and no energy transferred for heating or cooling so less biomass used in respiration 3 Contented animals gain biomass faster than stressed ones

Trigger interest

Remembering GM crops (5 mins)

Give students two minutes to write down as much as they can remember about genetically modified crops. Spend a few minutes sharing their ideas as a class and correcting any misconceptions.

Exposition

Sustainable fishing (20 mins)

Discuss the problem with overfishing and why it is necessary to conserve fish stocks. Present the restrictions used to increase fish stocks – fishing quotas, controlling the size of the holes in the nets, and bans on fishing during breeding seasons. Ask students to use information from the Student Book to explain how each method works.

Main activity

The role of biotechnology (20 mins)

Search for information about the production of mycoprotein to share with the class, such as diagrams and flow charts about how it is produced, and nutritional information. You can also show the class videos of adverts for mycoprotein meat substitutes. Ask students to use this information, plus that in the Student Book, to create an information leaflet on mycoprotein.

If students require support with this task, provide them with a pre-prepared template to show the kind of information that needs to be included in the leaflet.

Review and practice

Consequences spider diagram (10 mins)

Ask students to consider what might happen if the government decided that eating animals should be banned and the nation should eat more mycoprotein instead. Ask students to write 'ban animal farming' in the middle of a sheet of paper and write consequences around it that are linked (e.g., there would be fewer farms, which would lead to people losing their jobs).

As an extension task, students should distinguish between positive and negative consequences, for example, by underlining or putting a circle around the statements.

B18 Biodiversity and ecosystems

Introduction to checkpoint intervention

The checkpoint intervention provides suggestions for a lesson to follow up the *B18 Biodiversity and ecosystems* assessment. Depending on the outcomes of the assessment, these suggestions can be used to move a class towards secure, to target specific learning outcomes, or to consolidate knowledge for students achieving secure outcomes. Use students' outcomes from the checkpoint assessment to address general misconceptions from the content covered. Be prepared to re-cover content as required; three key concepts from the chapter are covered in more detail below.

Checkpoint secure learning outcomes

- Describe how acid rain is formed.
- Describe the importance of a good level of biodiversity to the stability of an ecosystem.
- Describe the effects of deforestation and peat removal in terms of atmospheric carbon dioxide concentrations and biodiversity.
- Calculate the percentage of biomass passed between trophic levels and explain how this loss of biomass affects the numbers or biomass of organisms at each trophic level.
- Describe projects designed to prevent more environmental damage and so maintain biodiversity.
- Describe the techniques used to conserve fish stocks.

Student reflection

Encourage students to reflect on whether there were any checkpoint questions they found difficult or straightforward as well as their preparation for the checkpoint assessment, including how they revised, what they revised, the time spent on revision, and what areas they could improve upon. For more strategies, see the Metacognition spread.

Intervention activity: Describe how acid rain is formed.

Provide students with a diagram demonstrating the formation of acid rain, such as Figure 1 from Topic 18.3 of the Student Book. Ask students to write a paragraph describing how acid rain is formed, using the diagram to prompt them if necessary. Their answers should include:

1. Acid gases such as carbon dioxide, sulfur oxide, and nitrogen oxides are formed when fossil fuels are burnt.
2. Sulfur impurities react oxygen to form sulfur dioxide gas.
3. Sulfur dioxide and nitrogen oxides dissolve in rainwater.
4. This reacts with the oxygen in the atmosphere to form dilute sulfuric and nitric acid.

Intervention activity: Describe the effects of deforestation and peat removal in terms of atmospheric carbon dioxide concentrations and biodiversity.

Ask students to complete the paragraph below by filling in the gaps using the following words:

photosynthesis, biodiversity, compost, decays, carbon dioxide, combustion

Destruction of peat bogs to produce garden ________________ or for fuel reduces the area of this habitat and therefore the ________________ associated with it. When the peat is burnt or when it ________________, this releases carbon dioxide into the atmosphere.

Deforestation decreases the volume of ________________ being taken in by plants from the atmosphere for ________________. Trees that have been cut down are often burnt to clear wooded areas for farming. This ________________ releases carbon dioxide into the atmosphere.

Answers: compost, biodiversity, decays, carbon dioxide, photosynthesis, combustion

Intervention activity: Describe the techniques used to conserve fish stocks.

Students match up each 'technique' to the correct 'explanation' in the table below to explain how each fishing technique helps to conserve fish stocks.

Technique
Larger holes in fishing nets
Fishing quotas
Fishing banned during breeding season

Explanation
This provides the opportunity for fish to reproduce to help increase population sizes.
This means younger, smaller fish are not caught and can live to grow and breed.
This means there are strict rules on the amount and type of fish that can be caught.

Answers

B1.1

1a Advantages: relatively cheap, can be used almost anywhere, magnification up to around ×2000 [1]; disadvantages: limited magnification and resolution [1]

b Advantages: high magnification (up to around ×2 000 000), high resolution, can give 3D images [1]; disadvantages: expensive, can only be used in temperature, pressure, and humidity-controlled rooms [1]

2 Magnification = $\frac{\text{size of image}}{\text{size of real object}}$

magnification = $\frac{800}{20}$ = ×40 [2]

3 Size of real object = $\frac{\text{size of image}}{\text{magnification}}$

capillary diameter = $\frac{5}{1000}$ = 0.005 mm = 5.0 μm [3]

4 Electron microscopes: magnify up to around ×2 000 000, [1] have a resolving power of about 10 nm (scanning electron microscope) or 0.2 nm (transmission electron microscope), [1] may be used to examine subcellular structures (e.g., chromosomes during cell division) [1]; light microscopes: magnify up to around ×2000, [1] have a resolving power of about 200 nm, [1] may be used to look at cells dividing (e.g., stained onion cells) [1]

B1.2

1a Nucleus, [1] cytoplasm, [1] cell membrane, [1] mitochondria, [1] ribosomes [1]

b Cell wall, [1] chloroplasts, [1] permanent vacuole [1]

2 Cell wall strengthens cell and provides support, [1] chloroplasts for photosynthesis, [1] permanent vacuole keeps cells rigid to support plant [1]

3 Nucleus: controls all cell activities, [1] contains instructions for making new cells or new organisms [1]; mitochondria: site of aerobic respiration, [1] releasing energy for the cell [1]

4 Any two from root cells – no exposure to light, cells in centre of tree trunk – no exposure to light, cells in flowers of plants – their function is not to photosynthesise [4] Any other valid suggestion.

B1.3

1a Genetic material in a prokaryotic cell isn't contained in a nucleus [1] and may include extra rings of DNA (plasmids) separate from main genetic material. [1]

b i Long protein strands that lash about [1]

ii Movement [1]

2 All cells have cell membranes [1] and cytoplasm [1] and both prokaryotes and eukaryotes can have a cell wall. [1] Prokaryotes have no nucleus [1] and no chloroplasts [1], whilst eukaryotes have no plasmids. [1]

3a Small animal cell length around 10 μm

$\frac{10}{6}$ = 1.7

Length of small animal cell is same order of magnitude as cell nucleus. [2]

b Large plant cell length around 100 μm

$\frac{100}{6}$ = 16.7

Length of large plant cell is an order of magnitude (10^1) bigger than cell nucleus. [2]

B1.4

1 Specialised cells form tissues and organs, that form a whole organism. [1]

2a Any one from lots of dendrites – to make connections to other nerve cells, long axons – to carry nerve impulse, synapses – to pass impulse to another cell or between nerve cell and muscle using transmitter chemicals [2]

b Any one from special proteins – that slide over each other to contract fibres, many mitochondria – to produce energy for movement, glycogen storage – for cellular respiration [2]

c Any one from long tail – to move sperm towards egg, many mitochondria – to produce energy for movement, acrosome – stores digestive enzymes for breaking down outside layers of egg, large nucleus – to contain genetic material to be passed on [2]

3 Transmitter chemicals [1] are required to pass impulse from cone cell to another nerve cell and then on to brain. [1] The many mitochondria [1] supply energy from cellular respiration needed to make transmitter chemicals [1]

4 Any three from mitochondria – number of mitochondria indicates how much energy cell uses, flagella or cilia – presence indicates whether cell moves around or moves substances such as mucus, nucleus – presence indicates whether cell is capable of reproduction, storage materials such as fat or starch – presence indicates whether cell stores materials it can use for respiration [6] Any other valid feature.

B1.5

1a Any one from increased surface area – to promote water uptake, large permanent vacuole – to speed up movement of water by osmosis, many mitochondria – to produce energy for active transport of minerals [2]

b Any one from cells die forming long hollow tubes – to allow water and mineral ions to move easily through them, lignin spirals formed – to help cells withstand pressure of water moving up plant and to support plant stem [2]

c Any one from cell walls break down to form sieve plates – to enable water and dissolved food to pass through, companion cells – to support phloem cells and provide energy to move substances up and down in phloem [2]

d Any one from chloroplasts – contain chlorophyll to trap light needed for photosynthesis, positioned in leaves and outer layers of stem – to absorb as much light as possible, large permanent vacuole – to keep cell rigid and support leaf and stem to capture light [2]

2 Cell not exposed to sunlight and therefore not adapted to photosynthesise (not photosynthetic). [2]

3 Any three from: chloroplasts – presence indicates whether cell is photosynthetic, large vacuole – presence indicates role in osmosis/rigid support, lignin spirals – presence indicates strengthening and transport of water, sieve plates/companion cells – presence indicates transport of dissolved food, mitochondria – presence indicates active movement of substances [6]

B1.6

1 Spreading out of particles of a gas, or a substance in solution, along a concentration gradient (from area of higher concentration to area of lower concentration). [1] This takes place as a result of random movement of particles. When particles are concentrated, there are more collisions. [1]

2 Folded membranes provide increased surface area. [1] The greater the surface area, the more diffusion of dissolved substances can take place across it. [1]

3 Heating makes particles move more quickly, [1] speeding up diffusion as particles collide more often and harder, [1] and spread out faster. [1]

4a Digested food molecules move from gut (high concentration) into bloodstream (low concentration) down a concentration gradient. [1] Large surface area of small intestine lining increases rate of diffusion. [1] Rich blood supply maintains concentration gradient. [1]

b Carbon dioxide moves from blood (high concentration) into air in the alveoli of the lungs (low concentration) down a concentration gradient. [1] Large surface area of alveoli increases rate of diffusion. [1] Rich blood supply maintains concentration gradient. [1]

c Chemicals produced by female moth spread out into air around her down a concentration gradient. [1] Chemicals more concentrated close to female moth (high concentration) than further away (low concentration). [1] Male moth flies up concentration gradient, following chemical to reach female moth. [1]

B1.7

1a Osmosis is the diffusion of water from a dilute solution to a concentrated solution [1] through a partially permeable membrane. [1]

b Osmosis takes place through a partially permeable membrane [1], diffusion takes place in any solution or gas. [1] Only water molecules are involved in osmosis [1], many different types of molecules/particles are involved in diffusion. [1]

2 Large intake of water dilutes the blood and body fluids so they are less concentrated than the cell cytoplasm. [1] As a result, water moves into brain cells by osmosis causing the brain to swell [1]. Red blood cells absorb water by osmosis, swell and burst [1] so can no longer carry oxygen around the body. [1]

3a i Solution with same concentration of solutes as inside of cell. [1]

ii Solution with lower concentration of solutes than inside of cell. [1]

iii Solution with higher concentration of solutes than inside of cell. [1]

b If solute concentration outside body cells is more dilute than cell contents, water will move into cells by osmosis – cells will swell and may burst. [1] If solute concentration outside body cells is higher than cell contents, water will leave cells by osmosis – cells will shrink and stop working properly. [1] Solute concentration outside body cells must be as constant as possible to minimise changes in size and shape of cells, [1] keeping them working normally. [1]

B1.8

1 Plants rely on osmosis to support their stems and leaves by keeping the cells turgid (firm) [1]. If a plant loses too much water from its cells then they become flaccid (soft) and the plant wilts. [1]

2 mass gained: surrounding sugar solution hypotonic to beetroot cells, [1] so water moved into cells by osmosis [1]; mass lost: surrounding sugar solution hypertonic to beetroot cells, [1] so water moves out of cells by osmosis [1]; mass constant: surrounding sugar solution isotonic to beetroot cells, [1] so no net movement of water into or out of cells by osmosis. [1]

3 Plant cells are turgid when the solution surrounding them is hypotonic to the cytoplasm. [1] Water moves by osmosis into the cells so the cytoplasm swells and presses on the cell walls and the cell is turgid. [1]

Plant cells are flaccid when the solution surrounding them is hypertonic to the cytoplasm. [1] Water leaves the cells by osmosis. They become flaccid (soft) as there is no pressure from the cytoplasm on the cell walls. [1]

If more water is lost by osmosis, the vacuole and cytoplasm shrink, and eventually the cell membrane pulls away from the cell wall. [1] This is plasmolysis.

Plasmolysed cells die quickly unless the osmotic balance is restored. It is rarely seen in living plants because if it happens, they rapidly die. [1]

B1.9

1 Useful molecule binds to transport protein in cell membrane. [1] Transport protein changes shape and moves useful molecule across membrane into cell [1] against concentration gradient. [1] Useful molecule is released, transport protein returns to original position. [1]

2a Active transport: substances moved against concentration gradient across partially permeable membrane, [1] process uses energy supplied by cellular respiration [1]; osmosis and diffusion: substances moved down concentration gradient, no requirement for energy from respiration [1]

b Cellular respiration in mitochondria releases energy needed for active transport. [1] Cells that carry out a lot of active transport often have many mitochondria to meet their energy requirements. [1]

3a Plants need to move mineral ions from soil into their roots. [1] Mineral ion solutions in soil are much more dilute than the solution in plant root hair cells, [1] so they are moved against a concentration gradient using active transport. [1]

b Marine birds are exposed to salt water in the sea. [1] They use active transport to remove excess salt from the body [1] against a concentration gradient. [1]

B1.10

1 For animals, any from: small intestine, lungs, gills. [1] For plants, any from root or leaf. [1]

2 Any two from: large surface area, being thin, having an efficient blood supply (in animals), being ventilated (in animals). [2]

3a Students should draw a large cube and a smaller cube, give them sensible dimensions and calculate the surface area to volume ratio of each. [3] They should compare the two cubes, showing that the ratio gets smaller as the cube (and therefore organism) gets larger. [2]

b Organisms with a large ratio can use diffusion alone. [1] Organisms with a small ratio cannot rely on diffusion alone so require a transport system/exchange surfaces. [1]

4 Any three from: large surface area – greater area over which exchange can take place, thin membrane/being thin – short diffusion path, efficient blood supply – maintains steep concentration gradient, being ventilated – maintains steep concentration gradient [6]

B2.1

1a Structures made of DNA found in pairs in the nucleus of cells, which contain inherited material. [1]

b Small packet of information (section of DNA) that controls a characteristic, or part of a characteristic, of the organism. [1]

c Chemical that makes up genetic material. [1]

2 Stage 1: chromosomes replicated and all sub-cellular structures such as mitochondria and ribosomes reproduced [2]; stage 2: (mitosis) nucleus divides to form two identical daughter nuclei [2]; stage 3: cytoplasm and cell membranes divide to make two independent cells [2]

3a New cells needed for growth and development, and worn out or damaged cells must be replaced with identical cells. [1] Mitosis produces cells with same chromosomes and identical genetic material, which fulfil same function as original cells. [1]

b Mitosis forms two identical daughter cells. [1] Two copies are made of each original chromosome before cell division and daughter cells each hold a copy of the full set of chromosomes. [1] Mitosis is important for forming identical cells needed for growth or tissue replacement. If chromosome number did not stay the same, new cells would not work properly and organism might die. [1]

B2.2

1a Stem cells are undifferentiated cells that divide to produce many identical cells which may differentiate to form different types of specialised cells. [1]

b All cells in an early animal or plant embryo are unspecialised (stem cells). [1] Differentiation fulfils organisms' requirements for different cells to carry out different roles (e.g., muscle cells, sperm cells, gut lining cells). [1]

2 In animals, differentiation occurs during embryo development and is permanent. [2] In plants, it occurs throughout life and can be reversed or changed. [2]

3 Order of magnitude = factor of 10 [1]
adult human has approximately 3.72×10^{13} cells [1]
fertilised ovum is one cell [1]
Adult human is around 13 orders of magnitude bigger than the original cell. [1]

4 In plants differentiation can be reversed and mitosis induced. Conditions can be changed to induce more mitosis. [1] Cells redifferentiate into different plant tissues needed to form a new clone plant, [1] so plants can be cloned relatively easily. [1] In animals differentiation cannot be reversed (cells differentiate permanently), [1] so clones cannot be made easily. [1] In order to make animal clones, embryos have to be cloned. [1]

B2.3

1a Stem cell: undifferentiated cell [1] with potential to divide by mitosis, differentiate, and form different specialised cells in the body [1]; normal body cell: specialised for a specific function, [1] division by mitosis can only form cells with same specialisation [1]

b Bone marrow, [1] embryos, [1] plant meristems [1]

2 Embryonic stem cells can make any type of adult cell [1] to repair or replace damaged tissues, [1] could grow organs for transplants as needed, [1] organs grown from stem cells could cause fewer rejection issues if right techniques are used [1] Any other valid advantage.

3a In research it is important as far as possible to change only one variable. [1] Ability to produce large numbers of identical plant clones enables researchers to change a variety of different variables and see effects on genetically identical individuals. [1] Any differences will be due to variables under investigation, not genetic differences between plants. [1]

b Single rare plant specimen may not reproduce or researchers may not understand conditions needed for it to thrive. [1] If specimen dies or does not make seeds, species will be lost. [1] Cloning allows scientists time to find out about the plant and to find other specimens to introduce long-term genetic variation. [1]

B2.4

1 Any three from: spinal cord damage, diabetes, heart damage after heart attack, eyesight in the blind, bone and cartilage repair, growing organs for transplant. [3] Any other valid suggestion.

2 For (any two): potential to cure many currently untreatable diseases/injuries, grow new organs for transplant, prevent organ rejection; against (any two): use of human embryos problematic for some, risk of side effects (e.g., cancer, viral infection), patients may need immunosuppressant drugs, development of treatments slow and expensive, difficult to control stem cells [4] Any other valid evidence-based argument.

3 Using embryonic stem cells from umbilical cord blood [1] and amniotic fluid [1] rather than embryos, using adult stem cells where possible, [1] developing therapeutic cloning [1] in which cells originate from patient so no reproductive embryo and no immune problems. [1] Any other valid point.

B3.1

1a Specialised cell (found individually) [2]

b Organ (several tissues working together) [2]

c Organ (several tissues working together) [2]

2a Collection of cells of similar structure and function all working together [1]

b Collection of different tissues working together to carry out a specific function [1]

c Digestive system [1] breaking down food into soluble substances [1]
Circulatory system [1] movement of blood around the body, which carries substances to and from cells [1] Any other suitable answer.

3 Layer of muscular tissue contracts to churn up food, [1] mixing it with digestive juices to help physical and chemical digestion. [1] Glandular tissue produces enzymes to break down food. [1] Folded lining provides increased surface area. [1] Tough epithelial tissue covers and protects inside and outside of organ. [1] Any other valid point.

B3.2

1 **A – 3**, [1] **B – 4**, [1] **C – 1**, [1] **D – 2** [1]
2 Breaking up (digesting) of large insoluble molecules into smaller, soluble molecules [1], that can be absorbed and used by the cells. [1]
3 Small intestine has villi for a large surface area/good blood supply [1] to increase diffusion of soluble molecules into the blood [1].
Liver produces bile [1] to help in the digestion of fats. [1]

B3.3

1a Molecule made up of long chains of amino acids [1]
b Structural components, [1] hormones, [1] antibodies, [1] enzymes (catalysts) [1]
2 Similarities: vital components of a balanced diet, contain carbon/hydrogen/oxygen, large molecules made up of smaller molecules joined together [3]; differences: carbohydrates made up of sugar units, lipids made up of fatty acids and glycerol, proteins made up of long chains of amino acids, lipids insoluble in water, proteins contain nitrogen [3] Any other valid point.
3a Iodine test (yellow-red iodine solution turns blue-black if starch present) [2]
b Ethanol test (ethanol added to solution gives cloudy white layer if lipid present) [2]
4 Lipids are made up of three molecules of fatty acids joined to a molecule of glycerol. [1] Different combination of fatty acids [1] determines whether lipid is solid (fat) or liquid (oil). [1]
5 Complex carbohydrates are made up of long chains of simple sugars (e.g., glucose, sucrose) [1] joined together. [1] Simple sugars are basic units of complex carbohydrates. [1]

B3.4

1a Substance that speeds up a chemical reaction but is not used up or involved in the reaction and can be used many times over [1]
b Large protein molecule that acts as biological catalyst [1]
c Area in structure of enzyme with unique shape that binds to specific substrate [1]
2a Substrate of reaction to be catalysed fits into active site of enzyme like a lock and key. [1] Once in place, enzyme and substrate bind together. [1] Reaction takes place rapidly [1] and products are released from active site. [1] Enzyme then ready to catalyse another reaction. [1]
b The independent variable is the source of the catalyst [1], the dependent variable is the rate of breakdown of hydrogen peroxide/rate of oxygen production/rate of rise of bubbles. [1] Variables to keep the same: temperature, concentration of initial hydrogen peroxide solution. [1]
3a Building large molecules from smaller ones, [1] changing one molecule into another, [1] breaking down large insoluble molecules into smaller soluble ones [1] Any other valid example.
b Chemical reactions needed for life could not take place fast enough without enzymes to speed them up. [1] Each reaction controlled by specific enzyme [1] so that many metabolic reactions can take place in same small space without interfering with one another. [1] Enzymes enable cells to perform basic reactions (e.g., respiration) [1] and specific reactions to carry out particular functions, [1] simultaneously. [1]

B3.5

1 Rate of enzyme-controlled reaction initially increases as temperature increases, up to optimum temperature. [1] Once temperature exceeds 40 °C, protein structure of enzyme starts to break down. [1] Rate of reaction slows and ultimately stops when enzyme is denatured (stops working). [1]
2 Both change shape of active site, [1] both change rate of enzyme-controlled reaction and ultimately stop it. [1] Temperature permanently denatures most enzymes over about 40 °C. [1] Different enzymes work best at different pH levels. [1]
3a If body temperature increases too much, proteins making up cell structure start to be damaged, which changes shape of active sites, [1] making enzymes less effective and ultimately denaturing them [1], preventing chemical reactions necessary for life. [1] Low temperatures cause reactions to slow down. [1]
b Low temperatures cause reactions to slow down, but these will return to normal once temperature returns to normal. [1] High temperatures permanently denature enzymes so they cannot catalyse reactions, even when temperature returns to normal [1]

B3.6

1a Amylase: salivary glands, pancreas [1]; protease: stomach, pancreas, small intestine [1]; lipase: pancreas, small intestine [1]
b Amylase: carbohydrate to sugar [1]; protease: proteins to amino acids [1]; lipase: lipids to fatty acids and glycerol [1]
c Amylase: mouth, small intestine [1]; protease: stomach, small intestine [1]; lipase: small intestine [1]
2a Bile emulsifies fats, [1] breaking large fat droplets into smaller droplets. [1]
b Fats do not mix with other liquids in digestive system, staying as large globules that are hard for lipase enzymes to act on. [1] Larger surface area produced by bile action [1] allows enzymes to reach more fat molecules and break them down more quickly. [1]
3 Large insoluble molecules in food cannot be absorbed into the blood [1]. They have to be broken down to form small insoluble molecules that can be absorbed. [1] Role of enzymes is to catalyse breakdown of food [1] so food can be digested at the right speed [1] and in the right region of the gut [1] to be absorbed and used by the body. [1]

B3.7

1a Acidic [1]
b Hydrochloric acid is made in glands in the stomach lining [1] to produce a low pH (acidic conditions). [1]
c Alkaline [1]
d Liver produces bile, stored in gall bladder and released when food enters small intestine. [1] Bile is alkaline, neutralising stomach acid to create a slightly alkaline environment in small intestine. [1]

2a pH 2 [1]

b pH 8 [1]

c Activity levels increase up to optimum pH, [1] then fall fast. [1]

d Increase in pH affects shape of active site of enzyme, [1] so it no longer bonds to the substrate. [1] Enzyme ultimately denatured [1] and no longer catalyses reaction. [1]

3 Any six from: Bread taken into mouth, broken up, and coated in saliva to start digestion of starch using amylase from salivary glands. Stomach muscles churn bread and mix it with pepsin and hydrochloric acid. Bread passed from stomach into first part of small intestine. Bile added from gall bladder via bile duct to emulsify fats, provide larger surface area for digestive enzymes to work on, and neutralise acid from stomach to provide alkaline pH for optimum enzyme action. Digestive enzyme amylase added from pancreas to break down carbohydrates (bread) to glucose. Semi-digested food squeezed on to rest of small intestine. Lining covered with villi giving large surface area so products of digestion can be absorbed into blood as efficiently as possible. Indigestible food remains pass into large intestine where water is absorbed. Faeces pass out of body through anus. [6]

B4.1

1 Red blood cells, white blood cells, plasma, platelets [2] for all 4 correct, 1 mark for any 1 correct

2a Any three from: transport of blood cells, transport of dissolved gases, transport of food, transport of hormones, removal of waste products, defence against infection, preventing blood loss through clotting [3]

b Carry oxygen from the lungs to the cells where it is needed [1]

c Biconcave (pushed in on both sides) [1] give increased surface area to volume ratio for diffusion. [1]
Packed with haemoglobin [1] that binds to oxygen. [1]
Have no nucleus [1] to make space for more haemoglobin. [1]

3 White blood cells: relatively large blood cells with nuclei, [1] protect against invasion of harmful microorganisms, [1] some (lymphocytes) form antibodies or antitoxins, [1] some (phagocytes) actively engulf and digest bacteria and viruses [1]; platelets: fragments of larger cells with no nuclei, [1] help with clotting to keep microorganisms out [1]

B4.2

1 Capillaries, arteries, and veins [1]

2a Arteries carry blood away from the heart. [1] Veins carry blood towards the heart. [1] Capillaries link arteries and veins/diffusion in and out of cells. [1]

b Arteries have a thick wall of muscle (to contain blood under pressure)/elastic fibres (to allow stretch as blood is forced through by heartbeat as pulse). [1] Veins have valves (to keep blood flowing towards heart). [1] Capillaries have very thin walls (to promote diffusion of substances in and out). [1]

3a Arteries carry blood from heart to organs, veins return blood to heart, [1] capillaries link arteries and veins. [1]

b Skeletal muscles squeeze blood towards the heart when they contract. [1] Valves in veins prevent backflow of blood. [1]

4 In fish blood leaves the capillaries of the gas exchange organ slowly, at low pressure, and without a pulse. [1] Active land mammals require lots of food and oxygen to supply muscles and organs for movement and warmth. They also produce a lot of waste materials such as carbon dioxide that must be removed. [1] Single circulation system would not be able to supply tissues or remove waste fast enough as blood would travel around body too slowly. [1] Double circulation system much more efficient because blood is pumped to gas exchange organ, then returns to heart and is pumped around body quickly at pressure. [1]

B4.3

1 A build up of fatty material on the lining of the coronary arteries. [1]

2a Prevent blood flowing backwards, [1] making heart more efficient [1]

b Supply heart muscle cells with oxygenated blood [1] for aerobic respiration and efficient contraction [1]

c Allows heart to pump blood around body very efficiently, [1] enables blood to leave heart at high pressure, [1] right ventricle only has to send blood to the lungs (where high pressure would be damaging) [1]

3 Pulmonary artery carries blood away from heart. [1] Blood carried from heart to lungs is deoxygenated blood from the body, [1] which is dark red until it picks up oxygen in the lungs. [1]

4a Metal mesh placed in artery and opened up by inflation of tiny balloon [1] to hold narrowed blood vessel open so blood can flow freely [1]

b One mark for each advantage and disadvantage for each of the two methods.

	Stent	Bypass surgery
advantages	general anaesthetic not required, relatively cheap, effective	very effective against severe blockages
disadvantages	ineffective against severely blocked or narrowed arteries	general anaesthetic required, expensive

[4]

B4.4

1 Group of cells in right atrium of heart [1] producing regular electrical signal that spreads through heart and makes it contract. [1]

2a Valves prevent backflow of blood in heart. [1] Leaky valve can allow blood to flow backwards, [1] which means full amount of blood does not leave heart [1] and blood coming into heart chamber mixes with blood that hasn't left, making heart less efficient. [1]

b i Advantage: work well/no medication needed [1]; disadvantage: limited lifespan (12–15 years) [1]

ii Advantage: lasts a long time [1]; disadvantage: lifetime medication required to prevent clotting [1]

3 Can be used to keep patient alive until suitable heart for transplant becomes available. [1] Can be used in some cases to rest patient's own heart and allow it to recover. [1] May be used to replace the natural heart in the long term. [1] Very expensive, [1] not effective over long periods, [1] could be overtaken by organs grown from stem cells. [1] Any other valid evidence-based argument.

B4.5

1 (Air), nose/mouth [1], trachea [1], bronchi [1] (lungs)
2 Intercostal muscles between ribs and diaphragm contract and relax, [1] changing chest volume and pressure [1] and forcing air in or out. [1]
3a Exchange of oxygen and carbon dioxide gases in lungs [1]
b Air moved into alveolus by breathing (ventilation) [1] is relatively high in oxygen, low in carbon dioxide [1]. Oxygen moves through thin alveoli walls/into red blood cells (blood) in capillaries [1] down concentration gradient by diffusion [1]. Carbon dioxide moves out of blood into air in alveolus/by diffusion down concentration gradient. [1]
4a Award marks for well-drawn bar chart correctly labelled. [3]
b Bar chart shows we breathe in air containing mainly nitrogen with oxygen and a tiny bit of carbon dioxide, and breathe out air containing mainly nitrogen but with less oxygen and more carbon dioxide. [1] We take oxygen from air breathed in and pass carbon dioxide into air breathed out, but we breathe in and breathe out air. [1]
c Ventilation and rich blood supply for steep concentration gradient, [2] clusters of alveoli/spherical alveolus shape for large surface area, [2] thin alveolus walls to minimise diffusion distances [2]

B4.6

1 Plant organs: Any three from: roots, flowers, stem, leaf [3]
Plant tissues: Any three from: epidermal, palisade mesophyll, spongy mesophyll, xylem, phloem [3]
2a Cover surfaces and protect them, may secrete waxy substance for waterproofing [1]
b Contains lots of chloroplasts for photosynthesis [1]
c Contains some chloroplasts for photosynthesis [1]
3 Tightly packed palisade mesophyll cells at top of leaf contain many chloroplasts for photosynthesis [1] and are protected by epidermis. [1] Spongy mesophyll cells also photosynthesise and have large air spaces and surface area to maximise gas exchange. [1] Xylem supply water for photosynthesis [1], phloem transport food from photosynthesis around plant. [1] Stomata can be opened or closed by guard cells to let gases in and out. [1]

B4.7

1 Transport of food made in leaves and water and mineral ions taken from soil to rest of plant [1]
2 Mature xylem cells are dead, phloem cells are living. [1] Xylem transport water and mineral ions from soil, phloem transport dissolved sugars from photosynthesis. [1] Xylem found on inside of vascular bundles, phloem on outside. [1]
3 Phloem in trees found in a ring just underneath bark. [1] Soft bark of young trees vulnerable to damage by animals. [1] If complete ring of bark is eaten, transport of water from roots and sugars from leaves stops and young tree will die. [1] Plastic covers protect young bark from animals. [1] Covers can be removed once trees are more mature and bark is harder. [1] If covers aren't used, most of young trees are likely to be destroyed and woodland will eventually die as old, mature trees are not replaced. [1]

B4.8

1a Small openings all over leaf surface surrounded by guard cells [1]
b Stomata open to allow air into the leaves to provide carbon dioxide for photosynthesis, [1] and close to control loss of water. [1]
2 Water vapour evaporates from cells lining air spaces [1] and diffuses out of leaf through stomata [1] down a concentration gradient. [1]
3 As water evaporates from leaf surface, more water is pulled up through xylem to replace it. [1] Water moves into roots by osmosis to replace water moving up xylem. [1] Transpiration stream is constant movement of water molecules through xylem from roots to leaves. [1]
4a Mean on lower surface: 257, mean on upper surface: 85.7 [1], Ratio of stomata per mm^3 lower surface : upper surface = 257 : 85.7 = 3:1 [2]. Accept other way round so ratio upper surface : lower surface = 85.7 : 257 = 1:3
b There are more stomata on the lower surface of the leaf compared to the upper surface. [1] This is because the lower surface of the leaf is often more sheltered from direct sunlight and wind [1], resulting in reduced evaporation and water loss. [1] By having more stomata on the lower surface, plants lose less water. [1]

B4.9

1 Light intensity, temperature, humidity and the amount of air movement. [2] for all 4 correct, 1 mark for any correct answer
2 Light: light increases rate of photosynthesis and so increases stomatal opening to take in more carbon dioxide. [1] At the same time this increases the loss of water by evaporation from the leaves. [1]
Humidity: high humidity reduces the concentration gradient between the inside and outside of the leaves [1] and so reduces transpiration. [1]
Air movements: more air movements [1] increases evaporation, diffusion, and water loss. [1]
Temperature: evaporation and diffusion are faster in higher temperatures [1] as particles are moving more rapidly. [1]
3a Standard conditions: 1.25 cm [1], with light: 3.50 cm [1]
b When the light was shone on the plant it photosynthesised at a higher rate [1] so opened stomata to take in more carbon dioxide [1], which meant water was lost through the stomata by evaporation. [1] Also the light would have increased the temperature [1], increasing the rate of evaporation [1] and reduced humidity [1] increasing the concentration gradient of water between the inside and outside of the leaf. [1]

B5.1

1 State of complete physical and mental well-being, not just absence of disease. [1]
2a Any three from: pathogens (bacteria, viruses), non-communicable diseases, poor diet, stress, lack of clean water [3] Any other valid factor.
b Pathogens: tuberculosis, flu; non-communicable diseases: heart disease, arthritis; poor diet: obesity, starvation, rickets; stress: heart disease, mental health problems; lack of clean water: diarrhoeal diseases, sickness [3] Any other valid example.

3a Bacterial infection with tuberculosis, viral infection with HIV, injecting drug use [2]

b Increases chances [1]

c HIV positive injecting drug users [1]

d $1.5 - 0.2 = 1.3$, $\frac{1.3}{1.5} \times 100 = 86.67$ [1], 87 times more likely [1]

4 Many diseases more prevalent in overcrowded situations as transmission is easier. [1] Poor cities in less developed countries have this problem. [1] Injecting drug abuse increases risk of transmission of many blood-borne diseases. [1] This problem is more common in wealthy countries. [1] Poor diet linked with non-communicable diseases (e.g., heart disease) in all countries [1] and to poor immune system and more communicable diseases in less developed countries. [1] Any other valid explanation.

B5.2

1a Microorganism that causes disease [1]

b Any two from: by air, direct skin contact, direct contact of body fluids, by contaminated water, by undercooked or contaminated food [2]

2a As a result of reaction to toxins produced, [1] or damage caused to cells [1]

b People often crowded together so pathogens pass easily from one to another [1] by droplet infection/direct skin contact. [1] Often no sanitation so disease spreads through contact with waste. [1] Lack of clean water supplies/water often contaminated with sewage so diseases spread through drinking or bathing in contaminated water. [1]

3a Common symptoms of infectious diseases result from the response of the body to the pathogens/infectious agents [1] including cell damage and destruction by dividing pathogens and/or toxins produced by pathogens. [1]

b Pathogens spread by air via air droplets (when someone coughs, sneezes or talks) [1] and other people breathe in the droplets. [1] These go into the gas exchange system and can cause infection there. [1] Pathogens can enter water when it is contaminated by sewage. [1] People drink the water [1] and pathogens enter the digestive system where they can cause infection. [1]

B5.3

1 Simple cell division [1] called binary fission [1]

2a Reduces risk of growing dangerous pathogens [1]

b To prevent contamination of culture by microorganisms from the environment [1] and to prevent release of potentially harmful microorganisms that might grow. [1]

c Higher temperatures enable microorganisms to grow much more rapidly [1] so that they can be identified sooner (e.g., in hospital labs) [1] or to produce products more quickly in industry (e.g., insulin-producing GM bacteria). [1]

3 Petri dish has limited supplies (nothing new added), [2] bacteria use up available food and oxygen (limiting growth), [2] Petri dish has no mechanism for removal of waste products (build-up of carbon dioxide and other toxins can poison bacteria and stop growth). [2] Any other valid limiting factor.

B5.4

1 Disinfectants are used to kill bacteria in the environment but are toxic to people, [1] antiseptics are disinfectants that kill bacteria and are safe to use on human skin, [1] antibiotics are chemicals that can be used inside body to kill bacteria or prevent them from growing. [1]

2 These places contain most bacteria [1] and are where people are most careful about hygiene. [1]

3a **A**: (radius of **A** = 2.5 mm) area = $3.142 \times 2.5^2 = 19.6\,\text{mm}^2$ [1]
B: (radius of **B** = 5.1 mm) area $3.142 \times 5.1^2 = 81.7\,\text{mm}^2$ [1]
C: (radius of **C** = 7.9 mm) area $3.142 \times 7.9^2 = 196.1\,\text{mm}^2$ [1]

b $\frac{196.1\,\text{mm}^2}{19.6\,\text{mm}^2} = 10.0$ [1]
Highest concentration 10 times more effective than lowest concentration (an order of magnitude more effective) [1]

B5.5

1 By air (droplet infection) – pathogens breathed in. [1] By direct contact – pathogens pass skin to skin or body fluids. [1] By water – pathogens taken in through drinking/eating. [1]

2 Any three from: wiping work surfaces with disinfectant, cleaning toilets, keeping raw meat away from food eaten uncooked, using tissues to blow nose, washing hands before handling food [3] Any other valid example.

3 Wiping work surfaces with disinfectant: destroys microorganisms present after preparing raw food; cleaning toilets: destroys bacteria from faecal material; keeping raw meat away from food eaten uncooked: prevents spread of pathogens that will not later be destroyed by cooking, allowing the pathogens into the gut; using tissues to blow nose: contains pathogen-filled mucus for disposal; washing hands before handling food: prevents spread of pathogens onto food [6] Any other valid explanation.

B5.6

1a Fever, [1] red skin rash [1]

b Improved living standards (less overcrowding) have reduced spread of disease by droplet infection, [1] vaccination programme for babies and children has reduced pool of infection [1]

2a HIV is the virus that causes AIDS. [1]

b HIV attacks immune cells, causing only a mild, flu-like illness initially. [1] Without treatment the virus remains hidden after initial mild illness, [1] damaging the immune system to the extent that it cannot defend the body against infections and certain cancers. [1] These diseases kill the patient. [1]

3a i 4,200,000 + 4,400,000 + 4,100,000 + 3,800,000 + 3,300,000 + 3,100,000 = approx. 22,9000,000 cases [3]

ii 800,000 + 800,000 + 550,000 + 700,000 + 500,000 + 600,000 = approx. 3,950,000 cases [3]

b 1980–1985: $\frac{22\,900\,000}{100} \times 5 = 1\,145\,000$ people [2]

2000–2005: $\frac{3\,950\,000}{100} \times 5 = 197\,000$ people [2]

4 Similarities: no treatment for either virus [1]; differences: measles spread by droplet infection [1], TMV spread by direct contact and insect vectors [1], measles prevented by improved nutrition/living standards [1], vaccination can be used to prevent measles [1], TMV prevented by good field hygiene and pest control [1]

B5.7

1 Most common causes are eating either undercooked food contaminated with *Salmonella* [1] or food prepared in unhygienic conditions that becomes contaminated by *Salmonella* bacteria raw meat. [1]

2a Kill bacteria/prevent growth of bacteria [1]

b Any three from: doesn't last long (only a few days), unpleasant but not serious for most people, antibiotics would be lost from system through sickness and diarrhoea, risks development of antibiotic resistance [3] Any other valid suggestion.

3a Sexually transmitted disease, [1] pathogens passed from one partner to another during unprotected sexual contact [1]

b Any three from: being celibate, having a single sexual partner, limiting number of sexual partners, always using barrier method of contraception (e.g., condom) [3]

c Any two comparisons e.g. *Salmonella* is spread by eating infected food [1], gonorrhoea is spread by unprotected sexual contact. [1] Gonorrhoea is treated by antibiotics [1], *Salmonella* is usually not. [1] *Salmonella* affects the digestive system [1], gonorrhoea affects the reproductive system. [1]

4a *Salmonella*: vaccinate poultry so they don't get infected [1], cook poultry thoroughly, to kill bacteria [1], keep raw chicken away from cooked chicken, to prevent contamination [1] Gonorrhoea: Use a barrier method of contraception, to prevent fluids containing bacteria moving from person to person [1], treatment with antibiotics to kill bacteria [1], treat all sexual partners to prevent community spread [1]

b Look for articulate writing and clear explanation. Main points could include: details of *Salmonella* poisoning and symptoms, *Salmonella* can be found on contaminated meat or mayonnaise made with raw eggs, bacteria killed by thorough cooking, salads often contaminated when prepared with chicken for barbecues, meat not always cooked thoroughly on barbecues, check meat is piping hot in middle, no pink juices. [4]

B5.8

1a Any three from: spores carried in air, spores in water splashes, contaminated plant material in soil, direct contact between healthy and diseased leaves [3]

b Through the bite of a female Anopheles mosquito [1] infected with malaria parasites from another blood meal. [1]

2 Fungus causes black spots on leaves that then turn yellow and fall early, [1] reducing leaf area for photosynthesis [1] and limiting food available to make flowers. [1]

3 Nets prevent mosquito bites, [1] insecticide kills mosquitoes [1]

4 Awareness: avoid malarial areas if possible, be aware if malaria is in intended region of travel [1]; bite prevention: precautions such as insecticide-treated mosquito nets to sleep under, mosquito netting at windows and doors, use of insect repellent, keeping skin covered [3]; chemoprophylaxis: take antimalarial drugs that kill parasites in blood [1]; diagnosis: early malaria treatment more likely to be effective, have blood test for early diagnosis if ill after travel to malarial area [1]

B5.9

1 Skin [1], mucus in airways [1], stomach acid [1]

2 To include detail on ingestion and digestion (destruction) of microorganisms, [2] production of antibodies to target and destroy particular pathogens, [2] production of antitoxins to counteract toxins released by pathogens [2]

3a Clotted blood prevents pathogens getting into body through cuts in skin. [1] If blood won't clot properly, pathogens may get in through open cuts. [1]

b White blood cells destroy pathogens. [1] If number of white blood cells falls, fewer pathogens will be destroyed, [1] increasing likelihood of infection. [1]

B5.10

1 Pathogens: bacteria, fungi, viruses [1] Any two plant pests e.g., aphids, nematode worms, insect larvae [1]

2a Plant growth stunted [1] as insufficient nitrate to convert sugars from photosynthesis into proteins for protein synthesis [1]

b Leaves yellow [1] as insufficient magnesium to make chlorophyll so plant can't photosynthesise effectively [1]

3 Plant attacked by fungal or bacterial pathogen, [1] or by nematode worms or other plant pest, [1] damaging roots [1] so they cannot absorb mineral ions from soil effectively. [1]

4 Basic table shown below – 1 mark for each correct set of symptoms, cause and example of disease up to 4 marks

Common symptoms	Pathogen/pest type	Example of disease
stunted growth	n/a	nitrate deficiency
areas of decay or rotting	fungi	black spot
growths	bacteria	crown gall
damaged roots, malformed stems/leaves	nematodes, aphids	root knot disease
yellow leaves (chlorosis)	n/a	magnesium deficiency
mosaic patterns	viruses	tobacco mosaic virus
presence of visible pests	aphids, caterpillars	n/a

[4]

B5.11

1a Any three from: cellulose cell walls, waxy cuticle, bark, leaf fall, antibacterial chemicals [3]

b Any three from: poisons, thorns, hairy stems/leaves, drooping or curling when touched, mimicry [3]

2 Cellulose cell walls: resist invasion by microorganisms; waxy cuticle: barrier to pathogens; bark: protective layer hard for pathogens to penetrate; leaf fall: pathogens removed when leaves fall; antibacterial chemicals: kill bacterial pathogens; poisons: make herbivores unwell so deter eating; thorns: make plant unpleasant/painful to eat; hairy stems/leaves: deter eating and egg laying; drooping or curling when touched: dislodges insects, frightens small animals; mimicry: mimic unhealthy plants or existing eggs [6]

3 Similarities: internal and external defences, antibacterial chemicals, protective layers [3]; differences: humans have immune system making specific antibodies to particular pathogens, humans have few defences against attack by predators (no thorns, little hair), humans do not produce poisons [3] Any other valid point.

B6.1

1a Protect the individual against serious diseases [1]; as part of herd immunity to protect entire populations [1]

b Protein made by white blood cells in response to specific antigens [1]

c Bacterial: tuberculosis, tetanus, diphtheria [1]; viral: polio, measles, whooping cough [1] Any other valid example.

2 Small amount of dead or inactive form of pathogen introduced into body, [1] stimulating white blood cells [1] to produce antibodies [1] needed to fight pathogen and prevent illness. [1] If same, live pathogen is encountered later, body can respond rapidly to make correct antibodies [1] just as if you had already had the disease, conferring protection. [1]

3 A large proportion of the population is vaccinated [1] so they are immune to a disease. [1] Spread of the pathogen in the population is reduced [1] because fewer people can get ill with the disease. [1] It protects vulnerable people. [1]

B6.2

1 Paracetamol is a very useful painkiller but will have no effect on the disease-causing pathogen (treats symptoms, not disease). [1] Penicillin is an antibiotic that kills the disease-causing bacteria (treats disease). [1]

2 Viral pathogens reproduce inside body cells [1], so it is very difficult to develop drugs that destroy them without destroying body cells too. [1] Bacteria exist outside body cells and are different from human cells, [1] so antibiotics can target bacterial cells without hurting body cells. [1]

3a $380 - 90 = 290$, [1] $\frac{290}{380} \times 100 = 76.3$, [1] 76% fall [1]

b Deaths fell sharply between 1940 and 1950 due to the introduction of antibiotics (infection could be treated and cured). [1] Hospital births increasingly common, so doctors were on hand to spot symptoms of infection and prescribe antibiotics. [1]

c Antibiotics have reduced rate of deaths from maternal septicaemia to very low levels. [1] Emergence of antibiotic resistant bacteria renders antibiotics ineffective. [1] Without effective antibiotics rate of deaths could return to 1920–1930 levels and thousands of women could die from infections after giving birth. [1] Any other valid point.

4 The larger the clear spot on Figure 3, the more bacteria the antibiotic has killed [1] so the more effective it is against that specific bacteria. [1] The right antibiotic needs to be chosen to kill the bacteria so the bacteria does not reproduce further [1] and the infection gets worse. [1]

B6.3

1 He noticed a clear ring in jelly around some spots of mould growing on bacterial culture plates. [1] He realised that a chemical in the mould must have killed the bacteria in the culture, [1] leaving the jelly clear. [1]

2 Plant extracts may be impure, or they may be found only in very tiny amounts in a plant, [1] meaning huge amounts of plant material would be needed to extract the drug. [1] Synthetic drugs can be made industrially in large amounts and chemists can change the molecules to change the properties of the medicine. [1] Chemists can make drugs that are more effective and have fewer side effects than natural products. [1]

3 Advantages: anecdotal evidence of medicinal properties saves trying many different plants, long and widespread usage suggests few side effects; disadvantages: may be difficult to source the living material, concentrations may be very low, may be difficult to isolate the chemical that has caused the antibiotic effect [6] Any other valid point.

B6.4

1a Flow chart should include: identify pathway in disease that needs cure → identify new chemical (in living organisms or lab) → preclinical testing on cells, tissues, and animals → phase 1 clinical trials check for side effects → phase 2 clinical trials check efficacy and safety → phase 3 clinical trials check efficacy, safety, and dosage → licencing → phase 4 clinical trials monitor safety throughout life of drug [8]

b A placebo is a tablet (treatment) [1] that does not contain a new drug (active drug) [1] being tested in a double blind clinical trial. [1]

2a Efficacy: whether a drug does the job it is designed to do [1]; toxicity: whether a drug is toxic and, if so, how much of it is has to be given to have a toxic effect [1]; dosage: dose of a drug that is effective in treating a disease [1]

b Involves other scientists working in the same field checking the results of a trial. [1] This helps prevent false claims because other experts look at the work. [1] Establishes agreement for which claims are valid (which drugs are potentially useful). [1]

3 Laboratory testing is used to look at the potential toxicity of new compounds [1] and also if they have the desired effect on the cells, tissues etc. (efficacy). [1] Healthy volunteers are used to check for potential side effects [1] as they are healthy, any problems will be the result of the new drug. [1]

B6.5

1a Group of genetically identical cells or organisms [1]

b Protein made by an immune system cell in response to presence of foreign antigen [1]

c Proteins produced from a single clone of cells to target specific body cells or chemicals [1]

2a Cell that divides to make large number of identical cells [1] that all produce specific antibodies, [1] formed by fusing a mouse lymphocyte with a tumour cell [1]

b To make hybridoma cells mouse lymphocytes are stimulated to make a specific type of antibody and then fused with special tumour cells. Lymphocytes can make antibodies but cannot divide, [1] tumour cells don't make antibodies but can divide rapidly. [1] Hybridoma cells both make required antibodies and divide to make clone of identical cells, meaning they can be used to make very large amounts of specific monoclonal antibodies. [1]

3 Monoclonal antibodies are made to bind to specific antigens [1] found on the pathogen. [1] They bind to the pathogen if present [1] producing a colour change signalling a positive test. [1]

B6.6

1 Monoclonal antibodies can identify and bind to specific cell types, [1] it was thought that they could be used to carry drugs to specific cancers and treat them, [1] producing much more targeted means of treating cancer than therapies such as radiotherapy. [1]

2 They only bind to the specific diseased or damaged cells that need treatment, [1] healthy cells are not affected at all. [1] They are specific [1] so could be used to treat a wide range of conditions. [1]

3 Monoclonal antibodies initially created more side effects than expected, [1] partly because mouse cells were used in formation of hybridoma cells, triggering an immune response in people. [1] Developments such as production of mouse/human hybrid cells [1] and use of fractions of antibodies to carry drugs to target cells are reducing side effects. [1] Doctors are now prepared for side effects and can treat symptoms. [1] Ongoing development of new skills and technologies to create monoclonal antibodies and attach them to drugs and other compounds means they are being used in more and different ways. [1] Any other valid point.

B7.1

1a Non-infectious disease that cannot be passed from one individual to another [1]

b Communicable diseases caused by pathogens (e.g., bacteria, viruses, fungi), [1] whilst non-communicable diseases affect people as a result of genetic makeup, lifestyle, or environmental factors. [1]

2a A correlation is when there is an apparent link between a non-communicable disease and an environmental risk factor. [1] A causal mechanism provides an explanation for how a risk factor affects a biological process. [1]

b Any two from: impact individual by preventing them working/preventing them taking part in family life/social activities/reducing their income so having a financial effect. [2]

3 Sampling allows scientists to discover similar patterns between particular risk factors and specific diseases. [1] These similarities may suggest a correlation, [1] which scientists can then use to discover a causal mechanism. [1] Any three from: large sample covering all the groups in the study population, random selection of people, collecting data at regular time intervals, taking account of differences in the population. [3]

B7.2

1 Mass of abnormally growing cells formed when normal control of cell cycle is lost and cells divide rapidly without growing and maturing [1]

2a A benign tumour grows in one place, is usually contained in a membrane, and does not invade other tissues. [1] A malignant tumour invades neighbouring healthy tissues and can also split, [1] releasing small clumps of cells that spread around the body in the blood and invade different healthy tissues. [1]

b A benign tumour can grow very large and compress/damage organs, which can be life-threatening (e.g., brain tumour). [1] A malignant tumour may split into pieces which are carried around the body in the blood or lymph [1] where uncontrolled cell division continues to form secondary tumours. [1] These tumours disrupt normal tissue and often cause death if untreated. [1]

3a Either stop cancer cells dividing, [1] or make them self-destruct [1]

b Cancer drugs are designed to target rapidly dividing cancer cells. [1] They also tend to affect other rapidly dividing cells. [1] Cells of hair follicles, skin, stomach lining, and blood-forming bone marrow are always dividing rapidly [1] so are more likely than other body cells to be affected by chemotherapy. [1]

4 Radiotherapy: cancer cells destroyed by targeted doses of radiation, [1] stops mitosis in cancer cells but can also damage healthy cells [1]; chemotherapy: chemicals used to stop cancer cells dividing or make them self-destruct, [1] aim to target cancer cells as specifically as possible [1]

B7.3

1a Nicotine, carbon monoxide, tar [3]

b Nicotine: addictive/produces sense of calm/increases heat rate [1] carbon monoxide: poisonous/taken up by blood instead of oxygen [1] tar: carcinogenic/damages lung tissue causing COPD [1]

2a Award marks for well-drawn chart correctly labelled. [4]

b The more cigarettes smoked, the greater the risk of dying from cardiovascular disease (CVD) [1]

c Smoking narrows blood vessels, which can increase blood pressure. [1] Nicotine in cigarette smoke increases heart rate. [1] Other chemicals in cigarette smoke damage lining of arteries, making atherosclerosis more likely, or increase blood pressure. [1] All of these effects increase the risk of CVD.

3 Smoking causes blood to carry poisonous carbon monoxide rather than oxygen, reducing oxygen-carrying capacity of blood. [1] In pregnant women this can lead to oxygen shortage for developing foetus. [1] Oxygen shortage can lead to premature births, [1] low birthweight babies, [1] and stillbirths. [1]

4 Scientists have several causal mechanisms that work together to explain link between smoking and lung cancer. [1] Cigarette smoke contains tar and other carcinogens. The higher the concentration of these chemicals in the lungs, the more likely they are to affect cells and turn them malignant. [1] Cilia that would naturally filter out tar are anaesthetised by chemicals in cigarette smoke so they do not work, [1] allowing tar and other carcinogenic chemicals to build up in delicate lung tissue. [1]

B7.4

1a Any three from: obesity, type 2 diabetes, heart disease, osteoarthritis, high blood pressure [3]

b Relative risk in men who exercise least: 4.5 [1]; relative risk in men who exercise most: 1 [1]

2 If excess food is consumed, it is stored as fat and may cause obesity. [1] Exercise uses up some of the energy [1] produced during respiration from food eaten. [1] As a result, less food is stored as fat, reducing risk of obesity. [1]

3 Relative risk of suffering cardiovascular disease is lower in people who exercise compared to people who don't. [1] Exercise builds cardiovascular fitness, so people who exercise are likely to have healthier hearts than people who don't exercise. [1] Exercise lowers blood cholesterol levels and reduces the risk of fatty deposits building up on the coronary arteries, lowering the risk of heart disease and other health problems. [1]

4 Risk of developing type 2 diabetes is much higher in women than in men with the same body mass index (BMI). [1] At BMI of just over 25 – just overweight – risk for men is 2.2% and for women 8.1%. [1] At BMI of over 35, risk for men is 42% and for women 93%. [1] Women may be less likely to do exercise whatever their BMI (need more data). [1]

B7.5

1a Agent that increases risk of cancer developing [1]

b Any three from: tar from cigarette smoke, sunlight, X-rays, alcohol. [3] Any other valid suggestion.

2a Alcohol destroys liver tissue/cirrhosis. [1] It is a carcinogen [1] so can cause liver cancer. [1]

b Alcohol damages brain tissue [1] leading to loss of function and possible death. [1]

3a Electromagnetic rays e.g. x-rays and gamma rays, [1] particles e.g. alpha and beta [1]

b Exposure to ionising radiation increases the risk of cancer [1] because it can penetrate cells [1] and damage the chromosomes [1] causing mutations in the DNA [1]

B8.1

1a Carbon dioxide from air, [1] water from soil [1]

b Carbon dioxide and water from water in which it lives, [2] light from Sun or electric light [1]

c Carbon dioxide + water $\xrightarrow{\text{light}}$ glucose + oxygen [2]

2 Carbon atom moves from air into air spaces in leaf, [1] into plant cells, [1] into chloroplasts, [1] joins with water to make glucose, [1] and is converted to starch for storage. [1]

3a $6CO_2 + 6H_2O \xrightarrow{\text{light}} C_6H_{12}O_6 + 6O_2$ [2]

b It needs bigger input of energy from environment [1] to break bonds holding carbon dioxide and water together [1] than is released when new bonds form in glucose and oxygen. [1]

4 Leaf kept in light photosynthesises, making glucose from carbon dioxide and water using energy from light. [1] Glucose converted to starch to be stored. [1] Iodine solution turns blue-black in presence of starch. [1] If leaf is then kept in dark for 24 hours it cannot photosynthesise so cells will use stored starch for respiration. Little or no starch left in the leaves so iodine solution unaffected. [1]

B8.2

1 Carbon dioxide, [1] light, [1] temperature [1]

2a As light intensity increases, so does rate of photosynthesis. [1] This indicates that light intensity is limiting factor. [1]

b Temperature acts as a normal limiting factor to begin with, increase in temperature increases rate of photosynthesis. [1] This is consistent with light and carbon dioxide concentrations (Figures 1 and 3). [1] Increasing light intensity and carbon dioxide concentration cause rate of photosynthesis to rise and then plateau when another factor becomes limiting. [1] Above a certain temperature, enzymes in cells are denatured so no photosynthesis can take place at all. [1]

3 Tropical rainforest: high light intensity, warm temperature, plenty of moisture, carbon dioxide from decaying material (relatively few limiting factors and rapid growth conditions available all year round, allowing plants and even individual leaves get very large); [2] UK woodland: low light intensity, short days in winter, cold temperature (most growth takes place in spring and summer with plenty of light and warmth but temperatures still lower temperatures than tropical rain forest, less time for growth so plants smaller); [2] Arctic tundra: no light all winter but lots of light in summer, lack of water due to frozen ground (low temperatures limiting factor on photosynthesis for most of year, so plants much smaller and slower-growing than tropical rainforest or UK woodland) [2]

B8.3

1a Leaves, [1] stems, [1] roots, [1] storage organs [1]

b Glucose is soluble and would affect movement of water into and out of plant cells [1] by osmosis. [1] Starch is insoluble and does not disturb plant's water balance. [1]

c Accept any appropriate suggestion involving a slice of potato and dilute iodine solution. [3]

2 Respiration to release energy for cell functions, [1] making amino acids for growth, [1] conversion into starch for storage, [1] making cellulose to form cell walls, [1] building up fats and oils for food store in seeds [1]

3 Bogs are wet and peaty and soil contains very few minerals, [1] especially nitrates. [1] Plants need nitrates from soil to make amino acids and build them into proteins. [1] Many plants cannot grow well in bogs. Carnivorous plants trap insects and digest their bodies, [1] which provide good supply of nitrates and other minerals. [1] These plants can grow and thrive in bogs as they do not rely on bog soil for minerals. [1]

B8.4

1 Can eliminate limiting factors and promote maximum rate of photosynthesis so plants grow as fast and as large as possible, [1] maximising profit [1] and allowing us to grow more crops in and out of season [1]

2a i Light levels low until sunrise – not enough light for photosynthesis, temperature falls overnight – low temperatures affect enzyme activity and slow rate of photosynthesis [3]

ii Carbon dioxide will limit photosynthesis – not enough carbon dioxide to make glucose as fast as other conditions would allow [3]

iii Days shorter in winter but no leaves on trees so light will reach woodland floor – light limiting factor for trees but not woodland floor plants, carbon dioxide levels normal or higher than normal as fewer leaves absorbing – not limiting, temperature colder – enzyme activity reduced by low temperature, so temperature most likely limiting factor [3]

iv Leaves on trees will limit light but light intensity high and days long, temperature warm, carbon dioxide levels most likely limiting factor – insufficient for maximum rate of photosynthesis [3]

b Each case is within natural environment [1] where light, temperature, and carbon dioxide levels change constantly and can interact, [1] meaning that any factor may be limiting factor for photosynthesis at any time. [1]

3 20 °C is a temperature that can be achieved easily in the UK during spring and summer with relatively little or no heating. [1] 30 °C is much higher than ambient temperature all year round so greenhouse would need almost constant heating. [1] Rate of photosynthesis with plenty of carbon dioxide is higher at 30 °C than at 20 °C but cost of constant heating very high (profit from one extra crop annually might not outweigh year-round heating cost). [1] Investing in removable insulation helps maintain a temperature of 20 °C or more for much of year with little or no heating. [1] When greenhouse is heated insulation retains warmth. [1] Minimising heating minimises carbon footprint. [1] Any other valid point.

B9.1

1 Glucose + oxygen → carbon dioxide + water (energy transferred to environment) [2]

2 Movement, [1] building new molecules, [1] heat generation [1]

3a $C_6H_{12}O_6 + 6O_2 \rightarrow 6CO_2 + 6H_2O$ [2]

b Muscle cells are very active and need a lot of energy. [1] They need large numbers of mitochondria to supply energy as aerobic respiration takes place in mitochondria. [1] Fat cells use very little energy [1] so need very few mitochondria. [1]

4 Accept any appropriate suggestion for practical investigation. [6]

B9.2

1 Heart rate increases with exercise. [1]

2a Resting heart rate before training: 72bpm, resting heart rate after training 63bpm. Change due to training = 9 bpm. % change = $\frac{9}{72} \times 100 = 12.5$ percent fall in heart rate. [3]

b The heart is a muscle with a good blood supply. [1] When people exercise regularly the heart muscle gets stronger with a better blood supply. [1] This makes the heart more efficient so it pumps more blood with each beat. [1] This means at rest it does not need to beat so often and the resting heart rate of a trained individual is usually slower than it was before they started training. [1]

c Blood pumped out at rest in one minute before training: $64 \times 72 = 4608\,cm^3$ per minute [1] Blood pumped out at rest in one minute before training: $80 \times 63 = 5040\,cm^3$ per minute [1] 2 marks for accurately drawn graph

3a Converted from glucose [1] to supply cells with fuel for respiration [1]

b Glycogen can be converted rapidly into glucose [1] to provide fuel for aerobic respiration, providing body cells with energy. [1] Muscle tissue often needs sudden supply of energy for rapid contraction in a way that most other tissues do not, [1] so muscle needs a glycogen store. [1] Other tissues don't need energy in the same way so have not evolved to have glycogen stores. [1]

4a Award marks based on ideas presented and method produced. Look for sensible ideas, safe investigation, clear instructions, realistic expectations, appropriate methods of recording and analysis, awareness of weakness in investigation, and understanding of variables. [6]

b Results will depend on method chosen and fitness parameter investigated. Look for awareness that fitness will vary (can be affected by age, lifestyle, or health issues). [4]

B9.3

1a Cellular respiration without oxygen [1]

b Animals: glucose → lactic acid (energy transferred to environment) [1]

plants and yeast: glucose → ethanol + carbon dioxide (energy transferred to environment) [2]

2a They become tired/they start to burn/stitch [1]

b After a long period of exercise muscles become short of oxygen and switch from aerobic to anaerobic respiration, which is less efficient. [1] Glucose molecules are not broken down completely, [1] so less energy is released than during aerobic respiration. [1] End products of anaerobic respiration are lactic acid and a small amount of energy. [1]

3 Animals: anaerobic respiration takes place in muscles when not enough oxygen, produces lactic acid [1] and relatively small amount of energy, [1] allows animal to continue running even when they cannot breathe fast enough to supply the oxygen they need [1]; plants and yeast: anaerobic respiration forms ethanol and carbon dioxide, [1] allows plant/yeast to continue to respire in low-oxygen atmospheres, [1] rare in plants as they form oxygen during photosynthesis but more common in yeasts, where fermentation is used in baking and brewing [1]

4 Waste lactic acid produced during exercise as a result of anaerobic respiration has to be broken down to produce carbon dioxide and water. [1] This needs oxygen, and amount of oxygen needed to break down lactic acid is known as oxygen debt. [1] Even though muscles have stopped working, heart rate and breathing rate stay high [1] to supply extra oxygen until all lactic acid broken down and oxygen debt paid off. [1]

B9.4

1 Sum of all the reactions in a cell or in the body [1]

2 Any four from: conversion of glucose to starch, glycogen, and cellulose; formation of lipid molecules from a molecule of glycerol and three fatty acid molecules; use of glucose and nitrate ions to form amino acids used to make proteins; respiration; photosynthesis; breakdown of excess proteins to form urea. [4]

3 Liver involved in a number of processes requiring a lot of energy, [1] including removal of lactic acid produced during anaerobic respiration (blood carries lactic acid from muscles to liver, where it is converted back to glucose for aerobic respiration in cells), [1] conversion of glucose to glycogen when in excess, [1] and conversion of glycogen back to glucose when needed for respiration. [1] Liver consequently carries out a lot of aerobic respiration to provide energy needed for these reactions. [1]

B10.1

1 Regulation of internal conditions of cell or organism [1] to maintain optimum functional conditions [1] in response to internal and external changes [1]

2 Receptors: cells that detect changes in internal or external environment, may be part of nervous or hormonal control systems [2]; coordination centres: areas that receive and process information from receptors and coordinate body's response, may be part of nervous or hormonal control systems [2]; effectors: bring about response to signal from coordination centres, may be muscles or glands [2]

3a Any three from: temperature increase, temperature decrease, levels of sunlight increase, wind levels increase [3] Any other valid suggestion.

b Temperature increase – warming body, temperature decrease – cooling body, levels of sunlight increase – burning skin, wind levels increase – increasing cooling from skin surface, lack of food – hunger. [3] Any other valid suggestion.

B10.2

1a To take in information from external and internal environment [1] and coordinate body's response [1] so the organism can react to surroundings [1]

b Neurone: single nerve cell [1]; nerve: bundle of hundreds or thousands of neurones [1]

2 Eye: light [1]; ear: sound, pressure [1]; skin: touch, pressure, pain, temperature, vibration [1]; nose: smell [1]; tongue: taste [1]

3a *x*-axis is age (years) [1], *y*-axis is reaction time (secs) [1], suitable scale used for each axis [1], points plotted correctly [1] points joined up by straight lines [1]

b The reaction times get faster as students get older from 8 to teenage years [1] – then they stabilise for a few years [1] but over 18 they get slower again [1]

4 Light from fruit detected by sensory receptors in eyes, [1] impulse travels along sensory neurone to brain, [1] information processed in brain, [1] impulse sent along motor neurone [1] to muscles of arm and hand [1] so you pick up fruit and put it in mouth [1]

B10.3

1a They are automatic, rapid responses (to stimuli), [1] which do not involve the conscious parts of the brain. [1] They enable an organism to avoid damage and danger because they happen very fast. [1] They control many vital bodily functions, such as breathing, without need for conscious thought. [1]

b This would slow the process down, making it less effective at preventing damage. It would be very difficult to consciously control all bodily functions and still be able to do anything else. [1]

2 Stimulus (pinprick) → receptor → sensory neurone (electrical impulse) → synapse (chemical impulse) → relay neurone (electrical impulse) → synapse (chemical impulse) → motor neurone (electrical impulse) → muscles in leg lift foot [6]

3 Reflex actions (e.g., breathing, swallowing) need to happen constantly and operate automatically, even when you are asleep, [1] so cannot rely on conscious thought processes. [1] Speaking and eating require coordination of different areas of body [1] so need to be under conscious control. [1]

B10.4

1a Takes in all sensory information [1] and coordinates responses [1] as well as allowing conscious thought, intelligence, and emotions [1]

b Millions of interconnected neurones [1] arranged in different regions [1] to carry out different functions [1]

2 Cerebral cortex: consciousness, intelligence, memory, language [2]; cerebellum: coordinating muscular activity and balance [2]; medulla: unconscious activities (e.g., gut movements, breathing) [2]

3a Any two from: matching changes in behaviour/memory to the damaged part of the brain, electrically stimulating areas of the brain and seeing what happens, MRI scans which show what area of a brain is affected in a patient [4]

b Stimulating areas of the brain advantages: carried out on conscious people so can describe what they are experiencing/can pinpoint exact area of brain. [1] disadvantages: top of skull needs to be removed, [1] risk of damaging the brain or infection [1] MRI scan advantages: shows what part of the brain is damaged [1] so scientists can link region to loss of certain function [1] disadvantages: cannot pinpoint exact area [1]

B10.5

1 Muscles in the iris [1] contract and relax [1] to change the size of the pupil [1]

2a Any three from: sclera, cornea, iris, pupil, lens, suspensory ligaments, ciliary muscles, retina, optic nerve, blind spot, fovea [3]

b Answer should match components given in part **a**. Up to two marks per component.
sclera: white outer layer of eye, relatively tough and strong so eyeball not easy to damage; cornea: transparent area at front of eyeball that admits light, curved surface to help focus light coming into eye on retina; iris: muscular structure to control size of pupil; pupil: hole in iris through which light enters eye; lens: clear disc to help finely focus light onto retina; suspensory ligaments: attach ciliary muscles to eyeball, involved in changing shape of lens; ciliary muscles: contract and relax to change shape of lens and focus light on retina; retina: layer at back of eye containing light receptor cells, optic nerve: carries impulses from light-sensitive cells of retina to brain; blind spot: where optic nerve leaves eyeball; fovea: area of cones where light is brought into clear focus on retina [6]

3 Light enters eye through cornea, [1] which focuses it towards the iris and retina. [1] Light passes through the iris, then lens. [1] Lens changes shape to focus light onto retina. [1] Light-sensitive cells of retina stimulated by light and send impulses along neurones. [1] Impulses travel along optic nerve and are translated by brain into visual images. [1]

B10.6

1 Close objects: ciliary muscles contract, [1] suspensory ligaments loosen, [1] lens becomes thicker and refracts light rays strongly [1]; far away objects: ciliary muscles relax, [1] suspensory ligaments pulled tight, [1] lens pulled thin and only slightly refracts light rays [1]

2a When ciliary muscles contract, [1] suspensory ligaments loosen [1] and lens becomes thicker. [1] When ciliary muscles relax, [1] suspensory ligaments pulled tight [1] and lens pulled thin. [1]

b Myopia: can see close objects in clear focus but distant objects look blurred, [1] as lens too curved or eyeball too long [1]; hyperopia: can focus clearly on distant objects but close objects appear blurred, [1] as lens too flat and thin or eyeball too short [1]

3a When doing close work ciliary muscles are contracted [1] to keep lens thick [1] and able to focus light from close objects. [1] When looking at distant objects ciliary muscles relax. [1] Looking up into distance regularly when doing close work rests ciliary muscles. [1]

b Glasses: advantages: relatively cheap, very effective, don't damage eyes, last a long time; disadvantages: can be nuisance or get lost, some people don't like look of them [3] laser eye surgery: only available to adults once eyes have stopped growing and vision has stabilised; advantages: vision permanently corrected (nothing to take on and off/in and out), good for playing sport and general activities; disadvantages: risk of errors and infections during surgery [3]

B11.1

1a Molecule made in one part of the body of an organism and then transported to another part of the organism where it has an effect [1]

b Gland that produces hormones and secretes them directly into the blood [1]

2 Hormones: chemicals that control the processes of the body, released from glands into blood, some act quickly but many act more slowly, all are slower than nervous control [3]; nervous system: electrical impulses passed from receptors to brain, transmitted along neurones, can be very fast (especially reflexes), transmission from one neurone to another involves chemical substances [3]

3a Child would not grow properly/would be short, may look younger than their age, and may have delayed tooth development. [2] Any other valid suggestion.

b Adult would have acromegaly (chin, nose, ears, hands, and feet continue to grow). If it happened before puberty they would become unusually tall. [2] Any other valid suggestion.

4 Pituitary gland is small endocrine gland in brain that controls many different body processes as well as controlling secretions of many other endocrine glands in the body. [1] It produces a number of different hormones that give it this control. [1] Pituitary gland called the master gland because so many other endocrine glands rely on it to function. [1]

B11.2

1a Hormone made in pancreas that causes glucose to pass from blood into cells, where it is needed for respiration [1]

b Condition under which pancreas cannot make enough insulin to control blood sugar or body cells stop responding to insulin made by pancreas [1]

c Insoluble carbohydrate stored in liver [1]

2 Type 1: condition where pancreas does not make enough/any insulin, treated by insulin injections to help control blood glucose levels, needs carefully controlled diet with regular meals and careful monitoring of carbohydrate intake, usually appears in children and young adults [3]; type 2: condition where body cells do not respond properly to insulin produced by pancreas, linked to obesity/old age/lack of exercise, treated by improving diet/increasing exercise/losing weight as well as by insulin injections [3]

3 Pancreas detects rise in blood glucose levels and secretes insulin. [1] Insulin triggers conversion of glucose to glycogen by liver, [1] which causes glucose to move out of blood into body cells, lowering blood glucose levels. [1]

4a Pancreas detects fall in blood glucose levels and secretes glucagon. [1] Glucagon triggers conversion of glycogen to glucose by liver, [1] increasing blood glucose levels. [1]

b Insulin: secreted and released by pancreas when blood glucose levels rise above ideal range, enables glucose to move from blood into body cell where it is needed, triggers conversion by liver of excess glucose into glycogen stored in liver and muscles, lowers blood glucose levels [3]; glucagon: secreted and released by pancreas when blood glucose levels fall below ideal range, triggers conversion by liver of glycogen into glucose, glucose released back into blood, raises blood glucose levels [3]

c Glucose is needed for cellular respiration, [1] releasing energy for body's metabolic reactions. [1] Too much or too little glucose in blood causes problems with respiration. [1]

B11.3

1 Their pancreas does not make insulin [1] so they have to inject it after meals so cells take up glucose [1] for respiration. [1]

2 A meal is eaten at 7am, 12 noon and 6pm [1] and so glucose is released into the blood from digestion of the food causing blood glucose levels to rise [1]. In people with diabetes blood glucose levels decrease rapidly when insulin is injected at 7 am and 12 noon [1] because the insulin promotes cells to take up glucose. [1] In people without diabetes blood glucose levels are kept fairly constant [1] because they release insulin naturally from their pancreas throughout the day. [1]

3 Type 1: regular meals and monitoring of carbohydrate intake to keep blood glucose levels steady, insulin injections to keep blood glucose steady – as body produces insufficient insulin to control blood glucose levels; [3] type 2: carefully controlled balanced diet to restore normal blood glucose balance – as body cells do not respond to insulin produced by body; weight loss – as type 2 diabetes is linked to obesity; regular exercise – as type 2 diabetes is linked to lack of exercise; drugs to help insulin work better/help pancreas make increase/reduce glucose absorption from gut – to improve effectiveness of insulin on body cells [3]

4 Pancreatic transplant: complex surgery, high risk, expensive, patients have to be on immunosuppressant drugs for the rest of their lives, not enough donors [2]; insulin injections: widely available, self-administered, relatively cheap [2]

B11.4

1 If a factor in internal environment increases, [1] changes take place to reduce it and restore original level. [1] If a factor in internal environment decreases, [1] changes take place to increase it and restore original level. [1]

2 Thyroxine: controls body's basic metabolic rate (in adults) and growth and development (in children), controlled in negative feedback loop by thyroid stimulating hormone (TSH) from pituitary gland [2]; adrenaline: increases heart and breathing rates, triggers conversion of stored glycogen to glucose, stimulates pupil dilation, heightens mental awareness, diverts blood from digestive system to limb muscles, secreted by adrenal glands upon nervous stimulation (no negative feedback loop) [2]

3a Points may include: how iodine levels affect thyroxine production, effects of lack of iodine on health (stunted growth/developmental problems in children, problems including fatigue/immune system damage in adults), effect on people's ability to fight other communicable diseases and work/care for family, impact of adding iodine to diet (allows body to make thyroxine and overcome health problems related to thyroxine deficiency) [4]

b Add iodine to diet through use of iodised salt [1]

B11.5

1 Females: oestrogen [1], males: testosterone [1]

2 Follicle stimulating hormone (FSH) causes eggs to mature in the ovaries, [1] luteinising hormone (LH) stimulates release of egg at ovulation from ovaries, [1] oestrogen and progesterone stimulate build-up and maintenance of uterus lining. [2]

3 Any three similarities from: adolescent growth spurt (slightly later in boys), growth of pubic and body hair, external genitalia grow and skin darkens, brain matures. [3] Any three differences from: boys develop muscles and male body shape/ girls develop female pattern of fat deposits, testes grow and start producing sperm in boys/ovaries start to form mature ova monthly in girls, boys develop facial hair, girls develop breasts, male larynx grows and voice breaks, uterus grows and becomes active in girls, menstruation begins in girls [3]

B11.6

1a Follicle stimulating hormone (FSH), [1] luteinising hormone (LH), [1] oestrogen, [1] progesterone [1]

b To provide protection and food for developing embryo if pregnancy occurs [1]

2a 0–5 [1]

b FSH levels rise in first part of cycle, stimulating eggs to mature in ovary and stimulating ovary to produce more oestrogen. [2] LH levels rise sharply around ovulation, triggering release of mature ovum, then fall again rapidly. [2]

c Oestrogen [1]

3 Oestrogen stimulates thickening of uterus lining. [1] FSH stimulates egg maturation in ovaries. [1] LH stimulates release of mature egg from ovary (ovulation) after about 14 days. [1] Progesterone maintains uterus lining for about 14 days. [1] Hormone levels then drop and uterus lining and egg are shed in monthly period. [1] Pregnancy is most likely if egg is fertilised between days 14 and 28 while uterus lining is ready to receive developing embryo. [1]

B11.7

1 Control of fertility to prevent pregnancy by preventing egg and sperm meeting or preventing implantation of fertilised egg in uterus [1]

2a All are forms of hormone-based contraception. [1]

b Progesterone-only pill must be taken very regularly; contraceptive implant is placed under the skin; contraceptive implant lasts for up to three years but can be removed at any time the woman wants to get pregnant; contraceptive patch applied to skin every seven days. [3]

c Contraceptive implant [1] lasts up to 3 years [1], hormones not lost through illness e.g., vomiting [1]

3 Hormone-based: relatively effective (0–10% failure rate per Figure 2), open to human error [1] barrier: less effective (12–18% failure rate per Figure 2), open to human error or damage [1] surgical: very effective, not open to human error following successful surgery [1]

B11.8

1 A problem with conceiving naturally [1]

2a Artificial FSH can be used to stimulate egg maturation [1] and oestrogen production [1] in ovaries so uterus lining builds up. Artificial LH can then be used to trigger ovulation if necessary. [1]

b Artificial FSH can be used to stimulate maturation of many eggs in ovaries. [1] Artificial LH can then be used to bring them to the point of ovulation [1] for harvesting for IVF. [1]

3 For: people have the right to have children when they want them (ageism), people still considered young at 40 now life expectancies have increased, may not meet right person until later in life, career demands for women make it difficult to have children earlier; against: chances of successful pregnancy very low, cost may exceed benefit, older parents will be old as children grow up, stress and physical risk associated with pregnancy increases with age [6] 1 mark each for up to 3 pros and 3 cons. Also consider any other valid argument.

4 Advantages: ability to control when to have a family or how many children to have, ease of starting and stopping contraception, ability to help infertile couples have children; disadvantages: use of female hormones in contraception or infertility treatments may have side effects, can give false sense of security (may forget to use other methods of contraception if sick), can lead to false expectation that you can have a baby whenever you want, can lead to multiple births if dosage not controlled properly. [6] Any other valid point. 1 mark each for up to 3 advantages and 3 disadvantages.

B11.9

1a Response of plant to light [1]

b Response of plant to gravity [1]

2 Plants need to respond to surroundings [1] so that roots grow down into soil when a seed germinates and shoots grow up towards light for photosynthesis. [1] Shoots need to be able to grow into light if something shades them, whilst roots need to be able to grow towards water. [1] Trophic responses enable plants to respond to light and gravity. [1]

3 In light from one side only, exposed shoots would grow and bend towards light. [1] Auxin moves to unlit side of shoots, stimulating cells on unlit side to grow faster and causing shoots to bend towards light until they are effectively illuminated evenly again. [1] Sensitivity to one-sided light is in shoot tip, so covered shoots would continue to grow straight up [1] as there is no change in auxin concentration. [1]

b Newly exposed shoots would start to bend towards light [1] because shoot tips would be exposed and auxin would move to unlit side. [1] Newly exposed shoots wouldn't bend as much as permanently exposed shoots [1] as they would have had less time to grow. [1]

B11.10

1 Auxins, gibberellins, ethene [1]

2a Hormone rooting powders contain auxin, plant hormone that stimulates root growth in cut stem. [1] Use of hormone rooting powders by gardeners make it more likely that cuttings will develop roots and grow well, making plant propagation easier. [1]

b When sprayed onto leaves, [1] hormones such as auxin can stimulate uncontrolled rapid growth that kills plants. [1] More spray is absorbed by broad-leaved plants (e.g., weeds) [1] than by narrow-leaved plants (e.g., cereals and grasses). [1]

3 1 mark for one of the possible uses of each of the three hormones: auxin, giberellins, and ethene. In each case 1 mark for first statement which describes the use and second mark for explaining how this helps feed the world. For example: auxin used in hormone rooting powders to help cuttings grow into new plants [1] making crop plant propagation easier [1]. Used in weed killer to stimulate uncontrolled growth and plant death [1] – reducing competition to cereal crops from weeds [1]. Used in plant tissue cultures to stimulate growth and cell division [1] – promoting plant cloning to produce many identical crop plants. [1] Any other valid point.

B12.1

1 37 °C [1]

2 Blood flows through thermoregulatory centre in brain, which monitors temperature and sends impulses to lower body temperature. [1] Blood vessels supplying capillaries in skin dilate, increasing volume of blood in capillaries and increasing cooling by radiation from skin surface. [1] More sweat produced by sweat glands, cooling skin surface as sweat evaporates. [1] Thermoregulatory centre monitors blood temperature and stops cooling mechanisms when it returns to normal. [1]

3 It must not rise above 40 °C because at these temperatures enzymes denature [1] and respiration stops causing death. [1] It must not fall below 35 °C because at these temperatures the rate of enzyme-controlled reactions is too slow [1] and eventually cells will die. [1]

4 The thermoregulatory centre in brain recognises loss of body heat to cold environment and sends impulses to raise body temperature. [1] The bird fluffs up feathers [1] to trap an insulating layer of air around the body. As insulating layer of air is warmed, less cooling will occur as temperature gradient between skin and surroundings will be reduced. [1] Blood vessels close to skin surface constrict, [1] reducing volume of blood in capillaries to minimise cooling by heat loss from skin surface. [1] As insulating layer of air is warmed, less cooling will occur as temperature gradient between skin and surroundings will be reduced. [1]

B12.2

1a Exhalation, [1] sweating, [1] excretion in urine [1]

b Controlled: excretion in urine [1]; uncontrolled: exhalation, [1] sweating [1]

2a Waste product formed from breakdown of excess amino acids in liver [1]

b Urea is poisonous to body cells [1] in high concentrations [1]

3a Excess amino acids are carried to liver in bloodstream. [1] Liver removes amino group from amino acids by deamination. [1] This initially forms ammonia, [1] which is converted into urea. [1]

b Balance of water, mineral ions, and nitrogenous waste (urea) in the body is important because of the effect solute concentrations in cells, tissue fluid, and blood have on osmosis. [1] If solute concentration outside body cells is higher than cell contents, [1] water will leave cells by osmosis and cells will shrink and stop working. [1] If solute concentration outside body cells is lower than cell contents, [1] water will enter cell by osmosis and cells will swell and may burst. [1] Balance enables body cells to retain shape and work efficiently. [1]

c Water loss from lung surfaces during breathing or from skin surface in sweating is uncontrolled. [1] It is vital that there is a system by which loss of water and mineral ions can be controlled to maintain correct osmotic balance between body cells and body fluids [1] and remove poisonous urea, [1] regardless of body's water needs. [1] Urea, excess water, and excess mineral ions are removed from body via kidneys and excreted in urine. [1] This process is very tightly regulated to maintain water and mineral ion content of blood within narrow limits, and to remove urea. [1]

B12.3

1a To filter out urea and balance water and mineral ion levels in blood [1]

b Sugar, mineral ions, amino acids, urea, and water are filtered out of blood into kidney tubule, [1] then selectively reabsorbed back into blood as liquid travels along tubule. [1] All sugar is taken back into blood but amounts of mineral ions and water vary with body's needs. [1] Some urea returns to blood down concentration gradient. [1]

2 Kidneys remove excess salt and excrete it in urine. [1] Urine will contain high levels of salt. [1] Hot day will cause a lot of uncontrolled water loss through sweating [1], so kidneys will need to conserve water. [1] Urine will contain less water/be more concentrated. [1] Up to a maximum of 4 marks.

3a After drinking large volume of water there is a slight delay and then volume of urine produced increases steadily for more than 60 minutes. [1] Over the next 60–90 minutes urine volume returns to normal. [1] After drinking, concentration of salt in urine falls below normal levels [1] before gradually returning to normal over same time period. [1]

b Drinking a large volume of water dilutes blood and body fluids. Body detects this and amount of water passing through kidneys without being reabsorbed increases. [1] Body's mineral ion concentration, including salt levels, has not increased. Body reabsorbs same amount of mineral ions as before. [1] This means that urine contains more water but same amount of mineral ions, so concentration of mineral ions in given volume of water falls. Body removes excess water but maintains levels of mineral ions needed in blood. [1]

4a Blood would become diluted. [1] Receptor cells in brain would detect reduced blood concentration [1] and pituitary gland would release less ADH into blood. [1] Kidney tubules would absorb less water, so kidneys produce lots of very dilute urine. [1]

b Salt and water would be lost in sweat and water would not be replaced. [1] Receptor cells in brain would detect increased concentration of urea in blood (even with loss of mineral ions in sweat) [1] and pituitary gland would release more ADH into blood. [1] Kidney tubules would absorb more water, conserving both salt (being lost in sweat) and water, so kidneys would produce small quantities of urine containing concentrated urea but a relatively low concentration of mineral ions. [1]

B12.4

1a When the kidneys stop working well enough to survive without intervention [1]

b Dialysis [1] transplant [1]

2a Blood passes out of artery via pump. [1] Blood thinners added to prevent clotting. [1] Blood passes between dialysis membranes where excess salt and urea are removed by diffusion down a concentration gradient. [1] Clean blood passes through bubble trap before returning to vein. [1]

b Advantages: cleans the blood [1] so keeps people with kidney failure alive [1] disadvantages: any two from: must follow a very carefully controlled diet, [1] have to spend regular, long sessions connected to a dialysis machine, [1] only a relatively short time when the balance of substances in the blood is at the right level and so may feel tired and unwell as the next dialysis session approaches, [1] over many years, the balance of substances in the blood becomes more difficult to control. [1] Up to a maximum of 4 marks.

3a No urea ensures steep concentration gradient between blood and dialysis fluid [1] to maximise removal of urea from blood into fluid by diffusion. [1] Normal plasma levels ensure that any excess salt, glucose, and mineral ions in blood diffuse into fluid down concentration gradient to give ideal blood concentration upon completion of dialysis. [1]

b Any excess salt and urea [1] are removed during dialysis [1]

B12.5

1 Transplanted kidney takes on functions of failed kidneys, balancing blood chemistry and removing urea [1]

2a Live organs have no tissue damage, [1] family donors often a close tissue match [1]

b Taking organs from a healthy, living person can threaten their health. [1] Surgery always carries a risk. [1]

3 Dialysis advantages: machines available, [1] no problem with tissue matching, [1]disadvantages: time-consuming, [1] lifetime commitment, [1] expensive in long-term, [1] diet must be carefully controlled, [1] transplant advantage: post-transplant relatively cheap, [1] permits relatively normal life, [1] disadvantages: shortage of donors, [1] need tissue match, [1] limited transplant lifespan, [1] personal choice must be justified by rational argument. Maximum of 6 marks.

B13.1

1a No fusion of gametes, only one parent, no variation [1]

b Fusion of gametes from two parents, variation [1]

c Sex cell containing only a single set of chromosomes, made as a result of meiosis (cell division) [1]

d Differences between individuals as a result of their genetic material [1]

2 Asexually produced offspring are formed by mitosis from cells of one parent, [1] with no joining of gametes and no mixing of genetic information, hence no variation. [1] Sexually produced offspring are formed by meiosis, [1] with fusing of male and female gametes from two parents, [1] hence offspring genetically different from both parents and from each other. [1]

3 Points may include: asexual offspring identical to parents so if conditions change and one individual cannot survive, none of them can; sexual reproduction introduces variation so if conditions change some offspring may survive; change in conditions may make it more difficult for sexually reproducing organisms to meet and mate, so may be less likely to survive than asexual organisms which can reproduce regardless. [6] Any other valid point.

B13.2

1a 19 [1]

b 38 (19 pairs) [1]

2 Sexual reproduction involves fusing of two gametes, bringing genetic information from two individuals together and introducing variation. [1] Gametes result from meiosis, [1] during which each gamete receives random mixture of original chromosome pairs, [1] so gametes are all different from original cells as well as from each other. [1]

3a Meiosis, [1] after chromosomes are copied, [1] cell divides twice, [1] resulting in sex cells each with single set of chromosomes [1]

b Reproductive organs (ovaries or testes) [1]

c It halves the number of chromosomes in gametes so that when two gametes fuse during sexual reproduction, new individual has full set of chromosomes. [1] It introduces variation as each gamete receives random mixture of original chromosomes. [1]

B13.3

1 Strawberries, daffodils, malarial parasites [1]

2a Flowering plants commonly reproduce sexually but can reproduce asexually (e.g., strawberry runners), [1] enabling formation of new plants even when sex organs (flowers) destroyed or eaten. [1] Fungi commonly reproduce asexually but can reproduce sexually when conditions are unfavourable (e.g., too dry). [1]

b Malaria parasites reproduce differently at different stages of life cycle – asexually in human liver and blood cells, [1] sexually inside red blood cells in mosquito. [1] Both sexual and asexual reproduction are part of life cycle [1] (asexual reproduction is not an alternative for bad conditions, as it is in fungi). [1]

3 Points may include: if organism can reproduce asexually in favourable conditions (e.g., fungi), variation is not needed – organism can reproduce fast and safely with genetic makeup that works; if conditions become more difficult, reproducing sexually can introduce variation to increase chances of survival (e.g., to produce spores better adapted to adverse conditions); reproducing sexually as fixed part of life cycle guarantees continuous variation, which can be an advantage to a pathogen (e.g., malaria parasite) – makes it harder for body to identify and reject it, and harder to make vaccines against it [6] Any other valid point.

B13.4

1a Entire genetic material [1]

b Major project to sequence entire human genome [1]

2a Genes are made out of sections of DNA. [1] Each gene codes for a specific protein. [1] Chromosomes contains thousands of genes. [1]

b Any three from: To find out as much as possible about human genome, to identify common differences and similarities in sequences between individuals, to identify genes associated with particular diseases, to discover more about human evolution, to find similarities and differences between humans and other organisms, to identify evolutionary relationships between different organisms, to identify organisms with genetic similarities to people that may be useful in models of disease [3]

3 All living humans originate from ancestors who migrated from Africa around 60 000 years ago. [1] Genome analysis enables scientists to map genetic markers in modern humans [1] to trace human migration patterns throughout history. [1] DNA samples are collected from different ethnic groups [1] and genetic differences are identified. [1] More closely related groups will have fewer genetic differences. [1] Any other well-researched point.

B13.5

1a One of four different compounds known as A, C, G, and T that are key to DNA structure [1]

b Single unit of DNA made up of sugar, phosphate, and base [1]

2 When the protein chain is completed on the ribosome, [1] it folds up. [1] Shape made depends on sequence of amino acids. [1] Unique shape enables it to carry out specific job as an enzyme, hormone, or structural protein. [1]

3 Template is produced that reflects sequence of bases in DNA of a single gene. [1] Template is small enough to leave nucleus through pores in nuclear membrane, and binds to surface of ribosome. [1] Carrier molecules in cytoplasm, each attached to specific amino acid, attach themselves to template in order given by DNA until template completed. [1] Amino acids join together to form specific protein. [1] Protein detaches from carrier molecules, which detach from template and return to cytoplasm to pick up more amino acids. [1] Once protein chain is complete molecule folds up to form unique shape that will enable it to carry out its functions in cell. [1]

B13.6

1a A section of DNA that codes for a specific protein [1]

b A change to the sequence of bases in a strand of DNA [1]

2 Points may include changes in order of amino acids/shape of protein may have bad effect – active site of enzyme changed so it no longer fits normal substrate, structural protein may lose its strength; mutation may cause changes in molecules controlling rate of cell cycle – causing cancer; mutation may mean organism doesn't compete well – leading to death of affected organism; mutation may cause production of enzyme with different-shaped active site – enabling it to produce different colour/better camouflage; may cause production of enzyme that allows breakdown of different food source – enabling survival when normal food is scarce; may cause production of enzyme that is faster or more efficient catalyst – improving cell function; mutation in bacteria may alter biochemistry to cause antibiotic resistance – advantage to bacteria if not to people; mutation may affect gene expression rather than changing product of particular gene – effect may be neutral, positive, or negative. [4] Two marks for positive impact and valid example, two marks for negative impact and valid example.

3 Gene expression is when a gene codes for a protein that is made in the cell. [1] Each gene controls the synthesis of many different proteins [1] depending on how much of the gene gets turned on or off [1] or which other genes are switched on or off at the same time [1] by the non-coding parts of DNA. [1]

B13.7

1a The alleles present regarding a particular characteristic [1]

b The physical appearance regarding a particular characteristic [1]

2a A dominant allele controls the development of a characteristic even when present on only one chromosome. [1] A recessive allele controls the development of a characteristic only if present on both chromosomes. [1]

b Homozygous: both alleles for particular characteristic are the same; [1] heterozygous: individual has two different alleles for particular characteristic [1]

3

Gametes	A	a
A	AA	Aa
a	aA	aa

probability 75% dangly earlobes, 25% attached. 1 mark for gametes, 1 mark for each successful offspring genotype shown and 1 mark for probabilities. [6]

4a i

Gametes	B	B
B	BB	BB
b	Bb	Bb

Gametes [1]; 1 mark for each correct offspring genotype shown. [5]

ii

Gametes	B	b
B	BB	Bb
b	Bb	bb

Gametes [1]; 1 mark for each correct offspring genotype shown. [5]

b Black fur in mice is dominant phenotype. Black mice may be BB or Bb. [1] Brown mouse has recessive phenotype with genotype bb. [1] Crossing black mouse with brown mouse can help work out genotype of black mouse. [1] Any brown mice in litter show black adult mouse is heterozygote. If offspring are all black, black mouse may be homozygous but not proven as all offspring could be black by chance. [1]

B13.8

1a XX [1]

b XY [1]

2a People think that the more times you have a girl, [1] the greater the chance that the next child will be a boy. [1]

b Every time an egg or a sperm meet it is a random event. [1] Whether the sperm is carrying an X or a Y chromosome is completely random, [1] so the chances of any baby being a boy or a girl are 1:1 or 50% that it will be a boy (XY) or a girl (XX). [1] Also accept Punnett square demonstrating this.

3 Gametes meeting is a random event. [1] Lots and lots of genetic crosses are needed for patterns to emerge, to overcome the random element of mating. [1]

4a

Gametes	G	g
g	Gg	gg
g	Gg	gg

1 mark for correct gametes for each individual; 1 mark for each correct offspring genotype. [6]

b Proportion of genotypes 50% Gg: 50% gg , [1] ratio 1:1, [1] random mixture of genes means offspring don't always follow expected ratios, [1] need lots of offspring before the ratios show and tigers don't have many cubs [1]

B13.9

1a Inherited disorder that causes extra fingers or toes [1]

b Polydactyly phenotype [1] is dominant [1]

2a A Pp, [1] individual had polydactyly but had unaffected child (must have recessive allele) [1]

B Pp, [1] individual had polydactyly but had two unaffected children (must have recessive allele) [1]

C PP or Pp, [1] both parents have polydactyly so each could have passed on dominant allele P, and both parents are heterozygotes as have unaffected children, so each could have passed on recessive allele p [1]

b Carriers have an allele for the normal dominant phenotype, [1] so their body works normally. [1]

c Cystic fibrosis caused by recessive allele, [1] which must be inherited from both parents for child to have disease. [1] If both parents had the disease themselves they would almost certainly be infertile, [1] so parents must be carriers (heterozygotes). [1]

3 Couple must both be carriers (heterozygous):

Gametes	A	a
A	AA	Aa
a	Aa	aa

1 mark for each correct parent genotype, [2] 1 mark for each correct offspring genotype, [4] 25% chance of a child (aa) affected by sickle cell disorder [1]

B13.10

1 Any one from: risk of miscarriage, false positive/negative result, mean that people have to make a decision about whether to terminate a pregnancy, it is expensive [1]

2 Amniocentesis: taking some fluid from around the developing fetus at around 15–16 weeks of pregnancy [2]; chorionic villus sampling of embryonic cells: taking small sample of tissue from developing placenta at around 10–12 weeks of pregnancy [2]

3a Ultrasound used to show position of fetus and needle, which is inserted through body wall into amniotic fluid. [1] Amniotic fluid sample containing fetal cells drawn up into syringe. [1] DNA isolated from fetal cells and tested for particular genetic disorders. [1]

b Early embryo is tested before implantation in mother. [1] DNA isolated from embryonic cells and tested for particular genetic disorders. [1]

4 Advantages: screening could reduce high societal cost of health care and support for children with genetic problems, screening could avoid children being born into pain/suffering caused by genetic disorders; disadvantages: collecting cells from fetus increases risk of miscarriage, screening can give false positive or negative result, screening necessitates decisions about termination, screening is expensive, screening could give rise to 'designer babies', no current cures for genetic disorders identified in screening, number of children born with genetic disorders relatively low so screening every fetus may not be cost-effective [6] Any other valid point.

B14.1

1a Any one from: eye colour, nose shape, sex, dimples. [1]

b Any one from: scarring, language, suntan, religion. [1] Any other valid example.

2a For comparison with normal population. [1] Studying identical twins reared together and reared apart allows comparison of impact of different environments on genetically identical humans. [1] Identical twins reared together are subject to same environmental conditions, meaning environmental variables are controlled. [1]

b Height seems mostly controlled by genes [1] as there is least difference in height between identical twins whether reared together or apart. [1] Mass [1] seems mostly affected by environment as identical twins brought up apart are no more identical than ordinary siblings. [1]

3 Award marks for any sensible suggestion demonstrating awareness of good experimental design. [6]

B14.2

1 Individuals within particular species may show wide phenotypic and genetic variation. [1] individuals with characteristics most suited to environment more likely to survive to breed successfully [1]. Alleles (variables) enabling survival passed on to next generation. [1]

2a New variant/form of a gene/new allele [1] resulting from change in organism's DNA. [1]

b Mutations occur naturally through mistakes made in copying DNA when cells divide. [1] Mutation may produce adaptation making organism better suited to changed environment. [1] Competitive advantage makes organism more likely to survive and breed, new variant rapidly becomes more common in population. [1]

3a Mutations gave some deer larger antlers, making them more successful in battles with other stags and more attractive to females. [1] Stags with largest antlers mated most successfully, [1] spreading alleles for large antler size within population. [1]

b Mutations gave some seals thick fur and thick blubber, making them more likely to survive and breed in very cold habitat. [1] Seals with thickest fur and blubber mated most successfully, [1] spreading alleles for thick fur and blubber within population. [1]

c Mutations gave some moths colouration like birch twigs that provided better camouflage from predators, [1] making them more likely to survive and mate successfully. [1] Moths with this colouration passed on alleles for it and eventually formed new species. [1]

B14.3

1a Animals or plants with desired characteristic chosen [1] and used as breeding stock [1] to produce offspring with same desirable characteristic over time [1]

b Any suitable example of animal use of selective breeding e.g. animals that produce more meat or milk, domestic dogs and farm animals with a gentle nature. [1] Any suitable example of plant use of selective breeding e.g. large, unusual, brightly coloured, heavily scented flowers, disease resistance. [1]

2 To produce animals and plants with useful or desirable characteristics [1] such as: increased yield or disease resistance in crops, [1] increased milk, meat, and wool yield in animals; [1] gentle nature in domestic animals and pets [1]

3 Award marks for explanation using any appropriate example. [6]

4a Breeding for particular characteristics [1] means only individuals with alleles for chosen characteristics are allowed to breed, [1] reducing variation in alleles for characteristic. [1]

b If conditions change (e.g., new disease, climate change) lack of variation within population reduces the chance of at least some of the organisms containing useful variant allowing them to survive [1] and breed, [1] threatening population's survival. [1]

B14.4

1 Modifying organism's genetic material [1] to produce a new, desirable characteristic [1] by cutting gene for desirable characteristic out of one organism [1] and transferring it to genetic material of another. [1]

2 Any three from: pest resistance – reducing crop vulnerability to pests and increasing yield, improved food value – changing crop composition to make crop more nutritionally valuable, flood resistance – reducing crop vulnerability to flooding and protecting yield, disease resistance – reducing crop vulnerability to common diseases and increasing yield. [6] Any other valid example.

3 Flow chart should include: use enzyme to cut out short stem gene from plant → use enzyme to remove plasmid from bacterium → insert short stem gene into bacterial plasmid → insert bacterial plasmid containing short stem gene into target plant [4]

B14.5

1a Individual produced asexually [1] that is genetically identical to parent [1]

b Only one parent, no joining of gametes, [1] offspring genetically identical to parent [1]

c Use of few cells from desirable plant to make big mass of identical plant cells, [1] each of which can produce tiny identical plant [1]

d Splitting apart developing embryo cells before specialisation [1] to produce several identical embryos [1]

2a Allows production of far more calves from best cows, [1] allows transport of high-quality embryos worldwide, [1] allows rapid replication of genetically engineered animals [1]

b Flow chart should include: cow given hormones to produce large number of eggs → cow inseminated with sperm/eggs collected and fertilised in laboratory → embryos collected and taken to the lab → embryos split to make more identical embryos → embryos transferred to host mothers [5]

c Economic: only wealthy farmers/countries can afford cloning technology; [1] ethical: cloning decreases cattle biodiversity; [1] how far should humans be able to interfere in production of new life [1] Any other valid suggestion.

3a Use mixture of plant hormones [1] to make small group of cells from desirable plant. [1] These cells produce mass of identical plant cells (callus). [1] Use different mixture of hormones [1] and conditions [1] to stimulate production of many identical plantlets. [1]

b Commercial use to produce identical plants for homes and gardens quickly and easily. [1] Conservation use to preserve rare plant species [1] by using single specimen to produce many new plants. [1]

B14.6

1 Production of new whole animal from cell of another adult animal [1]

2 Nucleus removed from unfertilised egg cell [1] and nucleus taken from adult body cell, [1] nucleus from adult cell inserted in empty egg cell, [1] new egg cell given tiny electric shock, [1] new egg cell divides to form embryo cells, [1] ball of cells (embryo) inserted into womb of adult female to continue development [1]

3 Any three comparisons e.g.,
in adult cell cloning an exact copy of an adult animal is produced, [1] in embryo cloning copies of one embryo are formed [1]; adult cell cloning involves transfering the nucleus from an adult body cell to an egg cell, [1] embryo cloning involves dividing up an embryo into smaller groups of cells; [1] adult cell cloning faces significant ethical concerns, [1] embryo cloning is ethically more acceptable [1]

4 The organism will have the same genes [1] but environmental factors affect the characteristics of the organism [1] e.g., the amount of food the animal receives. [1]

B14.7

1 Advantages - any one from: improved growth rates of plants and animals, increased crop yield, crops that grow well in dry, hot, or cold, crops that make their own pesticide or are resistant to the herbicides, crops containing vitamins and other nutrients; [1] disadvantages – possible effect on human health, insects may become pesticide resistant, expensive, genes might spread onto wild populations [1]

2 Put healthy versions of genes [1] into affected cells [1] so the cells work properly [1] or the cells of an early embryo is edited. [1]

3 Points may include: cost of genetically engineered products may be too high for those they are designed to benefit (e.g., golden rice), genetically modified crops have previously been made infertile forcing farmers in poor countries to buy new seed each year, fears about human engineering – people might use it to have 'designer' babies [6] Any other valid point.

B15.1

1a Pea [1]

b Seeing chromosomes through a microscope, [1] observation that chromosomes and units of inheritance behaved in similar ways, [1] working out the structure of DNA/mechanism of gene function [1]

2a He found that characteristics were inherited in clear and predictable patterns, [1] that some characteristics were dominant over others, [1] and that they never mixed together. [1]

b He kept records of everything he did [1] and analysed his results. [1]

3a People couldn't see units of inheritance, [1] so no proof of their existence. [1] People were unused to studying careful records of results. [1]

b Wilkins and Franklin [1] examined structure of DNA using X-rays. [1] Watson and Crick [1] worked on 3-D DNA model to explain how it works. [1] Watson and Crick ultimately developed double helix structure for DNA molecule [1] based on Franklin's X-ray results. [1]

B15.2

1a Changes that occur in an organism during its lifetime can be inherited [1]

b Organisms show variation. [1] The organisms that are most suited to their environment survive and breed. [1] They pass on these characteristics to their offspring. [1]

2a Different types of same bird lived in slightly different areas, [1] making Darwin question how they came about. [1]

b It gave Darwin time to collect many specimens [1] then make observations and consider theories. [1]

c It gave Darwin time to review specimens, drawings, and observations [1] before developing theories and ideas [1] and gathering evidence to support them. [1]

3 Points may include: helped pave way for Darwin's ideas, stimulated Darwin's thinking, first theory to challenge creation of world by God, instigated debate on origins of life, flaws in Lamarck's theory made Darwin's theory more plausible [5] Any other thoughtful point.

B15.3

1 Finches had adapted to differing environments [1] by natural selection. [1] Adaptations evolved for particular environments [1] made finch species look different. [1]

2 Breeding experiments with pigeons [1] to show how features could be artificially selected, as evidence for natural selection seen in Galapagos finches. [1] Examination of different barnacle types and their habitats to show adaptation of organisms to environment. [1]

3a Darwin observed variation in characteristics in his pigeons. [1] This was the result of inherited genes. [1] By selectively breeding pigeons with specific characteristics, [1] Darwin showed that characteristics could be inherited. [1] Through selective breeding, he showed how certain characteristics became more common in the population over generations. [1] This showed that the same could happen in natural populations where characteristics were selected for because they were the most 'useful' for survival. [1]

b Points may include: Darwin's ideas challenged popular and longstanding belief that God made animals and plants, Darwin's evidence was insufficient to convince some scientists, Darwin's theory lacked obvious mechanism for inheritance [6] Any other valid point.

B15.4

1a The formation of a new species [1]

b Geographical isolation: populations physically isolated by geographical feature, [1] such as new mountain range, river, or island [1]; environmental isolation: organisms separated by climate change in a particular area [1]

2 Part of population becomes isolated under new environmental conditions. [1] Alleles for characteristics enabling organisms to survive and breed successfully in changed environment will be selected for. [1] These are likely to differ from the alleles selected for in original environment. [1] Selection of different alleles for success in new environment causes characteristics of isolated organisms to change. [1] Eventually, isolated organisms can no longer interbreed with original organisms [1] and new species forms (speciation). [1]

3 Similarities: both avid naturalists, both made many observations in the wild in isolated areas of the world, both developed basic model of evolution by natural selection, both published initial papers simultaneously; differences: Darwin rich whilst Wallace relatively poor, Wallace had bad luck and lost specimens whilst Darwin didn't, Darwin formulated and refined his ideas over many years whilst Wallace wrote idea down immediately and sent it to Darwin [6] Any other valid point.

B15.5

1a The remains of organisms from millions of years ago, [1] any suitable example [1]

b No one there to see it, [1] early organisms soft-bodied so few fossils formed and most early fossils destroyed by geological activity [1]

2a When animal or plant does not decay after death [1] as preserved in ice or peat [1]; when hard parts of animal or plant replaced by other minerals, forming part of rock (rock fossil) [1]; when impression of organism made in mud becomes fossilised (mould fossil) [1]; when mould is filled in by minerals (cast fossil) [1]; when traces (footprints, burrows, droppings) are preserved [1]

b Any three from: soft-bodied organisms do not form fossils, many fossils destroyed by geological activity, most organisms did not become fossils, there are many fossils still to be found [3]

3a Animal or plant buried in ice [1] where conditions for decay are absent (temperature too low), [1] so does not decay after death [1]

b Animals and plants preserved almost intact, giving clear insight into what organism looked like [1]; very rare and often thousands of years old [1]; can show what animal had been eating or colour of long-extinct flower [1]; DNA can be extracted and compared to DNA of modern organisms [1]

B15.6

1 A complete fossil record gives a clear picture of how species evolves over time [1] in response to changing environmental conditions. [1] Some organisms have changed little – fossil sharks very similar to modern sharks as they evolved early into a form almost perfectly adapted to their environment. [1] Records not always complete. [1]

2 Shows how tall they were, [1] what their feet were like, [1] what terrain they walked on, [1], how fast they moved [1]

3a New predators: prey population unable to adapt quickly enough [1] to avoid new predators; [1] new diseases: population decline especially where whole population is close together (e.g., on island), [1] as disease spreads more rapidly than population can adapt to survive it; [1]
successful competition: competitive advantage (from mutation or due to newly introduced species) [1] rapid population decline if species unable to adapt quickly enough to compete effectively [1]

b As environments change, less adapted organisms are more likely to go extinct, [1] so there is less competition between species [1] allowing the more adapted ones to survive, breed and pass on their advantageous genes. [1]

B15.7

1a When many/most of species on Earth become extinct [1]

b Any two from: massive volcanic eruption, collision of giant asteroids with the Earth, major climate change [2] Any other valid point.

2 Any environmental change plus explanation e.g., a drop in temperature [1] means species that are not adapted to the cold [1] cannot survive and become extinct. [1]

3a Crater, [1] layer of rock debris, [1] mineral formed when massive force hits rocks, [1] age of rocks suggest impact happened immediately before dinosaur mass extinction event [1]

b i Asteroid impact would have blasted lots of dust and debris into atmosphere [1] and triggered fires, earthquakes, and landslides [1] generating more smoke and dust. [1]

ii Levels of light reaching Earth would have been much reduced, [1] which would have stopped plants growing [1] and caused very low temperatures. [1] This global winter would have caused mass extinctions through starvation and freezing. [1]

B15.8

1a Bacterium [1]

b MRSA has developed resistance to many antibiotics, including methicillin. [1]

2 Flow chart should include: colony of bacteria treated with antibiotic A → 5% survive through mutation → surviving bacteria form new colony resistant to antibiotic A and no longer affected by it → colony of bacteria treated with antibiotic B → 5% survive through mutation → surviving bacteria form new colony resistant to antibiotics A and B and no longer affected by them [6]

3a Antibiotic resistance caused by increased use of antibiotics/ lower hygiene standards in hospitals/spread of bacteria via hands and clothing of medical staff [1]

b Any four from: reduction in prescription of antibiotics, treating specific infections with specific antibiotics, reminders to medical staff to wash hands/use alcohol gel between patients, reminders to patients and visitors to wash hands/use alcohol gel on entering and leaving medical facilities, increasing hygiene standards in hospitals, treating patients affected by antibiotic resistant strains of bacteria in isolation [4] Any other valid point.

4 There is the presence of genetic variation within populations, [1] some bacteria are resistant to antibiotics, [1] the resistant bacteria survive and reproduce, [1] the number of resistant bacteria in the population increases. [1]

B15.9

1 Organisation of living organisms into groups according to their similarities and differences [1]

2a i Animals, [1] plants [1]

ii Archaea, [1] eubacteria, [1] protista, [1] fungi, [1] plants, [1] animals [1]

b Fewer organisms known when Linnaean system developed, [1] all classification based on observation of organism appearance. [1] Many more organisms have since been discovered observed using new techniques. [1]

3a System by which organisms are named, [1] with two names indicating genus and species [1]

b It enables scientists globally [1] to discuss huge variety of living and extinct organisms [1] by common names, preventing confusion. [1]

B15.10

1a Archaea, bacteria, eukaryota [1]

b Archaea: archaea [1] bacteria: bacteria [1] eukaryota: protista, fungi, plants, animals [1]

2a Domains are higher level of organisation than kingdoms – organisms they contain have fewer things in common. [1] Domains based on biochemistry of ribosomes and method of cell division. [1] Domain may contain several kingdoms. [1]

b Woese decided that there were too many differences between different types of bacteria for them all to be in the same group. [1] His ideas have been accepted because they have been backed up by evidence (e.g., DNA analysis), [1] and the new model works well in practice. [1]

3a Identify similarities and differences, [1] traditionally these were in physical appearance, including internal structures such as the skeleton, but biochemistry and cell function is now considered; [1] modern technology has made big changes, with use of microscopes to examine cells [1] and DNA analysis to identify precise relationships between organisms [1]

b Classifying organisms and seeing how closely they are related to other organisms allows us to build up evolutionary trees. [1] DNA technology makes it easier to recognise relationships between organisms that look very different but have evolved into separate species only relatively recently, [1] or between organisms that look similar but are not closely related. [1]

4 Giant pandas and red pandas share a common ancestor. [1] Giant pandas share a common ancestor with other bears. [1] Red pandas share a common ancestor with members of the weasel and raccoon family. [1] Evolutionary trees model evolutionary relationships between organisms, [1] determining how long ago they divided away from a common ancestor. [1] They promote understanding of evolutionary pathways and relationships between species. [1]

B16.1

1a All populations of interdependent different species [1] living in a habitat [1]

b Ecosystem made up of communities [1] of organisms in a habitat [1] and their interactions with abiotic elements of habitat [1]

2 Any five from: food production, feeding, pollination, predation, building nests and shelters, nutrient recycling [5] Accept any other valid example.

3 Look for evidence of good research, stable community chosen, example of plant and animal and their role in community, understanding of stability, ability of stable community to absorb change and act as reservoir of many different species. [6]

B16.2

1a Any three from: availability of food, new pathogens, new parasites, new predators, competition [3]

b Any two from: temperature, moisture levels, soil pH and mineral content, wind intensity and direction, availability of oxygen and carbon dioxide [2]

2 Light intensity limits photosynthesis, [1] which affects distribution of plants and any animals dependent on plants for food. [1] Accept any valid explanation of factors outlined in **1b**. [4]

3 New predator will compete with existing predators in community for prey, [1] reducing number of prey animals available to existing predators. [1] If new predator very competitive or if prey have no defences against new predator, [1] number of prey animals will be significantly reduced, causing big fall in existing predator numbers or even local extinction of existing predators. [1] If new predator not very competitive, number of prey animals and existing predators will be reduced only slightly. [1]

B16.3

1 Abundance is the number of individuals of a particular species in a given area. [1] Distribution is where organisms are found within a given area. [1]

2a Position quadrat (same size quadrat each time) [1] either using random number generator [1] or by having person hold quadrat, close eyes, spin round, open eyes, and walk 10 paces before dropping quadrat to frame sample area. [1] Sample as many areas as possible. [1]

b To give representative and unbiased sample [1] of organism number and distribution [1]

c Mean: 6 (6.3), median: 6.5, mode: 6, 7, 8 [4]

3 Quadrats are often used along transect, combining techniques. [1] Quadrats used for random measurements [1] to determine overall picture of population or distribution of organism/variety of organisms. [1] Transects used for specific study of particular section of habitat, [1] measuring zonal changes. [1]

B16.4

1 Food [1], territory [1], mates [1]

2a Fighting, [1] display [1]

b fighting: animal could be hurt or killed, needs lots of body resources [2]; display: needs lots of body resources, vulnerable to predation during display [2]. Accept any other valid point.

3a Any two from: very good eyesight – see prey, good hearing – hear prey, good sense of smell – track prey, binocular vision – pounce on prey [4]

b Any two from: special teeth – grind grass and break open cells, speed – avoid/escape from predators, good eyesight/peripheral vision – detect predator approach, good hearing – detect predator approach [4]

c Any two from: good camouflage – escape predator notice, good eyesight/peripheral vision – detect predator approach, good hearing – detect predator approach [4]

d Any two from: teeth and gut adapted to eating plants – grind leaves and break open cells, ability to reach top of trees (long neck or good at climbing) – reach food source, ability to grip branches – reach food source and remove leaves from branches. [4]

B16.5

1a Any three from: light, water, mineral ions, space. [3] Accept any other valid point.

b Any valid example with explanation e.g., grow and flower early in the year [1] before trees are in full leaf, removing competition for light. [1]

2a To avoid competition both between parent plants and seedlings [1] and between seedlings themselves [1]

b Any three from: fluffy/winged seeds to float on air, seeds in berries/fruits to be eaten and distributed by animals, explosive seeds to fire seeds away from parent plant, sticky/hooked seeds to be carried on animals, floating seeds to be distributed in currents. [3] Any other valid suggestion.

3a The crowded plants were taller after one month [1] but shorter after 6 months [1] (compared to the spread out seedlings). This happened because the spread out seedlings did not have to compete with each other so much [1] for resources such as light, water, and mineral ions. [1]

b The crowded plants had lower wet biomass after 6 months [1] (compared to the spread out seedlings). This happened because the spread out seedlings got more resources [1] so were able to carry more photosynthesis [1] to produce biomass. [1]

B16.6

1a Feature that makes it possible for organism to survive in particular habitat, [1] even in extreme conditions [1]

b Award marks for any three valid examples. [3]

2 Award marks for a sensible and explanation of the adaptations given in **1b**. Two marks per adaptation. [6]

3 Any four valid points e.g. Other organisms cannot survive in these environments. [1] High temperatures denatures the enzymes of other organisms. [1] Extremophiles have specially adapted enzymes that are not found in other organisms. [1] Salty environments kills other organisms because water moves into their cells by osmosis. [1] This causes the cells to swell up and burst. [1] Extremophiles have adaptations to their cytoplasm so this does not happen. [1]

B16.7

1 Structural, behavioural, functional [1]

2 Any three from: small ears – reducing surface area of thin-skinned tissue to reduce cooling, thick fur – trapping insulating layer of warm air close to skin to reduce cooling, layer of fat/blubber – providing insulation to reduce cooling, furry feet – insulating against contact with ice to reduce cooling, large size – reducing surface area to volume ratio to reduce cooling [6] Any other valid point.

3a Any valid example e.g., an elephant's loose wrinkled skin [1] plus explanation e.g., increases surface area to volume ratio [1] increasing energy transfer to the environment [1] increasing cooling [1]

b Any two valid comparisons e.g., Structural adaptations are visible adaptations to a part of the organism, behavioural and functional adaptations are not visible. [2] Behavioural adaptations are actions of the organisms, structural and functional are adaptations to the characteristics of an organism. [2] Plus, at least two examples for each type of adaptation. [3]

B16.8

1 Any three from: Photosynthesis, [1] cell support, [1] tissue support, [1] transport of substances around plant [1]

2a Evaporation from leaf cells into air spaces, [1] diffusion down concentration gradient through stomata into surrounding air [1]

b Any three from: curled leaves – reduced surface area, traps moist air to reduce evaporation, small leaves – reduced surface area, thick cuticle – reduced rate of evaporation, stem-like leaves – reduced number of stomata to minimise water loss by diffusion into air. [6] Any other valid adaptation.

3a Extensive root system covering wide area – maximising water uptake from soil, [2] long roots going very deep – maximising water uptake from soil, [2] water storage tissues – to store water in fleshy leaves/stems/roots after rain [2]

b Any four points from: cacti do not have the adaptations to survive in a cold climate, [1] frost can damage their tissues, [1] not enough light for the cactus to photosynthesise and grow properly, [1] soil conditions are not suitable, [1] different specific seasonal cues for flowering times, [1] lack of water in frozen soils [1]

B17.1

1a Organism that makes its own food [1]

b Because they produce glucose by photosynthesis [1]

2 As prey numbers go up, there is more food for predators, so predator numbers increase after an interval. [1] As predator numbers increase, prey numbers start to fall as more are eaten. [1] After an interval, fall in prey animals means fewer predators survive and breed, so predator numbers fall too. [1] As predator numbers fall, fewer prey animals are eaten and prey numbers begin to increase again (cycle repeats). [1]

3a Demonstrate relationships between producers [1] and primary, secondary, and tertiary consumers [1]

b Food chains suggest that each type of consumer eats only one other type of organism [1], when most consumers eat a variety of different foods. [1]

c Prey numbers may be affected by factors other than predators: weather conditions, disease, or competition affecting food supply; disease affecting prey; population balance between different types of prey for a particular predator; predator numbers may be affected by factors other than prey; disease affecting predator; competition balance between different types of prey [5] Any other valid point.

B17.2

1 Organism that breaks down dead and waste material, [1] one example e.g. beetles, maggots, bacteria, fungi. [1]

2a Precipitation, [1] respiration [1] transpiration, [1] evaporation, [1] condensation [1]

b Precipitation: water falls to land as rain, snow, hail, or sleet; [1] respiration: water released from animals and plants during life and after death during decay; [1] transpiration: water released into atmosphere by plants; [1] evaporation: water turned from liquid to water vapour as Sun heats Earth's surface; [1] condensation: water vapour condensed back to liquid as moist air rises [1]

3 Living organisms remove materials from the environment constantly. [1] If mineral ions taken from the soil by plants were not replaced, Earth's resources would be depleted quickly. [1] Decomposers break down plant and animal waste and dead animals and plants and return mineral ions to soil to be taken up by plants again, [1] and carbon to the atmosphere as carbon dioxide to be used by producers in photosynthesis. [1]

4 Animals produce droppings, some plants shed leaves and plants and animals die. [1] Detritus feeders e.g. maggots eat dead organisms and produce waste material. [1] Bacteria and fungi then digest everything – dead organisms, detritus feeders, and their waste. [1] The nutrients produced are used by the decomposers to grow and reproduce. They release carbon dioxide, water, and mineral ions as waste products. [1] Decomposers return mineral ions, including nitrates, to the soil. Plants take them up through their roots and use them to make proteins and other molecules in their cells. [1] Decay returns carbon to the atmosphere as carbon dioxide to be used by producers in photosynthesis. [1]

5 Any six sensible points that indicate an understanding of humans of part of the water cycle and the water cycle as an ongoing continuous process over time e.g., animals, including humans, are part of the water cycle in nature; animals take in water from food/drink; plants take up water from the soil; water passes through the bodies of animals and plants, released during respiration and decay; animals also release water in urine, and faeces; plants release water during transpiration - this water condenses as it rises into cooler air and forms clouds; water returns to the Earth as precipitation; water percolates through the soil and rocks to underground stores; animals and plants take up water by drinking.

B17.3

1a Cycling of carbon between living organisms and the environment [1]

b Photosynthesis, [1] respiration, [1] combustion [1]

2 Students should produce a reasonable representation of the carbon cycle with all the major processes shown. [6]

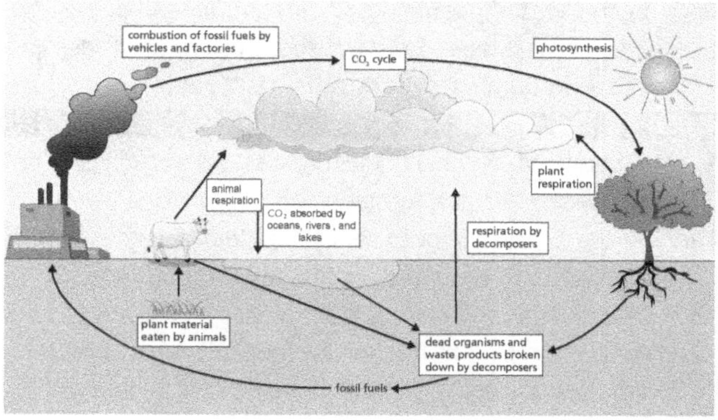

3 Photosynthesis: process by which plants use carbon dioxide and water to make glucose and oxygen using light, removes carbon dioxide from environment; [2] respiration: process by which plants and animals break down glucose and oxygen to make carbon dioxide and water, returns carbon dioxide to environment; [2] combustion: process by which organic material is burned in oxygen to make carbon dioxide and water in uncontrolled reaction, returns carbon dioxide to environment. [2]

B17.4

1a Detritivores, [1] decomposers [1]

b Plants remove minerals from the soil constantly. When animals eat plants these minerals are passed on through food chains and webs. [1] If minerals were never returned to soil, soil would soon be infertile. [1] Decay enables minerals that have become part of bodies of living organisms to be recycled and returned to environment. [1]

2a Rate of chemical reactions in microorganisms acting as decomposers [1] speeds up with increase in temperature. [1] Average temperatures much warmer in summer than in winter, [1] so rate of decomposition increases. [1]

b Moisture makes it easier for microorganisms to digest food and prevents them from drying out, [1] so they grow better in moist conditions. [1] Rate of decomposition slows down in dry conditions, even if temperature is high. [1]

3 Food decays as a result of decomposers. [1] Preserving food involves preventing the action of decomposers. [1] High temperatures preserve food by killing bacteria/fungi that would cause decay. [1] Freezing preserves food by slowing down or stopping the chemical reactions in the organisms that cause decay. [1] Most decomposers require oxygen. Packaging that excludes air preserves food because there is no oxygen available for decomposers to respire and so they die. [2]

B18.1

1 A measure of the variety of all the different species of organisms on Earth, or within a particular ecosystem [1]

2a We have discovered how to grow more food than we could ever gather from nature [1], learned to cure or prevent many killer diseases [1], have no natural predators. [1]

b Any four sensible answers, e.g: using it to build on, using it to grow crops, digging up land for quarrying/mining, polluting it with waste. [4]

3 Points may include: increased waste – including bodily waste, industrial waste, packaging, uneaten food, and disposable goods; exhaust gases from transport; use of pesticides and fertilisers in farming. [6] Any other valid point.

B18.2

1 Any two from: sewage, industrial waste, weedkillers/herbicides, pesticides, fertilisers [2]

2a Flow chart should show a series of at least five events in the correct order to show the process of eutrophication [5]

b If sewage enters a lake/the sea it causes high levels of nitrates [1] which can lead to the death of aquatic organisms. [1]

c Using data loggers [1], chemical analysis [1], bioindicators. [1]

3 Bioaccumulation refers to the gradual buildup of a substance within an organism's body over time [1]. Levels of DDT in insects were low. [1] Birds of prey and herons, consumed the insects and the DDT accumulated within their bodies. [1] Biomagnification is the process by which the concentration of a substance increases as it moves up the food chain. [1] The birds consumed multiple contaminated insects over time, [1] so accumulated higher concentrations of DDT in their bodies than those found in each insect. [1]

B18.3

1a Carbon dioxide/sulfur dioxide/nitrogen oxides [1] produced when fuels burn [1] dissolve in rainwater [1]

b Falls onto plants and kills them, [1] soaks into soil and destroys roots, [1] lowers the pH in lakes [1]

2 Flow chart should include: cars/power stations burn fossil fuels → acidic gases (e.g., sulfur dioxide, nitrogen oxides) formed → gases carried in the winds in atmosphere → sulfur acidic gases dissolve in rainwater and react with oxygen to form dilute sulfuric acid and nitric acid → acid rain falls [5]

3 Any three valid suggestions plus explanation e.g., low sulfur petrol and diesel used in vehicles, less sulfur dioxide produced. [6]

4 Sulfur dioxide and nitrogen oxides can be carried high in air by winds, [1] acidic gases can be blown from a country that does not control its sulfur emissions to a country that has strict emission controls [1] and fall as acid rain [1]

5a 1970: 6.5 million tonnes, 2020 0.1 million tonnes [1] 6.4 million tonne decrease [1] (accept +/– 0.2 million tonnes)

b Less sulfur dioxide being produced results in less acid rain [1] and less damage to plants and animals [1] leading to healthier ecosystems [1] and an increase in biodiversity. [1]

B18.4

1a Removal of large areas of forest by felling/burning [1] without trees being replaced [1]

b Any valid point e.g., reduction in rate of removal of carbon dioxide from the atmosphere, loss of biodiversity, destroying sources of new medicines or foods, species extinctions [1]

2a To grow crops for food, [1] to rear more cattle, [1] to grow crops that can be used to make biofuels [1]

b Carbon dioxide produced by burning of trees, [1] carbon dioxide produced by decomposition of dead vegetation, [1] less carbon dioxide removed from atmosphere by growing plants [1]

3a Use of peat as fuel [1] and by gardeners as compost [1]

b Carbon dioxide is released into atmosphere as peat is burnt or used as compost, [1] increasing atmospheric levels of carbon dioxide [1] and depleting the carbon store, [1] destruction of peat bogs destroys habitats and reduces biodiversity [1]

4 Points may include: cheap compost needed by gardeners and horticulturists to improve soil properties, promote seed germination, and increase food production; alternative, 'peat-free' composts are available but are less popular; peat bogs and peatlands vital to biodiversity as they form habitat for many organisms adapted to live in acidic conditions; peat forms very slowly and under very specific conditions – cannot be replaced as quickly as it is being used. [6] Any other valid point.

B18.5

1a Greenhouse effect: the way the surface of the Earth is kept warm by greenhouse gases; [1] global warming: an increase in the average temperature of the Earth; [1] climate change: a change in climate due to global warming [1]

b Any valid response e.g., loss of habitat, changes in distribution, changes in migration patterns, reduced biodiversity. [1]

2a Correct *y*-axis label and scale [1], Correct *x*-axis label and scale [1], data plotted correctly [1]

b Carbon dioxide levels rising steadily [1] over time, [1] partly as a result of human activities [1]

3 Energy is transferred from the Sun to the Earth. [1] Much of this heat is reflected back into space, [1] but some is absorbed by greenhouse gases in the atmosphere and reradiated back to Earth. [1] Earth's surface and atmosphere are warmed (greenhouse effect), [1] maintaining conditions ideal for life. [1]

4 Atmospheric levels of greenhouse gases increasing as a result of human activity, [1] increasing greenhouse effect [1] and causing global temperatures to rise [1]. Plus any two from: loss of habitat – reducing biodiversity; [2] changes in distribution – some organisms may disappear from some areas as habitat changes; [2] changes in migration patterns – caused by changes in climates and seasons; [2] reduced biodiversity – some organisms will become extinct as climate changes. [2]

B18.6

1 Any suitable example e.g., rising temperatures have caused the distribution of Dartford warblers to extend to more areas of the UK. [2]

2a Seasonal change [1] e.g., the temperature, amount of rainfall, and availability of water, levels of dissolved gases in the water, and hours of daylight change between seasons. [1] Geographical change [1] e.g., differences in soil, structure or pH, altitude, saltiness of water, and availability of water of different areas. [1]

b Any negative example with explanation e.g., deforestation [1] removes habitats [1] forcing them to relocate. [1] Any positive example with explanation e.g., reducing water pollution [1] increases habitat availability [1] so organisms may expand their distribution into previously contaminated areas. [1]

3a Number of breeding pairs of Dartford warblers in UK increasing [1]

b Dartford warblers cannot survive very cold winters. [1] Number of breeding pairs dropped sharply in 1963, [1] suggesting there was a cold winter. [1] Number increased sharply in 2010, perhaps as a result of temperature increase. [1]

c Temperature changes; [1] changes in number of Dartford warblers indicate that numbers have increased as UK temperatures have increased but drop sharply after cold winters. [1] Drop in number of Dartford warblers in Spain indicates that too high a temperature also affects distribution. [1] Evidence suggests that change in mean UK temperature [1] and seasonal extremes of temperature [1] affect distribution of UK birds both within the country and between countries. [1]

B18.7

1 Any three from: breeding programmes for endangered species, protection and regeneration of rare habitats, reintroduction of field margins and hedgerows, reduction of deforestation, recycling resources. [3]

2a Important for environmental health, [1] offers potential source of new food crops, [1] offers potential source of new medicines [1] Any other valid point.

b Any four from: increased field margins limiting land available for food production, reduction of deforestation limiting land available for food production, taxes on landfill waste increasing costs to businesses [4] Any other valid point.

3a Levels have fallen steadily [1]

b Data suggests that Kyoto agreement drove UK government to take action to limit carbon dioxide emissions [1] and that government intervention via legislation [1] has been successful. [1]

c Points may include: reduce or prevent increase in greenhouse effect and global warming – maintaining biodiversity by preserving Arctic and Antarctic ice habitat, preventing rises in sea levels that will destroy biodiverse mangrove forests, preventing temperature increase that could dry out wetlands and cause bleaching/death of coral reefs [6] Any other valid point.

4 Data shows that as landfill tax increased, amount of material going into landfill fell. [1] Taxes on waste and emissions [1] can be used as an effective incentive to change human habits [1] and promote research into less damaging alternatives (e.g., recycling rather than landfill). [1] Any other valid point.

B18.8

1 The positions of organisms within a food chain [1]

2a Award marks for accurate pyramid correctly labelled. [3]

b i $\frac{5000}{100\,000} \times 100$ [1] = 5% [1]

ii $\frac{30}{5000} \times 100$ [1] = 0.6% [1]

c Amount of material contained in biomass of organisms at each stage of a food chain less than it was at the previous stage; [1] not all of organism at one stage eaten by organism at next stage (only leaves of grass eaten by sheep, only blood of sheep eaten by ticks); [1] some materials taken in passed out and lost as waste products; [1] much of materials taken in used in cellular respiration to release energy for cell metabolism and some energy transferred to surroundings; [1] relatively small amount of material taken in converted to new biomass; [1] large percentage of biomass in grass cropped by sheep not converted to biomass in sheep and large percentage of biomass in sheep not converted into biomass in sheep ticks [1]

3 The pyramid of numbers for this example would not be a pyramid [1] because there will be many more ticks than sheep [1] so it does not accurately reflect the energy content of each organism. [1] A pyramid of biomass provides a more accurate representation of the energy content of the different trophic levels. [1] Biomass integrates both the number and size of organisms, reflecting the amount of energy available for consumption at each trophic level. [1]

B18.9

1 Some of the biomass an animal eats cannot be used, [1] biomass is lost between trophic levels in waste products, [1] biomass is used for temperature regulation in birds and mammals [1]

2a Flow chart should include: producers → primary consumer → secondary consumer → tertiary consumer → decomposers [4]

b Biomass lost by consumers in faeces [1] – cellulose cannot be digested by herbivores, animal parts such as hooves/claws/bones/teeth cannot be digested by carnivores; [1] biomass lost by producers and consumers in waste [1] – excess protein and water lost in urine, carbon dioxide and water lost in respiration; [1] biomass lost in consumers in maintaining a constant body temperature [1] – energy transferred to surroundings [1]

3 Only about 1% of energy from Sun's light falling on Earth (incident energy) is used by plants for photosynthesis [1] to create biomass. [1] Plant biomass eaten by animals cannot all be digested. [1] Some is broken down and used in respiration to provide energy for living processes such as movement and regulating body temperature. [1] Only a small amount of biomass eaten by animals is used for growth to produce new biomass. [1]

B18.10

1 Having enough food to feed a population [1]

2 Any three reducing factors from: increasing birth rate, new pests and pathogens, changing diets in developed countries, environmental changes, cost of agricultural inputs, conflicts. [3] Any three increasing factors from: maintaining or improving soil quality, efficient ways to produce food (e.g., selective breeding or GM technology), managing fish stocks, educating populations [3] Any other valid point.

3 Quinoa staple food in Bolivia; [1] production of quinoa in Bolivia has increased since 1997; [1] quinoa has increased in popularity in other countries, leading to increased exports [1] as farmers can charge more to foreign buyers than to local consumers and make more money; [1] local consumption of quinoa has fallen as exports have increased; [1] food security in Bolivia reduced as population increasingly dependent on imported pasta [1]

4 Accept any well-researched opinion and explanation. [6]

B18.11

1 Advantages – any two from: birds are always kept warm, birds always have plenty of food and water, consumers get meat relatively cheaply, birds can be sold for meat very quickly; [2] disadvantages – any two from: birds have limited space, there is a risk of disease spreading quickly, constant monitoring of birds costs money, supplying high quality food and maintaining temperature costs money [2]

2 Limiting the movement of food animals – this reduces the need for respiration and so they have more biomass available from their food for growth; [2] controlling the temperature of the surroundings – so the animals do not have to use much biomass in respiration for keeping warm or cooling down; [2] fed high protein food – to help build tissues [2]

3 Advantages: indoor work, animals grow faster so ready for market sooner, control of movement and temperature maximises conversion of food to biomass, weather conditions don't matter, fish farming conserves wild stocks; disadvantages: initial cost of buildings, cost of light and heating, higher feed bills, animals may be stressed (gain biomass more slowly), ethical concerns [6] Any other valid point.

B18.12

1a Almost all breeding fish have been caught [1]

b Some species may become extinct/reduction in biodiversity/less food for humans [1]

2a 1969 (accept between 1967 and 1969) [1] 1.8 million tonnes (accept 1.7 to 1.9 million tonnes) [1]

b 0.1 million tonnes [1]

c $1800000 - 100000 = 1700000$, [1] $\frac{1700000}{1800000} \times 100 = 94\%$, [1] (accept correct calculation based on answers given in parts a and b)

d Control size of the holes in nets so only largest fish caught, [1] ban fishing during breeding season, [1] strict catch quotas [1]

e Points may include: people dependent on fishing for livelihood and wary of reduced income, difficult to obtain overall picture of global fish stocks, difficult to apply quotas across large oceans fished by multiple countries [3] Any other valid point.

3 Any six sensible points – for example for the individual: food developed can have a healthy balance of nutrients; the food can be produced in large amounts so plenty for everyone; the food can be produced cheaply; for society: reduce diet-related disease globally; support the global economy as people are healthier; removes risks linked to global warming; need fewer animals; reduces land use for growing food [6]

MS1

1 25 μm [1]

2a 3.8×10^{-5} [1]

b 25 years [1]

3a 1:3 [1]

b 266 [1]

c $\frac{1}{5}$ [1]

MS2

1a 2 [1]

b 1 [1]

c 2 [1]

d 4 [1]

e 3 [1]

f 1 [1]

g 4 [1]

h 3 [1]

2 177.5

3a $89 + 275 = 364$ [1]

$\frac{89}{364} \times 100 = 24\%$ [1]

b 76% / 0.76 [1]

4 $\frac{25 \times 30}{4} = 187.5$ [1]

$= 188$ [1]

5 Mean: 1.65 [1] Median: 1.64 [1], Mode: 1.64, 1.76 [1]

MS3

1a The enzymes of the stomach work best in a pH that is less than 7 [1]

b Rate of reaction is proportional to concentration of reactant A [1]

c 22 cm^3 is approximately equal to 24 cm^3 [1]

2 0.2 mm [1]

MS4

1a [1]

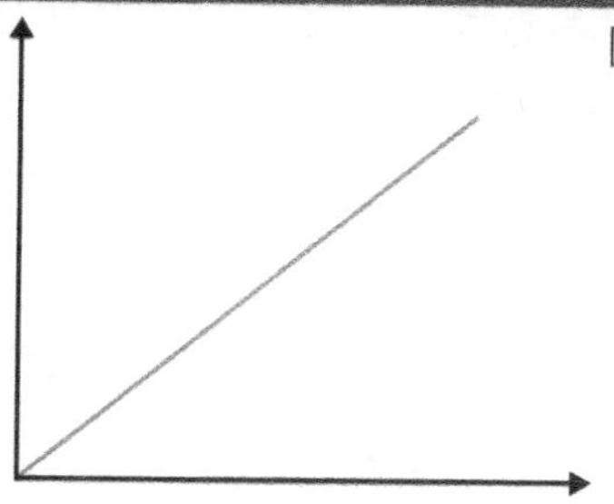

b [1]

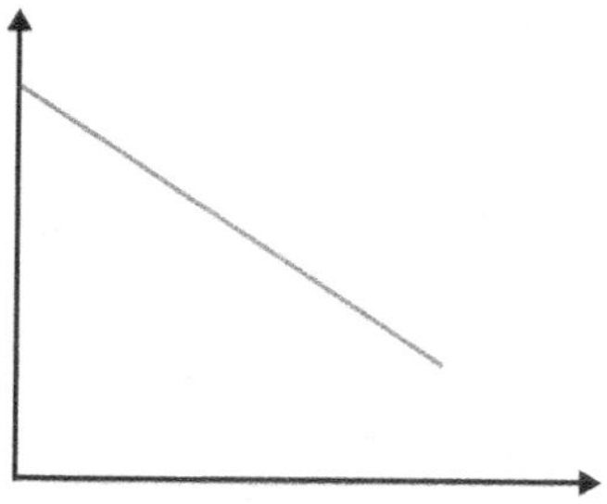

2a $y = mx + c$ [1]

b m: gradient, c: intercept [1]

3 $\frac{5500 - 3500}{400 - 150} = 8$ [1]

MS5

1 area of 1 cm cuboid (= 1 × 1 = 1) = 1 × 6 = 6 cm² [1]

Volume of 1 cm cuboid = 1 × 1 × 1 = 1 cm³ [1]

Surface area : volume = 6 : 1 [1]

Area of 2 cm cuboid (= 2 × 2 = 4) = 4 × 6 = 24 cm² [1]

Volume of 2 cm cuboid = 2 × 2 × 2 = 8 cm³ [1]

Surface area : volume = 24 : 8 = 3 : 1 [1]

Index

OXFORD
UNIVERSITY PRESS

Great Clarendon Street, Oxford, OX2 6DP, United Kingdom

Oxford University Press is a department of the University of Oxford. It furthers the University's objective of excellence in research, scholarship, and education by publishing worldwide. Oxford is a registered trade mark of Oxford University Press in the UK and in certain other countries.

First published in 2024

British Library Cataloguing in Publication Data
Data available

978-138-205142-2

10 9 8 7 6 5 4 3 2 1

The manufacturing process conforms to the environmental regulations of the country of origin.

Printed in Great Britain by Ashford Colour Press Ltd, Gosport.

Acknowledgements
A big thank you to my family and friends for their ongoing support in my new ventures, such as writing for OUP. With particular thanks to Becky for her encouragement in pursuing this career change and for her unwavering belief in me. Alex, thank you for your support and for patiently listening to me talk about writing book chapters every Thursday for months. And to Taylor Swift, whose playlists provided an epic soundtrack to keep me going. I also wish to express my gratitude to OUP for providing me with the opportunity to contribute to their esteemed publication.

The publisher and authors would like to thank the following for permission to use copyright material:

Cover: Sua Balac/closer&closer.

Artwork by Q2A Media, Michal Bednarski, Erwin Haya, Aptara, Jeffrey Bowles, Wearset Ltd., Peter Bull Art Studio, and Oxford University Press.